The Morals of Mea

The Morals of Measurement is a contribution to the social histories of quantification and of electrical technology in nineteenth-century Britain, Germany, and France. It shows how the advent of commercial electrical lighting stimulated the industrialisation of electrical measurement from a skilled labour-intensive activity to a mechanised practice relying on radically new kinds of instruments. Challenging traditional accounts that focus on metrological standards, this book shows instead the centrality of *trust* when measurement was undertaken in an increasingly complex division of labour with manufactured hardware. Case studies demonstrate how difficult late Victorians found it to agree upon which electrical practitioners, instruments, and metals were most trustworthy and what they could hope to measure with any accuracy. Subtle ambiguities arose too over what constituted 'measurement' or 'accuracy' and thus over the respective responsibilities of humans and technologies in electrical practice. Running alongside these concerns, the themes of body, gender, and authorship feature importantly in controversies over the changing identity of the measurer. In examining how new groups of electrical experts and consumers construed the fairness of metering for domestic lighting, this work charts the early moral debates over what is now a ubiquitous technology for quantifying electricity. Accordingly readers will gain fresh insights, tinged with irony, on a period in which measurement was treated as the definitive means of gaining knowledge of the world.

Graeme J. N. Gooday is Senior Lecturer in History and Philosophy of Science in the School of Philosophy at the University of Leeds. In 1989 he was joint winner of the Singer Prize, awarded by the British Society for the History of Science, for his paper 'Precision Measurement and the Genesis of Physics Teaching Laboratories in Victorian Britain'. He has written articles for numerous journals, including *The British Journal for the History of Science*, *History and Technology*, and *Osiris*.

The Morals of Measurement

Accuracy, Irony, and Trust in Late Victorian Electrical Practice

---

GRAEME J. N. GOODAY
University of Leeds

CAMBRIDGE UNIVERSITY PRESS
Cambridge, New York, Melbourne, Madrid, Cape Town, Singapore,
São Paulo, Delhi, Dubai, Tokyo, Mexico City

Cambridge University Press
The Edinburgh Building, Cambridge CB2 8RU, UK

Published in the United States of America by Cambridge University Press, New York

www.cambridge.org
Information on this title: www.cambridge.org/9780521187565

© Graeme J. N. Gooday 2004

This publication is in copyright. Subject to statutory exception
and to the provisions of relevant collective licensing agreements,
no reproduction of any part may take place without the written
permission of Cambridge University Press.

First published 2004
First paperback edition 2010

A catalogue record for this publication is available from the British Library

Library of Congress Cataloguing in Publication data
Gooday, Graeme.
The morals of measurement : accuracy, irony, and trust in late Victorian
electrical practice / Graeme J. N. Gooday.
p. cm.
Includes bibliographical references and index.
ISBN 0-521-43098-4
1. Electric measurements – History – 19th century.
2. Electric lighting – History – 19th century.   I. Title.
TK275.G66   2004
621.37′09′034 – dc21   2003053291

ISBN 978-0-521-43098-2 Hardback
ISBN 978-0-521-18756-5 Paperback

Cambridge University Press has no responsibility for the persistence or
accuracy of URLs for external or third-party internet websites referred to in
this publication, and does not guarantee that any content on such websites is,
or will remain, accurate or appropriate.

Contents

List of Illustrations		*page* viii
Abbreviations		ix
Preface		xiii

1   Moralizing Measurement: (Dis)Trust in People, Instruments, and Techniques   1
    1.1. William Thomson and the Limits of Measurement   2
    1.2. The Metrological Fallacy – Or What the History of Measurement Is Not   9
    1.3. Rival Narratives of Quantification: Networks of Trust versus Centres of Power   16
    1.4. Moral Economies of Trust and Quantification   23
    1.5. Trust and the Material Culture of Measurement   30
    1.6. Conclusion   39

2   Meanings of Measurement and Accounts of Accuracy   40
    2.1. Competing Rhetorics of Measurement: Comparison versus Reduction   42
    2.2. Uncertainties over the Identities of the Measurer and the Measured   50
    2.3. 'Reasonable Agreement' and the Multiple Meanings of Accuracy   57
    2.4. Responsibility for Accuracy and Error: The Politics of Ambivalence   65
    2.5. Reporting Accuracy: The Protocols and Languages of Error   72
    2.6. Conclusion   80

3   Mercurial Trust and Resistive Measures: Rethinking the 'Metals Controversy', 1860–1894   82
    3.1. Rethinking the 'Metals Controversy': Siemens versus Matthiessen   85
    3.2. The Mercurial Solution: Siemens' 1860 Proposal   90
    3.3. Matthiessen's Case for the Alloy: Trust in Solidity   94

	3.4. Controversy Begins: Challenging Accuracy and Metallic Utility	98
	3.5. The BAAS Committee's Contemplation of Mercury and Alloy Standards	103
	3.6. 'Dr Matthiessen Has Been Opposed to Mercury': The Acrimony of Commerce 1865–1866	110
	3.7. Resisting Mercury: The Unresolved Aftermath of the Metals Controversy	117
	3.8. Conclusion	124
4	Reading Technologies: Trust, the Embodied Instrument-User and the Visualization of Current Measurement	128
	4.1. 'Internalist' Histories and the New Historiography of Instruments	131
	4.2. Sensitivity versus Robustness: Galvanometer Accuracy in the Working Environment	141
	4.3. Temporal Characteristics of Current-Measurement Practices	148
	4.4. Proportionality versus Trustworthiness? Constructing the Direct-Reading Ammeter	153
	4.5. Ironies of Reading Instruments: Proportionality and Spot-Watching	160
	4.6. Conclusion	171
5	Coupled Problems of Self-Induction: The Unparalleled and the Unmeasurable in Alternating-Current Technology	173
	5.1. 'We Do Not Couple Machines': The Tribulations of AC Parallel Running	176
	5.2. The Problematic 'Inertial' Analogy: Maxwell's Account of Self-Induction	180
	5.3. Self-Induction as Momentum: John Hopkinson's Theory of AC Paralleling	186
	5.4. From Current Balance to Secohmmeter: Measuring Self-Induction at the STEE	192
	5.5. The Secohmmeter in Action: Gisbert Kapp and the Paralleling of Alternators	201
	5.6. Diagnosing Self-Induction: Mordey's Contested Analysis of Parallel Working	204
	5.7. Counting 'Ayrton's and Perry's Things': The Secohmmeter Further Contested	208
	5.8. Epilogue: The Lingering Marginal Career of the Secohmmeter	215
	5.9. Conclusion	216

6	Measurement at a Distance: Fairness, Trustworthiness, and Gender in Reading the Domestic Electrical Meter	219
	6.1. The Historiography of the Domestic Electrical Meter	222
	6.2. The Gas-Meter Paradigm of Measuring at a Distance	225
	6.3. The Dial-Less Meter: Edison's Technique for Measuring at a Distance	232
	6.4. Emulating the Gas Paradigm: Ferranti's Mercury-Motor Meter	239
	6.5. Fairness versus Expediency: Rival Interpretations of Electrical Consumption	244
	6.6. Meters and the Gendered Consumption of Electric Lighting	253
	6.7. Conclusion	261
Conclusion		263
Index		273

Illustrations

1. Werner Siemens' design for mercury column for resistance standard, c. 1890. — page 93
2. A Siemens electro-dynamometer, c. 1884. — 134
3. Integrated 'convenience' version of the Thomson mirror galvanometer, 1879. — 138
4. Willoughby and Oliver Smith stand over a mirror galvanometer, 1866. — 143
5. Ayrton and Perry's electric tricycle, 1882. — 157
6. Johnson & Phillip's hotwire ammeter, 1910. — 162
7. Ayrton & Mather version of the Deprez D'Arsonval galvanometer, 1902. — 163
8. James Clerk Maxwell's pulley and flywheel model for self-induction (n.d.) — 187
9. John Hopkinson's account of the parallel coupling of alternators, 1884. — 190
10. Ayrton & Perry's secohmmeter for measuring self-induction, 1887. — 198
11. Alternators running in parallel, Amberley Road Power Station, London, 1893. — 213
12. Gas meter with three contra-rotating dials, 1884. — 228
13. An Edison meter opened to show the electrolytic cells, 1888. — 235
14. Cross section of Ferranti DC mercury-motor meter, 1895. — 242
15. 'At the Door; or Paterfamilias and the Young Spark', *Punch* 1891. — 255

Abbreviations

BJHS *British Journal for the History of Science*
BAAS British Association for the Advancement of Science
ICE   Institution of Civil Engineers

The following three names applied successively to one and the same institution:

STE   Society of Telegraph Engineers (1871–80)
STEE Society of Telegraph Engineers and Electricians (1881–8)
IEE   Institution of Electrical Engineers (1889–)

Notes
Where no author is specified, publications were anonymous. The names of publishers are cited only for twentieth-century books.

One of the great difficulties experienced by people in mastering the *quantitative* science of electricity, arises from the fact that we do not number an electrical sense among our other senses, and hence we have no intuitive perception of electrical phenomena... an infant has distinct ideas about hot and cold, although it may not be able to put its ideas into words and yet many a student of electricity of mature years has but the haziest notions of the exact meaning of high and low potential, the electric analogues of hot and cold.
        William E. Ayrton, *Practical Electricity*, 1887, Preface

Every practice requires a certain kind of relationship between those who participate in it. Now the virtues are those goods by reference to which, whether we like it or not, we define our relationships to those other people with whom we share the kinds of purposes and standards which inform practices.
        Alasdair MacIntyre, 'The Nature of the Virtues', *After Virtue*, 2nd edition, 1985, p. 191

Preface

When electrification is produced by friction, or by any other known method, equal quantities of positive and negative electrification are produced... The electrification of a body is therefore a physical quantity capable of measurement... While admitting electricity, as we have now done, to the rank of a physical quantity, we must not too hastily assume that it is, or is not, a substance, or that it is, or is not, a form of energy, or that it belongs to any known category of physical quantities.
James Clerk Maxwell, *Treatise on Electricity and Magnetism*, 1873[1]

The first step is to measure whatever can be easily measured. This is O.K. as far as it goes. The second step is to disregard that which can't be measured or give it an arbitrary quantitative value. This is artificial and misleading. The third step is to presume that what can't be measured easily isn't very important. This is blindness. The fourth step is to say that what can't easily be measured doesn't really exist. This is suicide.
Daniel Yankelovich, interview with George Goodman, c. 1973[2]

As James Clerk Maxwell knew perhaps better than anyone else, dealing with electricity was no dull or easy matter. Like many contemporaries in industrial and academic spheres who sought to harness electricity to technological ends, he laboured extensively to comprehend its complex and occasionally shocking behaviour. Yet as Maxwell hinted early on in his famous *Treatise*, there was much uncertainty about what electricity actually *was*. Natural philosophers, electricians, and telegraphists could not agree among themselves about whether electricity was a form of energy, or constituted out of one or possibly two negatively or positively charged fluids (whether material or immaterial), or perhaps was even something hitherto altogether unknown.

---

[1] James Clerk Maxwell, *Treatise on electricity and magnetism*, 1st edition, 2 Vols., London: 1873. Unless otherwise specified, all quotations are from the 3rd edition (ed. J. J. Thomson), 1891, reprinted New York: Dover, 1954; quotation on p. 38.

[2] Daniel Yankelovich from interview quoted in 'Adam Smith' [pseudonym of George J.W. Goodman], *Supermoney*, London: Michael Joseph, 1973, p. 286. The context of Yankelovich's comment was a sharp critique of the U.S. government's approach to quantifying human losses in the Vietnam War.

They did at least share with Maxwell, though, the conviction that electricity was in fact measurable. Hence many adopted the pragmatic strategy of focussing on those manifestations of electricity that they thought could most easily be assigned stable numerical properties.

Or at least they supposed they could, for, in fact, efforts to measure even the most mundane electrical performance could be imbued with problems and ambiguities that arose for a wide variety of reasons. Not least among their concerns was an emerging question: what actually *constituted* the 'measurement' of a physical quantity? Was it necessarily a laborious activity of 'absolute' determination in terms of mass and length, or could it just be 'relative' comparison against a convenient calibrated commercial standard as was common in telegraphy from the late 1850s? Then again, could measurement be constituted by an instantaneous glance at the deflection of a needle or light-spot over the dial of a pre-calibrated instrument – as electrical-lighting engineers contended in regard to the new industrial instruments they developed in the 1880s? It is the vivid controversies that arose from these problems and ambiguities that form the substance of much of this book.

What then could the themes of 'morals' do to help the historical recovery of heterogeneous and contested technical practices of electrical measurement in the later Victorian period? My title is neither oxymoronic nor merely alliterative. It plays instead on the twofold signification of morals, drawing first on the historiographical *lessons* gleaned from attempting to reconstruct the diverse Victorian projects of quantifying electrical performance. On the one hand, it is about how physicists, chemists, electricians, and electrical engineers tried to measure what mattered to them for their very different reasons.[3] It is also about how they often disagreed interestingly about how to measure electricity and what could or *should* be measured. Much of their disagreements centred on the highly specialized technologies they deployed

---

[3] Here I allude to the wide range of scholarship on the complexity of the science–technology relationship in the nineteenth century: Donald Cardwell, *Technology, science and history: A short study of the major developments in the history of Western mechanical technology and their relationships with science and other forms of knowledge*, London: Heinemann, 1972; Ronald Kline, 'Science and engineering theory in the invention and development of the induction motor, 1880–1900', *Technology and Culture*, 28 (1987), pp. 283–313; Bruce J. Hunt, 'Insulation for an empire: Gutta Percha and the development of electrical measurement in Victorian Britain', in F. A. J. L. James (ed.), *From semaphore to shortwaves*, London: Royal Society of Arts, 1998, pp. 85–104; Sungook Hong, 'Historiographical layers in the relationship between science, technology', *History and Technology*, 15 (1999), pp. 289–311. These authors show how engineering endeavours furnished not only many of the theoretical and practical problems that occupied practitioners of physics and natural philosophy, but also the resources and personnel with which to solve them. In applying this approach to the history of measurement, my account can be contrasted with historiographies that present quantification as driven by a disembodied and culturally invasive 'spirit'. See, for example, Tore Frängsmyr, John Heilbron, and Robin E. Rider, *The quantifying spirit in the eighteenth century*, Berkeley: University of California Press, 1990.

Preface

to make electrical measurements beyond the bodily capacities or endurance of ordinary humans. Accordingly I focus much on the instruments through which my protagonists articulated their value-laden and thus often divergent practices of electrical measurement. This generally occurred in colourful discussions with instrument-makers and fellow instrument-users about which user-centred values should be incorporated into the construction of such instruments for their particular purposes and contexts.

And it is in this value-ladenness in the practices and instrumentation of measurement that my other moral theme lies. I explore the (limited) extent to which contemporaries construed judgements of trustworthiness in measurement to have some form of moral dimension. Accordingly I explore how judgements of fairness, fidelity, and honesty were used to decide which electrical practitioners and instruments should be trusted or distrusted in measurement work, and why. Historians really should not, of course, be squeamish about the intrusion of moral issues into their analytical discourse. Moralistic imperatives and judgements pervade even the most radical recent accounts in history and sociology of science.[4] Some sociologists have even argued that it is futile and counterproductive to seek neutrality when analysing scientific controversies.[5] And for the reflective historian, it is impossible to avoid making some sort of judgement – albeit often tacit – about the fair representation of the integrity (or self-interestedness) of historical actors. And if historians cannot avoid making morally loaded judgements of this sort in their everyday work, it is reasonable to suppose that past scientists and technologists might have interpreted and judged each others' actions in similarly evaluative terms.

---

[4] Schaffer and Shapin explicitly side with Hobbes against the many partisanly Boyle-centred accounts of late seventeenth-century science. Simon Schaffer and Steve Shapin, *Leviathan and the airpump: Hobbes, Boyle and the experimental life*, Princeton, NJ: Princeton University Press, 1985. Ashmore represents Blondlot's N-Ray experiments as a victim of unfair debunking experiments in Malcom Ashmore, 'The theatre of the blind: Starring a Promethean prankster, a phoney phenomenon, a prism, a pocket, and a piece of wood', *Social Studies of Science*, 23 (1993), pp. 67–106. In laying out a manifesto for a sociology of knowledge that treated claims for scientific truth and falsity with a 'symmetrical' impartiality, Bloor contends that sociologists' previous refusal to countenance such an approach amounted to a 'betrayal' of their disciplinary standpoint; David Bloor, *Science and social imagery*, London: Routledge, 1976, p. 1. For Brian Wynne's partisan role in opposing the nuclear power industry whilst undertaking a sociological analysis of it, see Brian Wynne, *Rationality and ritual: The Windscale inquiry and nuclear decisions in Britain*, Chalfont St Giles: British Society for the History of Science, 1982.

[5] Pam Scott, Eveleen Richards, and Brian Martin, 'Captives of controversy: The myth of the neutral social researcher in contemporary scientific controversy', *Science, Technology and Human Values*, 15 (1990), pp. 474–94; Dick Pels, 'The politics of symmetry', *Social Studies of Science*, 26 (1996), pp. 277–304; B. Wynne 'SSK's identity parade: Signing up, off-and-on', *Social Studies of Science*, 26 (1996), pp. 357–92. The last two papers appear in a volume of *Social Studies of Science* co-edited by Eveleen Richards and Malcom Ashmore titled 'The politics of SSK: Neutrality commitments and beyond'.

In seeking to identify the moral features of a technical practice such as electrical measurement I do not assume that all practitioners necessarily pursued moral agendas – whether morality be construed in terms of individual obligation (deontology), welfare maximization (utilitarianism), fair treatment (rights theory), or good conduct (virtue theory).[6] Victorian engineers and scientists certainly did not always act altruistically or impartially, nor were they systematically disinterested, communally oriented, or sceptical in all their claims. Indeed, on close inspection much of their conduct bears out Trevor Pinch's point that such norms of professionally virtuous behaviour (e.g., as identified by Robert Merton) are most obviously manifested in the post hoc justification of action rather than in its initial motivation.[7] But whatever the virtues or vices of the measurer(s) involved, their measurements could be intrepreted *by other* observers – especially critics – as bearing a significance that went beyond the merely technical or epistemological. The morals of measurement could be seen in at least four ways: in the *presuppositions* of a measurement; what was fair to assume about the integrity of previous measurers in the field? In the *performance* of a measurement; did its conduct instantiate trustworthy practices and appropriate experimental virtues? In the *reporting* of a measurement; was the written (published) account an honest and impartial summary of the performance? And in the *ramifications* of a measurement; what benefits – if any – might the quantitative information generated bring to others?

---

[6] Not all mutual 'obligations' of past practitioners can be simply represented as essentially 'moral' in character: see Steve Shapin, *A social history of truth: Civility and science in seventeenth century England*, Chicago: University of Chicago Press, 1994, pp. 310–11. For a general study of the relations between knowledge and obligations, see Morton White, *What is and what ought to be: An essay on ethics and epistemology*, New York/Oxford: Oxford University Press, 1981. For some interesting discussion of the moralization of science, see Robert Proctor, *Value-free science: Purity and power in modern knowledge*, London/Cambridge, MA: Harvard University Press, 1991; Anna Mayer, 'Moralizing science: The uses of science's past in the 1920s', *British Journal for the History of Science*, 30 (1997), pp. 51–70; and John Krige and Dominique Pestre, 'Introduction', in J. Krige and D. Pestre, (eds.), *Science in the twentieth century*, Amsterdam: Harwood Academic, 1997, pp. xxi–xxxv, discussion on pp. xxi–xxii.

[7] Robert K. Merton, 'The normative structure of science' [1942], in *The sociology of science: Theoretical and empirical generalizations*, Chicago: University of Chicago Press, 1973, pp. 267–78; Paul Feyerabend, *Against method*, London: New Left Books, 1975; Trevor Pinch, 'The sociology of the scientific community', in Robert Olby, Geoffrey Cantor, John Christie, and Jonathan Hodge (eds.), *Companion to the history of modern science*, London: Routledge, 1990, pp. 87–99, discussion on p. 89. Nevertheless, Rom Harré claims that the scientific community is 'morally superior to every other form of human association'; R. Harré, *Varieties of realism: A rationale for the natural sciences*, Oxford: Blackwell, 1986, pp. 1–2, 6–7. See further discussion in the next chapter; for contrasting studies of the widespread persistence of fraudulent science, see William Broad and Nicholas Wade, *Betrayers of the truth*, Oxford: Oxford University Press, 1985.

Preface   xvii

This last issue leads me to the moral–political point sharply delineated by the American social theorist Daniel Yankelovich in the quote in the opening of this Preface. Yankelovich's concern lies in what is *lost* and who are the *losers* when measurement is treated as the definitive arbitrating practice in creating representations of the world. He makes a stark observation about those measurers who are unreflexively enamoured of the apparently unique efficacy of measurement. He notes that, from the productivity and instrumental utility of their quantitative work, measurers can convince themselves and their allies that measurability is the only important feature of epistemology, and thence drift into self-serving circular arguments about the worldly understanding to be attained through measurement. Yankelovich identifies a three-stage slippage: That which cannot be (easily) measured can be at first disregarded, then treated as unimportant, and then indeed in the extreme case treated as if it does not really exist at all. He thus highlights a pernicious slide from a tendentious epistemological claim to a distinctly sinister moral claim and a rather nihilistic ontological claim. If accepted at face value, such claims enable measurers to have a monopoly of expertise in determining what can be considered to exist and what can be a legitimate matter of human concern – the only things that matter are those in which they have the predominant expertise.

If we follow the force of Yankelovich's observation, it is all too evident that if privileged significance is attached only to that which is easily measurable, then those people who cherish what cannot easily be thus quantified are likely to experience injustice or at least marginalization. Less extreme, but of great significance to this volume, is that such unfortunates may find their positions all too easily devalued by quantitative experts as deficient in (numerical) evidential support or even as grounded on mere speculation or delusion. Much has been made in this regard of a passing comment made in 1883 by William Thomson (later Lord Kelvin) that knowledge claims are 'meagre and unsatisfactory' unless based on the results of measurement. When extended to domains beyond that of the physical sciences – notably medicine and education – critics have condemned this partisan valorization of the easily measurable over the unquantifiable as the 'curse of Kelvin'.[8]

---

[8] Alvan R. Feinstein, 'Clinical biostatics XII: On exorcising the ghost of Gauss and the curse of Kelvin', *Clinical Pharmacology and Therapeutics*, 12 (1971), pp. 1003–16, esp. p. 1004. One critic of unreflective measurement invoked the antireductionist wisdom of Daniel Yankelovich as a poignant antidote to the 'Curse of Kelvin'; see letter from Sheldon H. White to the educationalist, Jerrold R. Zacharias, 7 November 1978, cited in Jack S. Goldstein, *A different sort of time: The life of Jerrold R. Zacharias*, London/Cambridge, MA: MIT Press, 1992, p. 287. By contrast, Ian Hacking notes that scientists seeking to defend contentious quantitative work against critical sceptics have often borrowed Thomson's words to bolster their position; Ian Hacking, *Representing and intervening*, Cambridge: Cambridge University Press, 1983, p. 242. For further discussion of Thomson's 1883 views on measurement, see Chapter 1 of this volume.

Who could deny that a politics of exclusion is readily facilitated by successful moves to promote a monopoly of expertise based on skill-intensive and resource-intensive practice of measurement? Subsequent chapters show that late Victorian debates on electrical matters were dominated by those physicists, chemists, electricians, telegraphists, engineers, or instrument-makers who had both the expert skill and access to resources to undertake sophisticated and lengthy electrical measurements. By contrast, there is an absence of voices from an older tradition of workers specializing in technologies of 'electrical display', as discussed by Iwan Morus. Whilst theirs was a form of practical virtuosity which by the last third of the nineteenth century still commanded public enthusiasm, it did not win such practitioners a place in debates at the Royal Society, Physical Society, or Institution of Electrical Engineers; nor did it win them many publications in technical journals of natural philosophy, such as *The Philosophical Magazine* or of the electrical trade, such as *The Electrician*. Significantly, though, there were some commercially important groups of consumers who were granted an indirect voice in some debates on how to quantify the behaviour of electricity, one prominent example being the tiny but growing elite of electric-lighting consumers in the 1880s and 1890s. Yet to learn about how customers' grievances at unreliable or misleadingly supplied electrical power affected the quantifying practices of electricians and engineers, we generally have to rely on the somewhat partisan testimony of the latter. And entirely absent from such debates were the voices of citizens who were not consumers of electrical products but whose quality of life was palpably diminished by the advent of noisy and polluting new generating stations and garish outdoor arc lighting.[9] *Their* disaffection was not something that the electrical community made any gesture towards quantifying: In Yankelovich's terms the electrical experts considered such disaffection as either not mattering or non-existent.

Ironically, however, we shall see that the experts discussed in this book kept encountering significant limits to their technologically enhanced capacities to quantify even straightforward electrical matters. In several instances, and in direct conflict with Thomson's identification of knowability with measurability, it proved extremely problematic to measure some electrical parameters that were considered not only to be real (despite being unquantifiable) but indeed of great technological importance. In Chapter 5 I examine the self-induction of a moving alternator as a case in point and in Chapter 6 the actual amount of light consumed by a household installed with electric illumination. In both cases we will see how debate shifted – albeit not without challenges, some moral in nature – from what arguably *should* have

---

[9] See 'Lines to the electric light at the G.W. Railway terminus', published by the *St James Gazette* in 1888, cited in Robert H. Parsons, *The early days of the power station industry*, Cambridge: Cambridge University Press, 1940, p. 42.

been measured to the kinds of electrical parameters that *could* easily and inexpensively be measured.

In a vein of Rortyian 'irony' this work explores the nature of limits in scholarly enterprises: both the limited power of interpretive themes, and the limits to which these themes can reasonably be taken. I show that practitioners acknowledged distinct limits to the power of electrical measurement to capture both the nature of electrical behaviour and the performance of electrical technology in relation to human demands. Thus although I argue for the importance of measurement in attempts to deal with the mysteries of electricity, I show also what a problematic and bounded enterprise it was. And although I argue for the importance of recognizing moral concerns in late nineteenth-century endeavours of science and technology, I also embrace the interpretive limitations of such a pursuit.[10] Whilst focussing often on the notion of trust as the important 'moral' dimension to measurement, I suggest that trust cannot be seen as an exclusively moral category, nor that moral considerations pertaining to trust are omnipresent in measurement practice. And accordingly I explore how practitioners used the complex qualitative and quantitative languages of 'accuracy' or 'degree of accuracy' to articulate the limits to which they considered they could trust – or should be able to trust – their measurements.[11]

As we shall see, the relationship between the limits of desirable accuracy and achievable accuracy was not stable. Although these converged when practitioners learned to live within the horizons offered by their instruments, they diverged when the demands of industrial efficiency or customer satisfaction called for greater robustness, celerity, transparency, sensitivity, or trustworthiness than extant instruments could be made to furnish. As Matthias Dörries has noted, it is the recurrent limitations of instruments rather than their 'successes' that lead to their users' and makers' trying to refine or adapt their construction and operation.[12] Throughout this book, I reiterate Dörries' insight and extend it to the ways in which problematic attempts to extend measurement practice into hitherto new domains of electromagnetism generated a reflexive awareness that understandings of what could or should constitute measurement were in need of re-examination. This reflexive awareness was not necessarily a matter of increased rigour: It was more often a matter of pragmatic compromise to abandon older practices to meet new desiderata.

---

[10] See Richard Rorty, *Contingency, irony and solidarity*, Cambridge: Cambridge University Press, 1989, pp. 73–95, esp. pp. 73–4. An alternative term that some readers might prefer to 'irony' here is 'finitism' or 'anti-reductivism.'

[11] A number of historians have shown that the meanings and significance of such terms are ineluctably embedded within the contingencies of cultural values. See M. Norton Wise (ed.), *The values of precision*, Princeton, NJ: Princeton University Press, 1995, and Chapter 2 of this volume for further discussion.

[12] Matthias Dörries, 'Balances, spectroscopes, and the reflexive nature of experiment', *Studies in the History and Philosophy of Science*, 25 (1994), pp. 1–36.

Put crudely, my overarching thesis is that electrical measurement underwent a form of 'industrialization' in the last two decades of the nineteenth century.[13] In the earlier period covered by this book, the 1850s to 1870s, physicists and telegraphic electricians generally relied to a great extent on their own manual skill and resourcefulness to manipulate instruments into the correct configuration to take a reading. They tended, moreover, to take great personal care and time to establish the propriety of calibrations, calculations, and results obtained in measurement activity. With the rise of electric lighting in the following two decades, however, such reliance on time-consuming and laborious care was displaced by the adoption of labour-saving and time-saving techniques of instrument operation and reading. Sophisticated automated (and fallible) instrument mechanisms were developed that displaced much of the interpretive skill from the human user to the ingenuity of designer and concomitantly shifted much of the all-important labour of calibration from user to instrument-maker. Measurement using 'direct-reading' ammeters and voltmeters thus became more like the 'minding' of automated factory machinery and less like the virtuosic skilled effort of a self-reliant expert. As we shall see, the shifting patterns of trust engendered in this new division of labour in measurement work generated debates which pitted the integrity of labour and virtues of mechanization in ways that echoed earlier controversies over factory mechanization.[14] Readers of this volume may wish to ponder the long amnesia about such debates. By reading what follows, they can recover how it became possible for them to trust readings they take in a glance from dials on their car dashboards or from rotating indices on their household electricity meters – a practice barely recognizable to mid-Victorian forbears as any kind of measurement, *sensu strictu*.

Following my detailed analysis of the role and character of trust in instrumental measurement practice in Chapter 1, I move in Chapter 2 to a deconstruction of the apparently simple notions of electrical measurements and accuracy to which late Victorians subscribed. Successive chapters are then devoted in a sense to considering the material, moral, and managerial problems of measuring specific electrical parameters: electrical resistance (Chapter 3), electrical current (Chapter 4), self-induction (Chapter 5), and domestic electrical consumption (Chapter 6). Each bears out my claim that problems of agreeing how to measure in a way were *not* solved by universalized use of metrological standards – these were neither necessary nor sufficient for this purpose. In each case, questions were raised about whether

---

[13] I use the term 'industrialization' here following Wolfgang Schivelbusch, *Disenchanted night: The industrialization of light in the nineteenth century* (trans. A. Davies) Oxford/New York: Berg, 1988.

[14] Maxine Berg, *The age of manufactures, 1700–1820: Industry, innovation and work in Britain*, 2nd edition, London: Routledge, 1994.

the parameter could be measured at all, and if so whether it could practicably be measured to within an appropriate 'degree of accuracy' in a suitably disinterested fashion. We shall thus see that William Thomson's advice to contemporaries to seek the safest knowledge of things through measurement was subverted by the difficulties of knowing which would be the most trustworthy methods, materials, and instruments to use in measurement.

In addition to these specific topics in the historiography of measurement, these chapters also address issues central to contemporary historiography of science and technology. A major theme in Chapter 2 is who could be the author of measurement – a question as contentious as attributions of *authorship* to texts. The complex division of labour that developed in the skilled design, manufacture, and pre-calibration of direct-reading instruments during the 1880s reduced the role of users to such a skill- and labour-free involvement that some traditionalists denied they were taking measurements at all. Chapter 3 explores how interpretive flexibility occurs in judgements of trust. In evaluations, so many measurements in constructing resistance standards were non-reproducible: According to the charity of their judgement, critics could impute various degrees of untrustworthiness to techniques, constitutive metals, or their human spokespersons. In Chapter 4, I explore the theme of how the bodies of measurers – and not just their eyes and brains – mattered in the measurement process. The various techniques of taking *readings* in measurement activities presupposed a particular kind of spatio-temporal configuration between instrument and reader, and this required different forms of bodily deportment – not all of which were equally acceptable to all practitioners. In Chapter 6, I consider briefly how the gendered identity of the putative measurer enters into considerations of the trustworthiness of measurement. I show how one female expert on electrical matters was technically adept at reading the electric meter but advised her readers to defer to male householders – as legally responsible for paying for the luxury of electric lighting – the general prerogative of taking such readings.

Although such familiar heroic masculine figures as William Thomson and James Clerk Maxwell often appear in my story it is not in their traditional role as abstract theorists, but rather as technologically informed experts on electrical measurement practice whose judgements impinged on the everyday lives of electricians and kindred workers. The foreground is filled by less-often-discussed but equally important characters such as the (physically disabled) chemist Augustus Matthiessen and his rival for expertise in electrical metals, the Prussian industrialist and telegraph entrepreneur Werner von Siemens. The telegraphic expertise of Robert Sabine, Harry Kempe, and Latimer Clark earns them a place as writers of major handbooks on electrical measurement techniques that were often read alongside Maxwell's *Treatise on Electricity and Magnetism* (1873) as a guide to refined techniques of laboratory measurement. Such works were familiar to William Ayrton as a telegraphic expert in India and Japan up to 1878 and continued to be sources of reference for him as City & Guilds Professor of physics and electrical engineering, first at Finsbury Technology in London then at South Kensington from 1884 to 1885. As a major teacher, instrument designer, and suffragete sympathizer, Ayrton takes up a major place in several chapters largely because of his work with collaborator John Perry in promoting the innovation of direct-reading instruments. A biographical piece in the *Electrical Engineer* for January 1889 argued that, although to the public Ayrton had played a major role in explaining the new technologies of electric light and power, for the 'practical man' his name more closely connected with 'simple and portable measuring instruments' than with anything else, as such instruments had largely been manufactured and used from 'his designs'.[15] Whilst thus aiming in part to illustrate Ayrton's important role in the history of measurement instruments, this book is not, however, a biography of Ayrton – an important enterprise that no one to my knowledge has yet attempted.

A recurrent critic of Ayrton at meetings of both the Institution of Electrical Engineers and the Institution of Civil Engineers was the aristocratic mechanical engineer turned freelance consultant electrician, James Swinburne. It was Swinburne who most often challenged Ayrton's pretensions as a protégé of Sir William Thomson to omniscience in matters of measurement, and Ayrton's somewhat quixotic agenda of attempting to render all things in the electrical domain subject to measurement. Appointed Professor of electrical engineering at King's College London in 1890, John Hopkinson was another eminence who crossed swords with Ayrton. As Cambridge-trained mathematician and UK consultant for Edison's direct-current supply system Hopkinson was reknowned for his fierce intellect and tongue, as well as for his controversial propensity for applying Cambridge mathematics to electrical machinery where other engineers considered it to have no place. An important part of the electrical community in which Ayrton, Swinburne, and Hopkinson worked was the network of instrument-makers who not only produced the instruments used by them on a daily basis, but regularly commented on the measurements produced with their devices. Accordingly I set into context the roles of Alexander Muirhead, Sydney Evershed, Kenelm

---

[15] Anonymous, 'Prof. W.E. Ayrton F.R.S.', *Electrical Engineer*, 3 (1889), p. 66. For further details of Ayrton's life, see Graeme Gooday, 'William Edward Ayrton', *New Dictionary of National Biography*, Oxford: Oxford University Press, forthcoming 2004. Biographical sources used include Phillip Hartog, 'Professor W. E. Ayrton: A biographical sketch', *Cassier's Magazine*, 22 (1902), pp. 541–4; 'Anonymous, 'William Edward Ayrton, F.R.S.', *Electrician*, 28 (1892), pp. 346–7; Evelyn Sharp, *Hertha Ayrton, 1854–1923: A memoir*, London: 1926; John Perry, 'William Edward Ayrton', *Proceedings of the Royal Society*, 85 A (1911), pp. i–viii; John Perry, 'Obituary: William Edward Ayrton, F.R.S.', *Electrician*, 62(1908), pp. 187–8; William Mordey and John Perry, 'Death of Professor W.E. Ayrton', *JIEE*, 42(1909), pp. 1–6; John Perry, 'William Edward Ayrton, F.R.S.', *Nature* (London), 79 (1908), pp. 74–5.

Edgecumbe, Marcel Deprez, Edward Weston, and Sebastian Ferranti, as well as such companies as Acme, Hookham, Elliots, and Siemens.

Elsewhere in the book, well-known characters appear in less familiar guises. Norman Campbell, more famous for his philosophical studies of science, is cast as a disaffected critic of accounts of measurement he encountered in his student period at *fin de siècle* Cambridge. H. G. Wells makes a cameo appearance as an unruly trainee science teacher, subverting his South Kensington training in measurement and instrument-making practices. Whilst male characters are very much in the foreground of this study of late Victorian science and technology, I show that several women were closely and importantly involved in the electrical work of their spouses and families – albeit in ways that are still a great challenge for the historian to recover. Eleanor Sidgwick, Elizabeth Muirhead, Gertrude Ferranti, and Hertha Ayrton thus appear in my narrative, and I devote considerable attention in my last chapter to the role of Alice Gordon – more familiarly known as 'Mrs J.E.H. Gordon' – in promoting domestic electrical lighting to British women and men in the early 1890s.

It was of course William Thomson who once suggested that the topic of measurement was simply 'teeming with interest',[16] and I owe substantial debts to the two people who first got me interested in the history of electrical measurement. Much of what I know about the history of electricity I learned from Andrew Warwick's undergraduate classes at the University of Cambridge in 1985–6. And it was Crosbie Smith at the University of Kent at Canterbury who introduced me to the history of measurement while supervising the doctoral thesis in 1986–9 that is the very remote ancestor of this book. Without their long-lasting inspiration this book would never have been written. Norton Wise too has exerted a most benign influence on my research, not least by inviting me to a workshop at Princeton 1992 at which I could develop my arguments on the moral aspects of late Victorian measurement. Sophie Forgan has been a cherished collaborator on a number of related historical projects on institutions and gender issues. I have also drawn much inspiration from the work, hospitality, and friendship of Bruce Hunt, Ben Marsden, Kathy Olesko, and Jack Morrell who over the past decade have shared their views on Victorian science and engineering with me. Without their help, this book would have been greatly impoverished.

As a postgraduate and postdoctoral researcher at the University of Kent at Canterbury from 1986 to 1992, I benefited greatly from the staff and students in the friendly discussions there, notably Jon Agar, Yakup Bektas, Ana Carneiro, Alex Dolby, Ian Higginson, Ben Marsden, and especially Crosbie Smith. At the University of Oxford from 1992 to 1994 I had the immensely valuable support and interest of Robert Fox, Bill Astore, Eileen Magnello, Cassie Watson, Roger Hutchins, Agusti Nieto-Galan, Vivane Quirke, Marten

---

[16] William Thomson, 'Scientific laboratories', *Nature* (London), 31 (1885), p. 411.

Hutt, Giles Hudson, Willem Hackmann, and Tony Simcock. I enjoyed conversations with these and the many other students who attended the graduate history of science and technology seminars at the Modern History Faculty at which some early ideas for this book were presented. My thanks go especially to Giles Hudson for his stalwart assistance in tracing the publications of Peter Willans.

Since joining the Division of History and Philosophy of Science at the University of Leeds in 1994, I have benefited greatly from the collegiality of Greg Radick, Helen Valier, John Christie, Adrian Wilson, Steve French, Peter Simons, Richard Noakes, Jon Topham, Otavio Bueno, Anna Maidens, Sean Johnston, and Geoffry Cantor for their encouragement and support in writing this book. My special thanks to Chris Megone, Mark Nelson, Jennifer Jackson, and Matthew Kieran in the School of Philosophy for guiding my initiation into the teaching of practical ethics and in clarifying my understanding of the complex nature of 'moral' issues.

I would like to thank the following people who patiently read through particular draft chapters and gave me the benefit of their constructive and thoughtful critical responses: Steve French, Hasok Chang, Sophie Forgan, Ben Marsden, Jack Morrell, Greg Radick, Phillip Good, Kathy Olesko, Bruce Hunt, Neil Brown, Andrew Warwick, and Brian Bowers. Early versions of chapters of this book were presented at seminars in the Universities of Kent, Oxford, and Leeds; the Royal Institution, the Massachusetts Institute of Technology, and the Institute of Historical Research in London. I thank all those who gave me critical feedback at those seminars and I owe a particular debt to Jed Buchwald for inviting me to the Dibner Institute at MIT in 1993 to present the paper that became the basis for Chapter 5.

For general interest in my work and for generous and useful exchanges of ideas I'd like to offer very warm thanks to Jed Buchwald, Jeff Hughes, Nani Clow, David Edgerton, Sam Alberti, Rob Iliffe, Colin Hempstead, Elizabeth Silva, Sasha Roseneil, Neil Brown, Bill Aspray, Anna Guagnini, Steve Johnston, Peter Reffell, Janet Cunniff, Jim Bennett, Nathalie Jas, Greg Morgan, John Krige, Neil Brown, Simon Schaffer, Frank James, Lenore Symons, Anne Barrett, Sungook Hong, Will Ashworth, and Steve Lax. There are many others with whom I have talked at many conferences and seminars over the years, and I hope that they can forgive me for being able to offer them only a general acknowledgement of their valuable collective input to my research.

I was helped enormously by the library staff at the Universities of Kent, Cambridge, Oxford, and Leeds; and the Science Museum/Imperial College Library. Thanks go particularly to Kirstyn Radford, Pippa Jones, and Kate Alderson-Smith, and all those in the Edward Boyle Library and Brotherton Library (Special Collections) at the University of Leeds who fetched so many obscure and dusty volumes up with great good will and efficiency. I am grateful to Tim Procter, formerly of the IEE Archives, for pointing out to me the substantial collection of electrical manufacturers catalogues in the

Preface

Silvanus P. Thompson Collection held at the Archives of the IEE in London. My thanks to Lenore Symons of the IEE Archives for granting permission to quote from this collection.

In the course of researching this book over the last ten years, I received funding from the Institution of Electrical and Electronic Engineers (New York), the British Academy, the Royal Society, and the Arts Faculty of the University of Leeds. I am very grateful to all these bodies for the invaluable opportunities that this funding has provided.

Thanks to Cambridge University Press: Fiona Thomson for suggesting that I write this book and subsequently to Alex Holzmann, Mary Child, and Frank Smith for gentle encouragement and tolerance of the lateness of its arrival.

Finally my undying gratitude goes to family, friends, and cats who have put up with all the nonsense and chaos for more than a decade. But all errors and omissions are still entirely my own responsibility.

Graeme Gooday, Leeds, August 2002

1

Moralizing Measurement: (Dis)Trust in People, Instruments, and Techniques

The scientific community is morally superior to every other form of human associ ation since it enforces standards of honesty, trustworthiness and good work against which the moral quality of Christian civilization in general stands condemned.

Rom Harré, *Varieties of Realism*[1]

The scientific laboratory is also populated by a wide variety of inanimate agents: experimental apparatus, oscilloscopes, measuring instruments, chart recorders and other inscription devices.

At any time, the culture of the laboratory comprises an ordered moral universe of rights and entitlements, obligations and capabilities differentially assigned to the various agents.

Steve Woolgar, *Science: The Very Idea*[2]

Whom and what should people trust or distrust? This question has long been a prominent concern not only in everyday human transactions but also in the most abstruse domains of science, commerce, and technology. Both Steve Shapin and Ted Porter[3] have shown the significance of this question in the complex relationship between trust and quantification. They demonstrate that, to a certain extent, Restoration natural philosophers and nineteenth-century engineers were able to win greater trust for their claims by giving them quantitative expression. At the same time, though, Shapin and Porter map some of the important historical contingencies of the subject. Quantification has not always been achieved to the satisfaction of all, nor has it necessarily made claims uniformly more highly trusted by all parties. Therefore, to avoid facile transhistorical generalizations about the relations between trust and numerical work, the historian has to ask questions rather more socio-historically specific in nature. Why did a *particular* group of practitioners come to trust or distrust particular strategies for quantification? Why did

---

[1] Rom Harré, *Varieties of realism: a rationale for the natural sciences*, Oxford: Blackwell, 1986, pp. 1–2, 6–7. See discussion of this passage in Theodore M. Porter, *Trust in numbers: The pursuit of objectivity in science and public life*, Princeton, NJ: Princeton University Press, 1995, p. 218.
[2] Steve Woolgar, *Science: the very idea*, Chichester, England: Horwood, 1988, p. 102.
[3] Steve Shapin, *A social history of truth: Civility and science in seventeenth century England*, Chicago: University of Chicago Press, 1994. T. M. Porter, *Trust in numbers*.

I

they come to trust or distrust particular means of achieving quantification for certain specific purposes? How did they come to judge the trustworthiness of particular individuals and instruments to quantify faithfully? What standards of honesty and openness were required for quantitative claims to be trusted? Addressing questions in this contextualist vein, this book has a principal aim to explore the particular modalities of trust and distrust that pervaded the tricky and relatively novel enterprise of measuring electricity in the latter part of the nineteenth century.

Whereas Porter and Shapin have focussed on the trust relations between individuals, I extend the exploration of the intricacies of trust into the material culture of quantification. I look back to the development of electrical measurement instruments in the late nineteenth century and consider how considerations of trust were unavoidably part of the complex division of labour in the business of designing, making, and using such devices. Because this was not simply trust in individual humans, a major concern is to show that the evaluation of measurements made with technologies involved considerable indeterminacy about the location and reference of the trust. In the last section of this chapter, I explore how the subject of trust or distrust might be non-human: It could also be the hardware itself, the materials out of which it was made, the techniques used to make or use it, or the theories employed in interpreting its performance.[4] To the extent that judgement of the trustworthiness of measurements was about the trustworthiness of individuals, we shall see in Section 1.4 that such evaluations of trust were only *in part* moral judgements concerning honesty, honour, and fidelity. In preceding sections I argue for the significance of 'trust' as being at least as important as more commonly discussed themes in the history of measurement, namely, 'power' and metrology. Accordingly, I discuss the historical literature on issues of trust, quantification, and electrical measurement to show how my approach both builds upon and goes beyond previous work. Before that, however, I reappraise the significance of William Thomson's well-known claims about the close relation between measurement and knowledge.

1.1. WILLIAM THOMSON AND THE LIMITS OF MEASUREMENT

I often say that when you can measure what you are speaking about and express it in numbers you know something about it; but when you cannot measure it, when

---

[4] For a related project on the location of trustworthiness in twentieth-century computing, see Donald Mackenzie, *Mechanizing proof: Computing, risk and trust*, Cambridge, MA: MIT Press, 2001. My perspectives on how practitioners determine the properties of material culture owe much to Mackenzie's piece 'How do we know the properties of artefacts? Applying the sociology of knowledge to artefacts', in Robert Fox (ed.), *Technological change: Methods and themes in the history of technology*, London/Amsterdam: Harwood Academic, 1996, pp. 247–63.

you cannot express it in numbers, your knowledge is of a meagre and unsatisfactory kind.

Sir William Thomson, 'Electric Units of Measurement',
Lecture to the Institution of Civil Engineers, 1883[5]

Towards the end of the nineteenth century, measurement work had unequivocally become a collective enterprise. It was premised on a shared trust in the efficacy of measurement to capture important characteristics of natural phenomena and machinery. It was also embedded within an ever-diversifying division of labour among designers, makers, and users of measuring instruments, all importantly supported by technicians and assistants. Some individuals nevertheless had a higher profile than others did in this enterprise, and one such was William Thomson, elevated to the peerage as Lord Kelvin in 1892. As Professor of Natural Philosophy at the University of Glasgow from 1846 to 1899, Thomson was uniquely wide ranging in his activity on electrical theory, metrological standards, submarine telegraphy, power generation, electrical lighting, and domestic supply meters.[6] Even when his theories lost favour, respect remained for the instruments he developed in collaboration with the Glasgow instrument-maker James White. Specifically important were the electrical-measurement devices for telegraphic signalling and testing developed from the late 1850s and those for measuring the electrical performance of lighting and power two decades later. When Thomson presented new instruments at the Society of Telegraph Engineers and Electricians (STEE) in spring 1888, his protégé William Ayrton declared criticism of them to be 'out of the question', coming as they did from one revered as if belonging in 'another universe'.[7] Indeed such Thomson–White measurement instruments as the 'current balance' were canonical for standardizing laboratories some decades thereafter (Chapter 4). Yet can we infer from such praise that Thomson's broader pronouncements *about* measurement were equally authoritative or even unproblematic?

Thomson's most famous remark about the epistemological significance of measurement was delivered in a lecture to the Institution of Civil Engineers (ICE) in London in 1883. It amounted to the claim that in order to have a satisfactory knowledge of physical properties it was necessary to be able to measure them. Thomson was doubtless heard with some deference when

---

[5] William Thomson, 'Electrical units of measurement', in *Popular lectures*, London: 1891, Vol. 1, pp. 73–76, quotation from p. 73.

[6] Crosbie W. Smith and M. Norton Wise, *Energy and empire*, Cambridge: Cambridge University Press, 1989, pp. 445–94, 649–722; Graeme Gooday, 'Precision measurement and the genesis of physics teaching laboratories', *British Journal for the History of Science* (hereafter *BJHS*), 23 (1990), pp. 25–51, esp. pp. 29–36.

[7] William Ayrton in discussion of W. Thomson, 'On his standard inspectional instruments', *Journal of the Society of Telegraph Engineers and Electricians* (hereafter *JSTEE*), 17 (1888), pp. 540–67, quote on p. 559. There was indeed no criticism voiced against Thomson's instruments at this meeting.

he related how he himself had nearly single-handedly established the importance of 'definite electric measurement' in 1858. He retold the already familiar story of how he had persuaded manufacturers of the first transAtlantic cable to take more care in measuring the resistance of copper cables and gutta-percha insulation so as to optimize the performance and profitability of their phenomenally expensive project. His audience was probably not surprised, though, to hear Thomson expounding on the virtues and efficacy of measurement: After all, civil engineers considered submarine telegraphy to belong to *their* professional territory. Given the ubiquity of the standard resistance coil that Thomson himself acknowledged, it is clear that in the intervening quarter century the world had hardly failed to notice the consequences of applying Thomsonian measurement techniques to the telegraph industry. Notably, though, Thomson's 1883 lecture was more than just an autobiographical recapitulation of the pragmatic benefits of quantification. He also forecast that the commercial requirements of electric lighting would bring a similar 'advance' in the practical science of electrical measurement – a point to which I shall return in my concluding chapter.[8]

Ironically, however, Smith and Wise have shown that, at the time of his ICE lecture, Thomson himself was anyway not unequivocally committed to measurement as the only possible grounding for knowledge. Nor was he irrevocably wedded to the view that all practical or epistemological disputes could decisively be solved by a measurement. As is well known, Thomson had challenged the late James Clerk Maxwell's claims for the existence of a 'displacement current' in electromagnetic propagation on the grounds that this theoretically constructed entity was in principle unmeasurable and thus unintelligible. Measurability of itself was not persuasive evidence for Thomson, however, not even the measurable similarity of numbers. When Maxwell pointed out that measurements of the velocity of light were very close to those of the velocity of telegraph signals and the theoretically important ratio of electrostatic to electromagnetic units, Thomson initially doubted this constituted definitive evidence that light was an electromagnetic wave phenomenon.[9] Smith and Wise show, in fact, that physical intelligibility was *more* important to Thomson than measurability. In theorizing the vortex construction of atoms in the ether he drew heavily on phenomenological analogies of steam engines, telegraph lines, and turbine vortices drawn from the Glaswegian landscapes of manufacture and marine technology.[10] Thus

---

[8] Thomson, 'Electrical units', pp. 82–6.

[9] Smith and Wise, *Energy and empire*, pp. 445–94. Simon Schaffer, 'Accurate measurement is an English science', in M. N. Wise (ed.), *The values of precision*, Princeton, NJ: Princeton University Press, 1995, pp. 135–72. Note, however, that, in his 1883 lecture, Thomson did concede that the speed of light and the speed of electromagnetic waves were 'probably connected physically'; Thomson, 'Electrical units', p. 90.

[10] Smith and Wise, *Energy and empire*, pp. 396–444. Smith and Wise have emphasized that Thomson's interest in quantification was strongly driven by both a pragmatic secular

it would be unhelpful to treat measurement as the unique key to unravelling Thomson's endeavours, and indeed Smith and Wise carefully avoid this reductionist trap in their wide-ranging study of his career.

Even if he was ambivalent about the definitive power of measurement, Thomson was certainly highly critical about the alternative(s): Any claim to knowledge that could *not* be expressed in numerical form was 'meagre' and unsatisfactory. Although Thomson named no specific targets for this criticism, his audience might have discerned two possible candidates. One was the plethora of semi-popular and educational books then being published which engaged in unresolved speculations about the nature of electricity; these described many venerable traditional experiments without quantitative interpretation.[11] Another possible allusion was to the showy public demonstrations of these or other qualitative electrical experiments by well-known public lecturers such as John Tyndall and Henry Pepper. They perpetuated an older tradition of entertaining audiences with spectacular demonstrations of electrical effects. As Iwan Morus has shown, William Sturgeon and others in the early to the middle part of the century survived on fees earned from such theatrical display[12]; and this qualitative culture of electricity was only gradually displaced from the 1860s when transoceanic telegraphy made manifest the more lucrative value of rigorously quantifying electrical performance. Because, on Smith's and Wise's account, Thomson

---

concern for financial economy and a personal moral imperative to minimize waste of divinely endowed resources. Ibid., 248–9, 255–6, and esp. p. 684. For further discussion, see Crosbie Smith, *The science of energy: A cultural history of energy physics in Victorian Britain*, London: Athlone, 1998.

[11] Examples of an entirely qualitative experimental treatment of electricity are to be found in such textbooks as Edmund Atkinson, *Natural philosophy for general readers and young persons*, London: 1872 (adapted from Ganot's *Cours Elementaire de physique*); John Angell, *Elements of magnetism and electricity*, London/Glasgow: 1879; Frederick Guthrie, *Magnetism and electricity*, 1876. Note: Guthrie's revised edition of 1884 included a supplementary chapter (constituting about 15% of this edition) by his assistant, Charles Vernon Boys, which covered electrical machines and measurements.

[12] For a discussion of John Tyndall see William H. Brock, N. D. McMillan, and R. C. Mollan (eds.), *John Tyndall: Essays on a natural philosopher*, Dublin: Royal Dublin Society, 1981. On Pepper, see Kenneth Chew and Anthony Wilson, *Victorian science and engineering portrayed in the Illustrated London News*, Stroud, England: Sutton/Science Museum, 1993, pp. 11, 95, 97. For a detailed discussion of the culture of electrical display see Iwan Morus, *Frankenstein's children: Electricity, exhibition and experiment in early-nineteenth century London*, Princeton, NJ: Princeton University Press, 1998; idem, 'Currents from the underworld: Electricity and the technology of display in early Victorian England', *Isis*, 84 (1993), pp. 50–69; idem, 'Telegraphy and the technology of display: The electricians and Samuel Morse', *History of Technology*, 13 (1991), pp. 20–40; and David Gooding, 'In nature's school' in David Gooding and Frank A. J. L. James (eds.), *Faraday rediscovered*, Basingstoke, England: Macmillan, 1985, pp. 105–36. I am grateful to Richard Noakes for pointing out that William Crookes was almost certainly not the target of Thomson's criticism. For details of Henry Pepper's career, see J. A. Secord, 'Quick and magic shaper of science', *Science* 297 (2002), pp. 1648–9.

avowed a divine obligation to maximize efficiency, we might infer that he discouraged qualitative speculation and entertainment because they incurred the expenditure of resources without necessarily solving important technological problems.[13]

A notable irony throughout this book, though, is that measurement techniques borrowed from Thomson were not always sufficient to furnish unproblematic means of quantifying the performance of electrical technology. An injunction to measure in order to acquire a better knowledge about electricity was of itself simply not sufficient for these scientific communities to know how to proceed. There was no single obvious answer for Thomson's contemporaries to the question of what even constituted a measurement. As we shall see in Chapter 2, the execution of a measurement could be construed in three distinct ways that were not self-evidently equivalent. Measuring a quantity could involve a *comparison* with a standard pre-calibrated unit of the same kind until equality or balance was reached. Alternatively, it could involve a *reduction* of the unknown quantity to be determined into (absolute) determinations of length displacement and or mass, with a theory-laden calculation used to produce a final result. Then again, in the mid-1880s, the electrical lighting fraternity rather radically extended the meaning of the term 'measurement' to a high-speed practice that involved neither direct comparison nor simple reduction. Instead, this new approach used complex electromagnetic–mechanical techniques to deflect a needle or light-spot over a certain length of a dial pre-calibrated in the relevant units, so that users could take 'readings' in volts or amperes at an instantaneous glance. This new 'direct-reading' technology embodied the industrialization of measuring instruments. To achieve a faster (if somewhat fallible) result, 'automatic' apparatus replaced both the human labour hitherto required in experimental manipulation and the human skill formerly used in theoretical interpretation of instrumental action. Importantly, Thomson embraced this latter approach in his own electrical engineering work alongside the two more traditional approaches to measurement. Unlike some contemporaries in natural philosophy, Thomson did not publicly attack this new approach as constituting a practice that was *less* than authentic measurement.[14]

There were further questions left open by Thomson's public exhortation to engineers to measure, answers to which required shared commitments to value-laden decisions of measurement practice: *What* should they

---

[13] The original electrician on the 1858 Atlantic cable expedition, Wildman Whitehouse, was certainly soon marginalized when he refused to acquiesce in the new regimes of quantitative instrumentation; see Bruce Hunt, 'Scientists, engineers and Wildman Whitehouse: Measurement and credibility in early cable telegraphy', *BJHS*, 29 (1996), pp. 155–69.

[14] See Chapters 1 and 2 of this book, and Graeme Gooday, 'The morals of energy metering: Constructing and deconstructing the precision of the electrical engineer's ammeter and voltmeter', in M. Norton Wise (ed.), *The values of precision*, Princeton, NJ: Princeton University Press, 1995, pp. 239–82.

measure? How should they conduct their measurements? With what sorts of instruments? To what degree of accuracy? How should they judge the reliability of measurements made by others, and how should they interpret the outcome of their own measurement activities? Thomson's 1883 lecture offered little explicit advice on all but the first of these questions. Even on that subject he raised significant ambiguities about what should constitute the working ontology of electrical measurement. Should measurements of electrical current represent metals fundamentally as conductors of electricity or as inclined to resist such conduction? For Thomson the question had a pragmatic 'instrumental' solution. He suggested, for example, that those working on lighting installations ought to measure not the resistance of circuit elements but its reciprocal – conductivity – as that was far more algebraically useful in quantifying the performance of Edison – Swan filament lamps connected in parallel.[15] Thomson's proposal for the 'mho' as a unit of conductivity as the reciprocal of the 'ohm', did not gain wide currency, however until the 20th century. In Chapters 3–4 and 5 we shall see that there were several major debates on the contentious points just raised about measurement practice that not even Thomson's techniques, nor even personal interventions, could resolve.

To be more specific, there were several important difficulties in extending Thomson's quantifying practices and imperatives into the enigmatic new domains of electrical technology. Thomson's support for a plan by the London chemist Augustus Matthiessen to use a stable metal alloy in constructing resistance standards was insufficient to defeat Werner von Siemens' rival arguments for the trustworthiness of mercury (Chapter 3). The legitimacy of extending Thomson's 'mirror' techniques of galvanometry to instruments used for dynamo testing was strongly challenged by the aristocratic mechanical engineer James Swinburne in the 1890s (Chapter 4). The most determined advocates of a Thomsonian agenda to render measurable electrical parameters hitherto difficult or impossible to quantify were his Glasgow protégés, William Ayrton and John Perry, whose fertile collaboration from 1875 to 1889 produced a number of novel instruments, discussed in Chapters 4 and 5. But the attempt of Ayrton and Perry to develop a 'secohmmeter' to give direct-readings of self-induction faltered when even its developers acknowledged that even this commercially significant and undeniably 'real' quantity could not in principle be measured *at all* when the machinery was in motion. This was a distinctly limiting blow to the Thomsonian programme of equating the measurability of electrical phenomena with their 'reality'.

---

[15] Thomson, 'Electrical units', pp. 133–4. The 'Siemens' was first proposed as the unit of electrical conductance in 1933 but was not formally adapted as an SI unit until 1971. H. G. Jerrard and D. B. McNeill, *A Dictionary of Scientific Units*, London/New York, 4th edition, 1980, pp. 127–8.

The writ of Thomson's agenda was limited not merely in scope, but also in time. When he stood down from his Glasgow chair in 1899, new preoccupations were already diverting the attention of many researchers and students in physics and natural philosophy away from a narrow programme of measurement. Thomson's successor at Glasgow was his former student Andrew Gray, and we can perhaps gauge something of this broader transition away from measurement from the successive editions of Gray's *Absolute Measurements in Electricity and Magnetism*. In the first edition of 1884, Gray echoed in a celebratory tone a Thomsonian account of how telegraphy had stimulated the interest and expertise in measurement since the 1850s. But in his greatly reworked edition of 1921, Gray complained bitterly that, such was his fellow physicists' obsession with X-rays and radioactivity, they were now only interested in measurements that pertained to the telegraph's ethereal successor:

> ...if it were not for the needs of Wireless Telegraphy, I question whether the theory and practice of absolute measurements would at the present time command serious attention... As it is, we now have an army of students and others talking glibly of Einstein and of quantum theory, whose attention to the fundamentals of dynamics and physics has been wo[e]fully slight.[16]

Gray nevertheless resigned himself to addressing a large part of his physics textbook to the theory and practice of measuring the constants of coils used in wireless work: A 'difficult and thankless task', he opined melancholically.[17] As one of only a small minority of physicists still interested in electrical measurement per se, Gray thus acknowledged the persistent reliance of this specialist group on the activities of electrical engineers to provide the problems, techniques, instruments, and audiences for their research. Accordingly, I devote most of my attention to those in the engineering community whose work made electrical measurement both possible and important – both before physicists were greatly interested in the subject and indeed after most of them had lost interest in it. In the next section I thus consider the role of physicists and engineers in the early development of standardized technologies for measurement.

---

[16] Andrew Gray, *Absolute measurements in electricity and magnetism*, 2nd edition, Glasgow: 1921. Notwithstanding its abbreviated title, this was an updated version of Gray's *The theory and practice of absolute measurement in electricity and magnetism*, Parts 1 and 2 (issued in 3 volumes), London: 1888–1893. Gray published an earlier volume titled *Absolute measurements in electricity and magnetism*, London/Glasgow: 1884, 2nd edition 1889, and in the preface to that Gray presented an upbeat Thomsonian historiography of the telegraphic origins of electrical-measurement practice.

[17] Gray, *Absolute measurements*, 1921, p. v–vi.

1.2. THE METROLOGICAL FALLACY – OR WHAT THE HISTORY OF MEASUREMENT IS NOT

> But more is necessary to complete the science of measurement in any department; and that is the fixing on something absolutely definite as the unit of reckoning ... The great house of Siemens [has ...] worked upon this subject in the most thorough and powerful way – the measurement of resistances in terms of the specific resistance of mercury – in such a manner as to give us a standard which shall be reproducible at any time and place, with no other instrument of measurement at hand than the metre measure.
>
> Thomson, 'Electrical Units of Measurement,' 1883[18]

Thomson and his fellow electrotechnologists paid much attention to the topic of metrology. They were undeniably preoccupied with promoting the universal adoption of measurement units and developing material standards in which to embody and reproduce those units at any place and time. In their analyses of nineteenth-century electrical-measurement practice, Schaffer, Hunt, Olesko, and others rightly emphasize the epistemological significance of metrology in attempts to attain objective universal knowledge through measurement. Proper units and standards were considered important means for ensuring – or trying to ensure – that measurement produced universally valid numbers that represented authentic properties of nature or technology in ways unpolluted by material contingency or cultural subjectivity. Accordingly these historians show how electrical standards committees and individuals in both Europe and the USA expended much effort towards such ends from the 1860s until well into the twentieth century. These efforts were directed to arguing for the merits of one system of measurement units against rivals and to securing a definitive form of ultimate standard [*Urmaass*] that reliably embodied fundamental units of electrical measurement and the means of copying it into easily manageable everyday form that could perform consistently at all sites and over the *longue durée*.[19] The general explanandum of such accounts is the way in which any suitably skilled

---

[18] Thomson, 'Electrical units', pp. 87, 94.

[19] Simon Schaffer, 'Late Victorian metrology and its instrumentation: A manufactory of ohms', in Robert Bud and Susan E. Cozzens (eds.), *Invisible connections: Instruments, institutions, and science*, Vol. IS09 of the Society of Photo-Optical Instrumentation Engineers (hereafter SPIE) Institute Series, Bellingham, WA: SPIE, pp. 24–55; Kathy Olesko, 'Precision, tolerance, and consensus: Local cultures in German and British resistance standards', Dordrecht, Boston/The Netherlands/London: Kluwer, 1996, pp. 117–56; Bruce Hunt, 'The ohm is where the art is: British telegraph engineers and the development of electrical standards', *Osiris*, 9 (1994), pp. 48–63; Arnold C. Lynch, 'History of the electrical units and early standards', *Proceedings of the Institution of Electrical Engineers* (hereafter *PIEE*), 132A (1985), pp. 564–73, see esp. p. 568; Larry Lagerstrom, 'Universalizing units: The rise and fall of the international electrical congress, 1881–1904', unpublished manuscript. Personal communication, November 1994.

person could use such units and standards to replicate any given measurement result. By implication, this replication would be to within a tolerable degree of uncertainty: Quite what constituted a tolerable degree of uncertainty for the late Victorians will be discussed in my account of the context dependence of 'accuracy' in the next chapter.

Nevertheless, historians of electrical metrology have also highlighted two themes that undercut any simplistic interpretation of it as a pure enterprise of epistemological universalization. With no small irony, they note that putatively universal systems for calibrating measurements drew on very particular culturally embedded values[20] to lend their metrological schemes meaning and legitimacy. Olesko has argued that, when Werner von Siemens promoted a unit of resistance measurement grounded on a metre column of mercury of cross section 1 mm² in the 1860s, he did so with a view to its congruence with, and easy integration into, existing length-based metrological schemes in the Germanic states. By contrast, (mostly) British Association (BA) physicists followed the agenda of Wilhelm Weber and William Thomson in articulating all units of electrical measurement in dimensions of mass, length, and time in the 'absolute' universal framework of energy transference articulated in the new thermodynamics. Accordingly they promoted the rival absolute unit of resistance which had the (somewhat counterintuitive) dimension of 'velocity'.[21] Whilst BA lobbyists thus criticized Siemens' resistance unit for its lack of intrinsic connection to other electrical units, the Prussian replied caustically that only a tiny constituency of energy-obsessed physicists needed to use the absolute unit, which was in any case very hard to realize in practice (Chapter 3). Siemens' important point was that most users of such resistance standards worked in the telegraphic sphere and would use them primarily for fault diagnosis and quality control rather than to make laboratory claims about the universal nature of electricity. I extend this theme by showing how similarly *localized* technological purposes stimulated the development of commercial units and standards for current (Chapter 4) and self-induction

---

[20] Jan Golinski, *Making natural knowledge: Constructivism and the history of science*, Cambridge: Cambridge University Press, 1998, pp. 173–7; Schaffer, 'Late Victorian metrology', p. 27; Olesko, 'Precision, tolerance, and consensus', pp. 117–56; Hunt, 'The ohm is where the art is', pp. 48–63; Lynch, 'History of the electrical units, p. 568.

[21] (The acronyms BA and BAAS refer to the same body: The British Association for the Advancement of Science. They are used interchangeably.) Within the BA's favoured 'electromagnetic' (vis-à-vis electrostatic) system of absolute units, resistance had the dimensions of 'velocity' rather than of length and was thus not straightforward to embody in a permanent material form. In this system the British Association for the Advancement of Science unit figured as 10^9 cm/s and, according to the most recent data available to Thomson in 1883, the Siemens unit was 9.413×10^8 cm/s. For Thomson's elaborate efforts to explain how resistance could be interpreted as velocity see Thomson, 'Electrical units', pp. 97, 130–3. On the unification of physics through considerations of energy, see Gooday, 'Precision measurement', pp. 36–7.

(Chapter 5); the grounding of electrical measurement in absolute energetic considerations became broadly significant *only* with the commercial advent of domestically metered supply in the 1890s (Chapter 6).

Insofar as historians have (rightly) focussed on the importance, problems, and ironies of developing such standards and units, they have written histories of *metrology* – what Thomson called the 'science of measurement'. These are not, however, strictly the same as the histories of measurement, for in a sense 'measurement' encompasses and yet also goes far beyond metrological topics. This differentiation has sometimes been obscured by use of the word 'metrology' as a loose synonym for measurement.[22] More unhelpfully, this elision has perhaps tempted some to come close to committing what I call the 'metrological fallacy'. This is the view that well-defined universal standards and units are somehow necessary and sufficient to facilitate the practice of measurement and thus that the history of measurement consists in explaining how past measurers overcome the lack thereof. Although no single historian has argued explicitly for this position, undertones of it are arguably apparent in some of the attempts to locate the development of standards and units at the forefront of narratives on the history of measurement.[23] I contend, by contrast, that the historiography of measurement cannot be simply about standards of measurement. This is because historical evidence shows that universally agreed standards of measurement are neither necessary nor sufficient for a particular quantification to be judged to be a proper measurement by expert commentators.

First I put the case that units and standards were not of themselves sufficient resources for making reliable electrical measurements. Performing an electrical measurement required much more than having access to a reliable and universal set of standards and units. Also needed were the relevant skills, measuring technologies, and discretionary (decision-laden) practices, and these by no means inevitably accompanied the use of such standards. As Harry Collins argued some time ago, and Otto Sibum has recently reiterated, the mere fact that two different experimenters possess identically calibrated and otherwise standardized apparatus does not generally enable them to attain quantitatively identical results. Much tacit skill and 'gestural knowledge' is also required for accomplishing even similar results, and this can normally be learned only face-to-face from skilled practitioners.[24] Even with such direct-emulation practical knowledge, it is entirely normal to find slightly discrepant results between prima facie identical measurements.

---

[22] This is a tendency for which I have been guilty – see Gooday, 'The morals of energy metering', p. 240, where I spoke misleadingly about 'metrological communities'.
[23] Golinski, *Making natural knowledge*, pp. 186–206.
[24] Harry Collins, *Changing order: Replication and induction in scientific practice*, London: Sage, 1985; Otto Sibum, 'Reworking the mechanical value of heat: Instruments of precision and gestures of accuracy in early Victorian England', *Studies in History and Philosophy of Science*, 26 (1995), pp. 73–106.

Indeed, ironically, slight discrepancies between experimenters' results elicit *greater* communal trust than claims for an exact digit-for-digit consonance between them: The latter is widely presumed to be the hallmark of fraud.[25] As Thomas Kuhn pointed out over three decades ago, experimenters habitually *expect* errors to arise; if the spread of results falls within their expectations of unavoidable error, this mild discordance is deemed to lie within the limits of 'reasonable agreement'. By appealing to such canonical standards for reasonable numerical (dis)agreement, non-identical outcomes of the 'same' measurement need not *necessarily* be seen to be in mutual conflict. By recovering the ways in which late Victorian practitioners arrived at and deployed thresholds for what constituted 'sufficient accuracy' or 'sufficient degree of accuracy', we will see in Chapter 2 how they coped with the insufficiency of metrological standards to guarantee agreement between otherwise identical measurements.

The insufficiency of putative standards to bring 'agreement' between measurements is most apparent when the criteria for reasonable agreement require measurements of a reliability or sensitivity that is close to the limits of what can be accomplished at the most sophisticated levels of contemporary practice. At such levels there can be a suspension of trust in innovative practices and techniques that makes the use of even the most precisely constructed standards uncompelling grounds for the acceptance of results. This was notably the case with attempts to standardize readings of current-reading devices and domestic meters during the 1880s and 1890s in which trustworthiness was only cautiously and selectively attributed to new devices (Chapters 4 and 6). The problem was more serious still when there was an outright breakdown of trust between practitioners engaged in measurements who used nominally the same standards. When practitioners did not trust the reports, techniques, or integrity of rivals engaged in measurements who used the same physical standards, the rendering of numerical results became unavoidably partisan. In the 1860s Augustus Matthiessen and Werner von Siemens (as well as their respective allies) repeatedly challenged each others' claims to be able to replicate their respective material resistance standards to within 0.1%. Matthiessen's vitriolic attacks on Siemens' claims regarding the trustworthiness of mercury columns were matched by Siemens' challenges to the trustworthiness of Matthiessen's alloy resistances; the reconciliation of these claims took decades to accomplish.

The use of widely distributed standards was thus insufficient for the consensual conduct of measurement with them. Following the lead given by Shapin and Porter, I contend rather that any discrepancies in the use of standards can become problematic if there is significant distrust among the

---

[25] This is arguably an obvious extension of what Thomas Kuhn, dubbed the 'fourth law of thermodynamics'; Thomas Kuhn, 'The function of measurement in modern science' in *The essential tension: Selected studies in scientific tradition and change*, Chicago: University of Chicago Press, 1977, pp. 178–224.

practitioners concerned. Such subtleties of trust in measurement equipment and in those who design, make, or use them have perhaps been understated in previous historical accounts. These subtleties arise from a complex of personal and commercial issues. For example, when individuals were trusted to undertake certain tasks competently and to report them honestly, this was not, by convention, a point that needed to be mentioned explicitly. Moreover, certain kinds of skill and specific dedicated characters seem to have been tacitly accorded particularly high levels of trust in this regard, most notably Lord Rayleigh and his collaborators, Arthur Schuster and Eleanor Sidgwick, in the early 1880s. Nobody contested their conclusion that the hitherto widely used BA 1865 unit was a full 1.3% adrift from its ideal absolute value.[26] And once they had vindicated Werner von Siemens' long-held claim that the resistance of his standard mercury column could be replicated to within 0.1% when 'necessary' precautions were taken, William Thomson at least contended that no further evidence was required for proving the point. Yet, as we shall see in Chapter 3, not all British practitioners accepted that mercury standards could *by themselves* be trusted to provide such a level of reasonable agreement, especially when used by those with rather less expertise than Rayleigh and Sidgwick.[27]

My next point is that universally sanctioned metrological standards were in other contexts not even strictly *necessary* for attaining trustworthy measurement. This was particularly so for everyday transactional purposes in which toleration of 'error' was much greater and or in which the results were merely local in import. The ubiquity of this localism is an important counter to an extreme view of the metrology dependence of society in Keith Ellis's popular treatise *Man and Measurement*. With confident counterfactuality he maintains that if societies did not use widely shared and uniformly regulated standards, everyday commercial transactions would become so contentious and prolonged that civilization would give way to anarchy as starving people took to the streets and rioted.[28] Yet clearly this apocalyptic vision is not borne out by historical evidence. Indeed, as Witold Kula and

---

[26] Thomson, 'Electrical units', pp. 93–7, referring to Lord Rayleigh, and Arthur Schuster, 'On the determination of the ohm in absolute measure', *Proceedings of the Royal Society*, 34 (1883), pp. 104–41, and to more recent results announced in 1882 by Lord Rayleigh and Eleanor Sidgwick, published as 'Experiments by the method of LORENZ, for the further determination of the absolute value of the British Association unit of resistance', summarized in *Proceedings of the Royal Society*, 34 (1883), pp. 438–9; the complete version was published in *Philosophical Transactions of the Royal Society*, 174 (1884), pp. 295–322.

[27] One might present this as an alternative to Harry Collins' point that controversies in matters of replication are ended by the fiat of a 'core set' of practitioners exercising sufficient power to decree what must be the definitive answer to a contested research question; Collins, *Changing order*, pp. 142–9. Judging by the response of fellow practitioners in the case of resistance standards, though, this core set did not simply or even necessarily have to be powerful: It had to be *trusted* by those over whom it claimed jurisdiction. See subsequent discussion of the relationship between trust and power.

[28] Keith Ellis, *Man and measurement*, London: Priory, 1973, pp. 1–2.

Sally Dugan[29] have shown, ordinary consumers at different locations have long used their highly localized standards of length and weight and interconverted their quantitative results in commercial transactions as easily as they could exchange foreign currencies – if not always without squabbles, challenges, or occasional fraud. Thus a locally reproducible standard could be *sufficiently* effective for many purposes, irrespective of its relationship to other sorts of measurement for other purposes and quantities. Although its integrity and universality might be challenged, societies did not permanently collapse into disorder as a consequence. If such disruption broke out in the electrical world – because of discordant experiments, failed cables, or extinguished lighting – many other possible causes might be suspected than a slight uncertainty in the value or definition of an electrical standard.

Much electrical activity in the nineteenth century proceeded with a proliferation of localized and fallible standards and units without major disaster ensuing. For example, in Chapter 3 we shall see that, even into the 1870s, telegraph companies still used a variety of different proprietary standards in the testing and laying of landlines. And despite the palpable variations between many resistance coils allegedly representing the *same* unit, submarine cables were tested, laid, and retested without calamitous inefficiencies or financial strain. In his 1883 ICE lecture, Thomson noted that, although Rayleigh and Sidgwick had shown the BA 1864–5 resistance unit to be 1.3% adrift from its absolute value, thousands of commercial coils copied from it were used with considerable efficacy in telegraphy and lighting as all copies agreed *with each other* to within 1/10%.[30] Given also that converting results to alternative systems was a relatively straightforward procedure, 'arbitrary units' of electrical resistance in miles of iron or copper wire were effective for many purposes. After all, throughout the period covered by this book, the quantitative value of standards was being contested. Yet the everyday measurement work of telegraph clerks, lighting engineers, and experimental physicists continued almost entirely unperturbed by the endless wrangling among national or international standards committees. The only significant aggravation they endured was when they were obliged to give up their cherished standards and either purchase new forms with fractionally different values or perform extra correctional calculations.

Thomson himself significantly sanctioned a pragmatic pluralism about measurement standards. He advised the assembled engineers at the ICE in 1883 that they could usefully express electrical resistance of a conductor in *either* the BA unit and the Siemens mercury unit. Notwithstanding his vocal advocacy of the absolute BA unit for the preceding two decades, Thomson

---

[29] Witold Kula (trans. R. Szreter), *Measures and men*, Princeton, NJ: Princeton University Press, 1986; Sally Dugan, *Measure for measure: Fascinating facts about length, weight, time and temperature*, London: BBC Books, 1993.
[30] Thomson, 'Electrical Units', pp. 93–8.

took particular pains to praise the Siemens product. This is not surprising given that, at the Paris international conference for the determination of units in October the previous year, form of the Siemens unit had been adapted to become the material embodiment of the theoretically absolute universal 'ohm' standard.[31] Although this form of standard was technically ratified as an internationally agreed form by all the major industrial nations represented at the meeting, this did not bring the end of metrological pluralism. As O'Connell has noted, this apparent unification in a universal system of measurement was largely rhetorical: Several European nations explicitly avoided upholding this compromise. Whilst France and Germany continued to use the mercury definition without explicit reference to the absolute system, Britain persistently avoided the use of mercury in its metallic standards – preferring a solid metal form instead (Chapter 3). As Lagerstrom has shown, the international committees that adjudicated universal standards of electrical measurement from the 1860s to the 1910s tried repeatedly and rarely with success to enact many diplomatic compromises – especially in the naming of units – to prevent their enterprise disintegrating into nationalistic factionalism.[32]

Given this sustained socio–cultural resistance to universal definitions, the principal explanandum of a history of measurement in the nineteenth century *cannot* be the universal adoption and implementation of a single unified system of units and standards. Whilst many of those involved indeed *aimed* for this universalization, to tell the story of electrical units and standards with the focus on only this narrative theme would be to tell an unacceptably teleological tale directed to explaining what happened only later (if at all) in the twentieth century. Thus, although my approach draws upon the rich extant literature on the history of metrology, I focus on the questions of trust in measurement that are quite distinct from those of defining a standard metrology. What sort of materials, methods, instruments, and people were trustworthy enough to employ in the task of making reproducible quantitative measurements? It is these sorts of quotidian issues in the *practice* of measurement that I shall pursue in later chapters.[33] In the next section I

---

[31] Ibid., pp. 93–8.
[32] Lagerstrom, 'Universalizing units'; J. O'Connell, 'Metrology: The creation of universality by the circulation of particulars', *Social Studies of Science*, 23 (1993), pp. 129–73; Golinski, *Making natural knowledge*, pp. 173–77.
[33] Golinski, *Making natural knowledge*, p. 9; Joseph Rouse, *Knowledge and power: Toward a political philosophy of science*, Ithaca, NY/London: Cornell University Press, 1987; Joseph Rouse, *Engaging science: How to understand its practices philosophically*, Ithaca, NY/London: Cornell University Press, 1996; Jan Golinski, 'The theory of practice and the practice of theory', *Isis*, 81 (1990), 492–505; Andrew Pickering (ed.), *Science as practice and culture*, Chicago/London: University of Chicago Press, 1992; Jed. Z. Buchwald (ed.), *Scientific practice: Theories and stories of physics*, Chicago/London: University of Chicago Press, 1995.

explain why accounts of measurement practice can more cogently appeal to considerations of trust than can narratives centred on power.

1.3. RIVAL NARRATIVES OF QUANTIFICATION: NETWORKS OF TRUST VERSUS CENTRES OF POWER

Practices must be distributed beyond the laboratory locale and the context of knowledge multiplied. Thus networks are constructed to distribute instruments and values which make the world fit for science. Metrology, the establishment of standard units for natural quantities, is the principal enterprise which allows the domination of this world.... Standardization is therefore an obvious concern for historians of science. The physical values which the laboratory fixes are sustained by the social values which the laboratory inculcates.

Simon Schaffer, 'Late Victorian Metrology and Its Instrumentation'[34]

Metrology-based histories of measurement typically place much explanatory weight on the centralized power of standardizing laboratories. To explain the (alleged) universalization of quantitative knowledge, they appeal to the efficacy of such metrological centres[35] to 'dominate' the rest of the world with their particular schemes of quantification. By contrast, I focus more attention on the persistent localization of measurement practices and the bonds of 'trust' between practitioners. Indeed, to understand the extent to which institutions or individuals have any effective 'power' over large-scale processes of quantification, I argue that we should consider the degree of trust that other practitioners accorded to their utterances and actions, not the capacity of eminent individuals to dominate their contemporaries. My approach is thus very much unlike Bruno Latour's provocative depiction of 'centres of calculation' in *Science in Action* and differs in important respects from Simon Schaffer's analysis of the Cavendish laboratory as a putative centre for electrical standards for late nineteenth-century Britain and its colonies.

Some critics of Bruno Latour's *Science in Action*, notably Lorraine Daston, see it as epitomizing a view of science premised on a particular notion of power.[36] In *Science in Action*, Latour indeed presents 'technoscience' as a

---

[34] Schaffer, 'Late Victorian metrology', p. 23

[35] Bruno Latour, *Science in action. How to follow scientists and engineers through society*, Milton Keynes, England: Open University Press, 1987, pp. 250–1. Ted Porter comments 'There is something deeply right about Bruno Latour's phrase "center of calculation" to describe the point from which empires are administrated and about his emphasis on this in a book about technology and science.' See Porter, *Trust in numbers*, p. 224.

[36] Lorraine Daston, 'The moral economy of science', *Osiris*, 2nd series, 10 (1995), pp. 3–24 (see p. 6, n. 12). For Latour's laudation of the political shrewdness of Eastman and Pasteur and the implicit scorn of Diesel and Koch, see Latour, *Science in action*, pp. 104–7, 110–11. For more subtle accounts of the role of power in science, see Stanley Aronowitz, *Science as power: Discourse and ideology in modern society*, Basingstoke, England: Macmillan, 1988; John Christie, 'Aurora, Nemesis and Clio', *BJHS*, 26 (1993), pp. 391–406. I thank Jeff Hughes for pointing out that Latour's book can be read as a normative rather than descriptive treatise as

quasi-Machiavellian game in which dominant figures manipulate networks of allies and resources to win credit for new discoveries and technologies. In his concluding account of 'Metrology', however, Latour presents a subtly different account of how quantitative knowledge is promulgated. There his explanation assumes the deference of whole populations to the power of a centralized institution. Latour invokes the agency of the 'centre of calculation' to account for the management not just of quantification but of *all* efforts at universalizing knowledge. Metrology for him is an exercise in bureaucratic power: It is the gigantic and 'frightfully expensive' enterprise of transforming the world outside the laboratory into a paper-administered form in which laboratory-generated facts and machines can safely survive. Infallible self-registering instruments that he labels 'immutable mobiles', previously calibrated at the centre, generate paper registrations which are then administered to a presumptively subordinate external world by a vast stable paper-based bureaucracy. It is these continuous long-range paper trails by which facts and machines are controlled and kept in calibration with the centre's standards. Whatever the plausibility of applying this 'big-picture' approach to science in the late twentieth century,[37] there are certainly at least two specific major problems raised in applying this sort of model to the account of nineteenth-century electrical measurement discussed in this book. The first problem is his emphasis on the medium of paper to bear the results of measurement practice – a point I shall be at pains to criticize in Chapter 4 and will accordingly discuss no further here.

The second such problem in a Latourian account is the identification of candidates for the so-called centres of calculation. Even at the very end of the nineteenth century, no country had an unequivocally authoritative 'central' laboratory for quantitatively researching and calibrating *all* electrical phenomena – notwithstanding efforts by their denizens to make them so. David Cahan has shown that the Physikalische-Technische Reichanstalt opened in a Berlin suburb in 1887 struggled hard to gain credibility and influence in its early years. Its diverse founders continually disagreed over the nature of the 'central' role it should play for the German State, and in the 1890s its

it is based on lectures that he gave to engineering students at the Ecole des Mines, Paris, on how to succeed in their professional careers. See particularly Latour's prescription of 'Rules of method', in *Science in action*, Appendix 1, p. 258. Latour's more recent publications have not focussed so exclusively on success; indeed, one notable discussion of his on 'failure' is *ARAMIS, or the love of technology*, Cambridge, MA/Landon: Harvard University Press, 1996. For a discussion on the historiography of failure see Graeme Gooday, 'Re-writing the "book of blots": Critical reflections on histories of technical failure', *History and Technology*, 14 (1998), pp. 265–91.

[37] See, Latour, *Science in action*, p. 251. For the late twentieth century, Latour implicitly cites as an example of such a centre of calculation the US National Bureau of Standards, which spends a large annual sum on maintaining the constancy of physical constants – the sort of work that is conventionally encompassed by the term 'metrology'. It is not at all clear that this sort of laboratory promotes metrology in Latour's extended used of the term.

18   *The Morals of Measurement*

director, Friedrich Kohlrausch, had great difficulty preventing its delicate measurement operations from being wrecked by the passage of nearby trams.[38] In Britain, various bodies vigorously promoted their electrical research programmes and development of standards; notable among these were William Thomson's laboratory in Glasgow and the Cavendish laboratory in Cambridge under James Clerk Maxwell and then Lord Rayleigh. Neither, however, became the dominant centre of national and imperial measurement activity. When the Board of Trade Electrical Standardizing Laboratory in central London was established in 1891, it served only strictly limited purposes in commercial testing and statutory ratification until such work was transferred to the National Physical Laboratory along with a new research role in 1900–1.[39] These British laboratories were not – *pace* Latour – the prerequisites of quantification, but rather a later response to the demands of practitioners in academia and industry who had *already* long been measuring without reference to an infallible 'centre of calculation' anyway. To capture the decentred power relations of electrotechnology in late Victorian Britain, we can borrow from Peter Galison's vision of late twentieth-century physics. Both consisted of disunified and dispersed sites of expertise in which no single academic institution was dominant, and yet their diverse practices still mutually interacted in a collective fashion through specific arenas of encounter which Galison labels 'trading zones'.[40]

---

[38] David Cahan, *An institute for an empire: The Physikalisch-Technische Reichaustalt, 1871–1918*, Cambridge, UK: Cambridge University Press, pp. 29–58, 126–75.

[39] In 1889 prominent members of the British electrical profession appealed to the Board of Trade for financial support for a standards laboratory, and soon afterwards this was set up in six rooms in the basement of No. 8 Richmond Terrace, No. 56 Whitehall. From 1891 to 1894 this is where the practical electrical standards required by the relevant 1889 legislation on electrical measurement were determined (and then implemented in further legislation in August 1894). The sole purposes of this lab were these: 'First, to obtain and preserve standards for the measurement of electrical quantities; second, for giving the standard measurements of those quantities; and, third, to enable the Electrical Adviser of the Board of Trade to make such tests of instruments and material as may be necessary for the performance of his duties. No scientific work outside these purposes can be undertaken.' 'The Board of Trade electrical standardizing laboratory', *The Electrician*, 33 (1894), pp. 665–6. For further details see also IEE, *Electrical handbook for London*, London: IEE, 1906, p. 2. On the National Physical Laboratory, see Edward Pyatt, *National Physical Laboratory*, Bristol: Hilger, 1983, and M Eileen Magnello, *A century of measurement: An illustrated history of the National Physical Laboratory*, Bath: Canopus, 2000.

[40] For discussion of the disunified nature of much of modern science and technology see Peter Galison and David J. Stump (eds.), *The disunity of science: Boundaries, contexts, and power*, Stanford, CA: Stanford University Press, 1996. For Galison's account of 'the trading zone', see Peter Galison, *Image and logic: a material culture of microphysics*, Chicago/London: Chicago University Press, 1997, pp. 781–844 and Peter Galison, 'Contexts and constraints' in Jed Buchwald (ed.), *Scientific practice, theories and stones of physics*, Chicago/London: Chicago University Press, 1995, pp. 13–41, esp. 35–9. In this context it is interesting to note that the Dynamicables Club was set up in the 1880s for manufacturers of dynamos, mica insulation, and telegraph cables to dine amicably together to ameliorate the divisive

How then are we to understand the relation of such subcentres to the practice of electrical measurement in other commercial and academic settings? An important distinction must be drawn between the exercise of power inside and that outside the institutionalized centre of measurement. As Schaffer notes in his study of Cambridge physics in the 1870s, Maxwell and his assistants exercised considerable power over the inexperienced students within the walls of the Cavendish laboratory through the supervision of their training. The acquisition of a wide range of intellectual and bodily skills was necessary to manipulate devices into orderly behaviour; to differentiate between artefactual and authentic results of measurements; and to discern the differences among 'kicks', 'shoves', and 'jerks' in a galvanometer's performance. Maxwell and his laboratory staff largely determined which instruments his 'captive' students would use, how they would use them, how they would report their measurement outcomes, and what standards of accuracy were appropriate in such reportage. From the testimony of such students, we know the efficacy of closely managed supervision of training in measurement, and that acquiescence in such training was normally a prerequisite for joining the community of expert electrical experimenters.[41] Whilst we can easily see how such power operated within the Cavendish laboratory, what of practitioners of electrical measurement who worked beyond its walls? What or who governed their decisions on these points when they were not under the jurisdiction of Maxwell *et al.*?

To answer this question, Schaffer takes as his explanandum the matter of how Maxwell and others tried to set up a uniform practice of electrical measurement throughout Britain and its colonies. To explain this, he appeals not to a Latourian bureaucracy of paper trails, but identifies five distinct features of the socio–political administration of experiment. The same types of instruments and standards apparatus as used at the Cavendish were widely deployed owing to the effective means of their commercial distribution. The efficiency and (arguable) uniformity of a widespread programme of laboratory training explains how all electrical practitioners in industry and academia practiced electrical measurement in a similar way. The authority of textbooks such as Friedrich Kohlrausch's *Leitfaden der Praktischen Physik* and *Practical Physics* by the Cavendish's own R. T. Glazebrook and W. N. Shaw acted as textual epitomes for those researchers needing guidance on the canons of measurement practice. The direct lines of administrative communication between laboratories and telegraph testing stations on ships facilitated a continuity of values and practice between the academic centre

---

effects of commercial competition; see Rookes E. B. Crompton, *Reminiscences*, London: 1928, pp. 110–11.

[41] See Schaffer, 'Late Victorian metrology', pp. 33–4, *A history of the cavendish laboratory, 1871–1910*, London: 1910, which is a collection of essays. For evidence of the effects of subverting a regime of physics laboratory teaching, see discussion of H. G. Wells in the next chapter.

and the commercial periphery. Finally, even for those who had not been trained in an academic laboratory and who were working outside the networks linking commercial practice to the labs, there was always the authority of eminent laboratory managers to prescribe what they should buy and do. Schaffer suggests that such individuals as W. E Ayrton in London, Andrew Gray in Glasgow and Bangor, and Glazebrook and Shaw in Cambridge 'told' their customers and extramural students where to buy the 'right' robust instruments, where to learn the 'right' techniques, and which 'morals' were appropriate for their work.[42] In this last claim we find echoed the power-laden assumption of Schaffer's opening claim that laboratory-based metrology enabled the 'domination' of the world beyond the laboratory walls.

Schaffer's account captures very evocatively how these laboratory professors would have *liked* to accomplish the universalization of electrical practice. He cautiously avoids, however, claiming that they succeeded in *achieving* this goal. Whilst some practitioners of electrical science and technology doubtless did learn their mode of electrical practice as outlined in Schaffer's account, I present evidence in this book that not all of them did so. This is especially the case with those who were trained not in the laboratory but through the mechanical engineering workshop apprenticeship in civil engineering, telegraph installation, or autodidactic learning. These characters did not feel compelled to subordinate their conduct to the grand pronouncements of self-appointed laboratory authorities. The figure of James Swinburne will feature prominently in this regard in Chapters 4–6 as a mechanically trained aristocratic engineer who persistently challenged the laboratory-based authority of Ayrton *et al.* As we shall see in Chapter 2, in the years after he left Cambridge, the disaffected Norman Campbell was also severely critical of the nature of the Cavendish training he received in measurement techniques in the 1890s. Moreover, the controversies documented in Chapter 3 reveal the persistent *impotence* of those laboratory-based professors in the British Association Electrical Standards Committee to determine completely the measurement practices and standards of those working beyond the Committee's networks of influence.

In the broad literature that expounds the notion of 'science as power',[43] there is a notable lack of emphasis of such significant *failures* in attempts to wield social power in science and technology. The ubiquity of failure should not be surprising, however, if we consider the implications of Marger's and Olsen's concise analysis of social power. These scholars have plausibly contended that social power consists in the ability to affect the actions and

---

[42] Schaffer, 'Late Victorian metrology', pp. 29, 38, 43. Schaffer highlights the irony of the way that practitioners spelled out such valorized particulars of laboratory material culture whilst also maintaining that the 'absolute system' of measurement depended on no particular instrument, technique, or institution.

[43] Aronowitz, *Science as power*; Christie, 'Aurora, Nemesis and Clio'.

beliefs of others, *despite* their resistance.⁴⁴ The obvious corollary is that the resistance of others defines the limits of individual and institutional power: Such 'authorities' learned the circumscribed scope of their influence when confronted by the opposition or indifference of others. Accordingly we cannot accept an 'impetus' theory of power to account for the characteristics of late Victorian electrical measurement: We cannot suppose that professorial pronouncements about proper techniques of measurement carried enough rhetorical 'momentum' to cause the capitulation of all those on whom they impinged. On this account, one would not prima facie expect the imperatives issued by Maxwell from the Cavendish Laboratory or by William Ayrton inside his London City and Guilds laboratories to have provoked deferential acquiescence of those outside their immediate spheres of influence.⁴⁵ Insofar as such external practitioners did accept professorial pronouncements about how to measure, we need instead to understand rather why there was an *absence of resistance* to their claims. This is more readily explained, I contend, in terms of trust rather than of centralized power. Those who trusted the judgements of Maxwell and Ayrton – those personally acquainted as friends, colleagues, collaborators, or former students – were most likely to give a charitable judgement of their trustworthiness on such matters of measurement. Conversely, the many other people who had not had the opportunity to judge their trustworthiness at such close quarters might be expected to exercise less immediate compliance to their authority, perhaps exhibiting even an outright indifference that constituted a form of social resistance.⁴⁶

Yet, to understand the central role of trust in science and technology, we cannot consider just the trustworthiness of individual testimony. The credibility of claims about measurement – whether about how to measure, or the authentic result of a particular measurement – was not adjudicated by reference to the trustworthiness of persons alone. Claims about the trustworthiness of measurement encompassed not just human judgement but also the appropriateness of an instrument, technique, or materials for achieving a particular kind of measurement in a given context, plus other auxiliary decisions about how reliable quantification ought to be accomplished. Thus if the results of a measurement result seemed suspect, perhaps by being difficult to replicate to within an acceptable degree of reasonable agreement, it was not always self-evident what or whom in the set-up should be considered problematic. By extending the Duhem–Quine thesis here to the topic of trust,

---

44 See Marvin Olsen and Martin Marger (eds.), *Power in modern societies*, Boulder, CO/Oxford: Westview, 1993, pp. 1–2.
45 For a critique of essentialist theories of power in relation to technology see John Law (ed.), *A sociology of monsters: Essays on power, technology and domination*, London: Routledge, 1991; see especially Bruno Latour, 'Technology is society made durable', in ibid. pp. 103–31.
46 Annette Baier has observed that the nature of trust in any interpersonal relationship is closely related to the relative balance of power within it. Annette Baier, 'Trust and antitrust', *Ethics*, 96 (1986), pp. 231–60, esp. 234–44.

we can see that in such situations it might not be a straightforward matter to determine whether the specific source of the untrustworthiness was the measurer, instrument, or some other factor in the experimental context. As I explore later in this chapter and in Chapters 2 and 3, the complex division of labour and distribution of responsibility in measurement recurrently creates such *ambiguities* (perhaps as interpretive flexibility) in how to make judgements of trustworthiness or attributions of untrustworthiness.[47]

I argue for the contingency and heterogeneity of the subject of trust. By problematizing the subject of trust in this way, my account contrasts with Ted Porter's analysis in *Trust in Numbers*. Porter argues that over the past two centuries the public has come to regard 'impartial' numerical evidence as more intrinsically trustworthy than the testimony of experts. He offers examples from the history of financial administration to show how sceptical publics sometimes tried to wield democratic power – not always successfully – to get auditable numerical facts to replace fallible human judgements.[48] By contrast, my study is not about the trustworthiness of numbers versus experts for neither ever became unproblematically trustworthy of themselves. Mine is a study of *which* experts, *which* measurements, and *which* techniques were proposed as trustworthy and why judgements on such matters could be accepted or rejected. Some sceptical electrical engineers, for example, persistently distrusted the potentially treacherous readings of unreliable electrical-measuring devices and diagnosed the condition of electric-lighting installations by using bodily skills to gauge the heat and brightness of light bulbs (Chapters 2 and 4).[49] In other cases, specifically the measurement of electrical resistance (Chapter 3) and electrical currents (Chapter 4), there was persistent dissent over which approach posed fewer risks of misleading results. Some researchers considered that certain electrical parameters were altogether beyond measurement, notably self-induction in moving machines (Chapter 5) and a household's actual consumption of electric lighting (Chapter 6).

---

[47] The Duhem–Quine thesis asserts that a single experiment can never definitely disprove a theory because an apparent disparity between experimental result of theoretical prediction might be attributable to a hitherto unidentified false theoretical assumption in the experiment's conduct and interpretation. See Sandra G. Harding (ed.), *Can theories be refuted? Essays on the Duhem–Quine thesis*, Dordrecht, The Netherlands: Reidel, 1976.

[48] According to Porter this move to quantification reflected 'weakness and vulnerability' on the part of the 'distrusted' expert community; *Trust in numbers*, p. xi. There could hardly be a greater contrast with Latour's claim that quantification follows the contours of centrally managed metrology.

[49] Corporeal 'self-evidence' of this sort is as historians of the body have recently noted, a hitherto neglected topic that can yield rich evidential rewards. See Schaffer, 'Self-evidence', *Critical Enquiry*, 18 (1992), pp. 327–62, and Christopher Lawrence and Steve Shapin (eds.), *Science incarnate: Historical embodiments of natural knowledge*, Chicago: University of Chicago Press, 1998.

For many situations, though, the prospect of trustworthy electrical measurement was considered prima facie plausible. And insofar as measurement was considered possible, my interpretation focusses on the way that experimenters tried to deal with the irreducibly material problems of assessing and managing fallible and perhaps previously unknown people, substances, instruments, and techniques. Rather than deferring to a definitive centralized authority, I suggest that such practitioners gauged the trustworthiness of people and devices that produced numbers by various means: Experimental testing, institutional debate, face-to-face encounters, and the reports of other trusted practitioners. Later I examine the way that the ever-increasing division and dispersion of labour in electrical manufacture and testing in the late nineteenth century made it ever more difficult to evaluate the trustworthiness of a measurement accomplished by the compounded labour of several individuals at different sites. Before considering that, let me take a step back now to address in more detail the issue of how the historian can construe the meaning of trust and particularly the *moral* connotations of trustworthiness.

1.4. MORAL ECONOMIES OF TRUST AND QUANTIFICATION

Although trust is an obvious fact of life, it is an exasperating one. Like the flight of the bumblebee or a cure for hiccoughs, it works in practice but not in theory. When we think about it, the obvious fact that, on the whole, we manage to live together in mutual confidence turns mysterious.

Martin Hollis, Trust within reason[50]

In examining how instruments and materials were the subjects of trust or distrust in the past, I am pursuing a relatively unexplored path. Most historical, philosophical, and sociological scholarship on the issue of trust has been almost exclusively devoted to the questions concerning trust among people.[51] There is broad assent to Sissela Bok's thesis that trust among human beings is a necessary (if not sufficient) condition for the flourishing of collective human activities.[52] In his *Social History of Truth*, Steve Shapin argues in this vein that there could be no 'science' *qua* universalized telling of authenticated 'truths' about the world unless there were a high degree of mutual trust among the many persons in the complex and dispersed division of scientific labour. Practitioners simply have to trust the secondary

---

[50] Martin Hollis, *Trust within reason*, Cambridge, UK: Cambridge University Press, 1998, p. 1
[51] Baier argues, for example, that 'Trust, the phenomenon we are so familiar with that we scarcely notice its presence and its variety, is shown by us and responded to by us not only with intimates but with strangers, and even with declared enemies.' Baier, 'Trust and antitrust', pp. 233–4.
[52] Sissela Bok, *Lying*, New York: Pantheon, 1978, p. 31.

testimony of others if they are ever to learn about how the world works beyond their own narrow area of expertise. On matters which they have not witnessed or otherwise participated in, scientists can make reliable judgements only by knowing how they could reliably judge *who* would not mislead them.[53] Thus for Shapin, as for Bok and Baier, the bond of trust among people is strongly imbued with moral significance. According to the perspective one adopts, trustworthy conduct is of moral value because it embodies such virtues as honesty and truthfulness, or because it is an absolute duty not to mislead others, or because collective trustworthiness is a prerequisite for the maximized welfare of the majority.

Before the force of the moral issues in trust is examined in more detail, it should be noted that interpersonal trust is not an entirely moral matter. Shapin notes that a standard distinction between two major components of trust is that between the normative and the predictive. Once this distinction is granted, it is obvious that some kinds of interpersonal trust have no obvious moral content whatever. The 'weak' enactment of trust is the inductive view that we can anticipate the future behaviour of people and things by assuming they will act as (reliably as) they have in the past. This fulfilment of inductively generated expectations is the non-moral sense of trustworthiness: A person cannot be 'blamed' if entirely reasonable expectations of regularity are confounded by unavoidable mishap or unforeseen contingency. Shapin extends this *amoral* notion of trust to the non-human domain too. He contends that it would make no sense to feel betrayed or wronged if our trust in the weather to turn out well were later confounded by experience. This purely epistemic form of trust is not particularly interesting for Shapin as it had been widely explored in methodological analyses of 'induction' in traditional philosophy of science. We shall see later, however, that there are greater subtleties to be explored in this matter, particularly in the way that late nineteenth-century scientists and engineers often used quasi-moralistic language in evaluating the trustworthiness of instruments and materials.[54]

Shapin is much more concerned with a rather stronger 'moral' notion of trust among people. This is the notion of trust as a relation that is both predictive *and* normative, indeed, it is predictive largely because it is normative. Trusting people in this sense is to have faith in the honesty and integrity of their personal conduct and in their fidelity in reporting of it, and this faith is premised on the assumed *obligation* of individuals to behave in this reliable way. If individuals did otherwise than they had promised, misled others

---

[53] Indeed, granted only a minor typographical alteration, Shapin's work could be seen rather as a *Social History of Trust*.

[54] Shapin, *Social history of truth*, pp. 7–8. See Hollis, *Trust within reason*, pp. 10–13, for a cognate discussion of the distinction between 'expecting of' and 'expecting that' and of the arguable artificiality of drawing a distinction between them.

about what could be expected of them, or willfully misreported their performance of a task, all would be breaches of normative trust. Failure to fulfil this trust is blameworthy, and Shapin moves quickly from the blameworthiness of betrayers of trust to the claim that normative trust is 'morally textured'. He declares that to trust people is to perform an act that is 'moral' in form: It is to always make a judgement about their moral character as trustworthy and how worthy they are of our confidence in their integrity.[55]

He immediately raises a challenge, however, to the traditional notions of the putatively moral nature of trust. Normative sanctions against, for example, lying, slander, or fraud are commonly assumed to draw their force from an appeal to transcendent moral imperatives, not merely from fiats of cultural convention. Yet, notes Shapin, cogent challenges have been made to this distinction. Some philosophers even concede it might just be anthropocentric projection or wishful thinking to maintain that universal ethical imperatives are anything but delusory reinventions of prudent civic regulation.[56] In this vein Shapin declines to differentiate between ethics and etiquette, that is between obligations that are based on morals and those grounded on manners. When he writes of the 'morally consequential' character of trust he means that trust is quite literally the 'great civility'. According to Shapin, this is the sharing of presuppositions about the self, the world, and others that enable communal discourse about a 'world-known-in common' in the 'moral fabric' of ordinary social interaction. For the purposes of this study I maintain the historian's strategic agnosticism about whether morality can be reduced to purely social considerations.[57] Nevertheless I consider a particular strength of Shapin's highly socialized notion of trust and morality is that it grounds a coherent (social) epistemology of trust. It enables him to provide one historically specific answer to the central question with which I started this chapter: Who should be trusted and why?

In Restoration Britain, Shapin argues, the answer to the question of whom to trust lay primarily in judgement of socio–political class. He contends that gentleman philosophers were the accredited arbiters of how the world was, and they trusted each other to offer only reliable testimony. Shapin offers a detailed account of how Robert Boyle and the network of Protestant gentlemen to which he belonged shared a well-established code of social engagement in which 'trust' in the honesty and openness of character was paramount in the conduct of natural philosophy as in all matters. Such Protestant gentlemen were expected not to intentionally mislead another, although they might sometimes be licensed to harbour doubts about the

---

[55] Shapin, *Social history of truth*, p. 38.
[56] Hollis, *Trust within reason*, p. 11. Whilst entertaining the plausibility of this view, Hollis himself is notably uneasy about giving up this distinction.
[57] Shapin, *Social history of truth*, p. 36. It is sufficient that the characters portrayed in what follows took trustworthiness to be an obligation in ways that went beyond mere self-interest or arbitrary convention.

testimony of those who they deemed to have questionable credentials as gentlemen, notably Catholics and mathematicians. By contrast, Shapin contends, those at the lower end of the social spectrum could not be trusted by gentleman at all in any circumstances. The testimony of artisans was specifically untrustworthy because, as paid servants, their independence of moral character could not be guaranteed; Robert Hooke's intermediate social status as the Royal Society's experimental servant clearly generated some painful equivocation on this matter.[58] And tellingly, in Shapin's account, there was no point in asking questions about whether artisans trusted the testimony of their social superiors – lower social orders had no power to constitute themselves as a tribunal in such matters.

Shapin's identification of an association of trustworthiness with high social status is avowedly a contextually specific claim. Whilst aristocrats in later Victorian Britain might well have expected to be treated with some deference in science and technology, this association was considerably weaker than in Boyle's time. Certainly the bashful Lord Rayleigh enjoyed the confidence of many physicists on matters pertaining to electrical measurement, and the prospective hereditary baronet James Swinburne certainly talked as if he had similar authority on many matters of engineering.[59] But this social status was not enough by itself to accord these figures the status of trustworthiness in all matters. In Chapters 4 and 5 we shall see that the reliability of testimony was also judged by reference to experience and *training*. Edison's senior British advisor, John Hopkinson, was, like Rayleigh, a product of the Cambridge Mathematics Tripos,[60] whereas Swinburne had passed through a mechanical engineering apprenticeship. The former group downplayed the value of judgements of the 'practical men' unversed in calculus, and the latter doubted that 'mathematicians' had adequate 'hands-on' industrial experience of electrical machines to understand the problems of quantifying their complex behaviour.

Distrust of technical testimony on the grounds of insufficient experience did not *necessarily* imply, however, any kind of moral judgement against the practitioners in question. Antagonists could and sometimes did question the 'predictive' trustworthiness of particular instances of each others' work or instruments *without* raising questions of personal integrity. There were important exceptions, however. Distrust of measurements could sometimes

---

[58] Shapin, *Social history of truth*, pp. 36 and 392-3.
[59] For a discussion of aristocratic engineers in this period see David Cannadine, *Decline and fall of the British aristocracy*, London/New Haven, CT: Yale University Press, 1990, pp. 396-7. I thank Frank James for this reference. For Swinburne's life and aristocratic pedigree, see F. A. Freeth, 'James Swinburne, 1858-1959', *Biographical Memoirs of the Royal Society*, 5 (1959), pp. 253-64.
[60] On the Cambridge Mathematics Tripos in the nineteenth century, see Andrew Warwick, *Masters of theory: Cambridge and the rise of mathematical physics*, Chicago: Chicago University Press, 2003.

arise from suspicions about the dishonesty, unfairness, or self-delusion of those making (putative) measurements. Such suspicions were typically aired when issues of institutional power, personal reputation, or commercial success were at stake. In such situations the trustworthiness of measurements could become a considerably more fraught and morally loaded business. Quantitative claims of those in the lucrative business of producing resistance standards (Chapter 3) or domestic meters (Chapter 6) could be attacked quite independently of their social status or training. The consequences of such moralistic attacks could also be long-lasting and complex, leaving a sustained inheritance of distrust or at least factional division. Accordingly, it is easy to see why trustworthiness – by whatever means it was adjudicated – was an important virtue in late Victorian electrical practice.

Before examining what it might mean for trustworthiness to be a virtue in electrical practice, let us consider the more general relation between moral virtue and practice. In *After Virtue*, Alisdair Macintyre depicts a practice – science, for example – as a cooperative creative activity which promotes 'goods' which are internal to that practice. As such, a practice is an activity which not only promotes worthy ends, but also extends the shared ability to achieve such ends in a morally virtuous manner, implicitly according to the Aristotelian canon of such virtues as courage, liberality, and truthfulness. In Macintyre's account, a practice of science or technology is thus inevitably a virtue-laden enterprise: It is by reference to the virtues that practitioners define their relationships to those with whom they share the 'kinds of purposes and standards which inform practices'.[61] Focussing specifically on the case of science, Lorraine Daston has argued similarly that the exercise of certain moral and intellectual virtues is a *prerequisite* of science as a collective 'moral economy'. Shared moral values maintain some form of equilibrium among a community of practitioners, and the communal valorization of certain specific virtues explains the coherence and collectivity of quantification far more cogently than do Latourian accounts of Machiavellian individualism. She argues that such putative intellectual 'virtues' as accuracy, precision, impartiality, and communicability can mould the practice of quantification in quite distinct ways in different contexts.[62] For example, whereas Ted

---

[61] Alisdair Macintyre, *After virtue: A study in moral theory*, 2nd edition, London: Duckworth, 1985, pp. 187–203, quote from p. 191. By Macintyre's criteria of creativity and virtue, science and history are both bona fide practices, whereas planting turnips and bricklaying are not. Macintyre's account of moral virtues and the contrast with 'intellectual' virtues draws heavily on Aristotle's *Nichomachean ethics* (available in many different editions).

[62] Daston, 'The moral economy of science'. Daston stresses that a moral economy is not a set of neo-Mertonian 'norms', nor is it synonymous with 'ideology': moral economies 'moralize scientists' into a stable practice, whereas ideologies 'moralize nature' to serve social interests; ibid. pp. 6–7. Shapin similarly alludes to the moral economy of science in Shapin, *Social history of truth*, pp. 289–90. E. P. Thompson originally used the term 'moral economy' to refer to the communal action by the eighteenth-century British agrarian poor to force

Porter's research shows that bureaucratic precision often aims at impersonal objectivity in the public interest, Kathy Olesko's study of physics teaching in nineteenth-century Germany shows that a major concern of precision measurement can be the integrity of the individual measurer.[63]

We can easily accept Daston's general point about the importance of shared virtue in a self-identifying community of practitioners engaged in quantification. Nevertheless, the scholarship of Shapin and Golinski shows that we cannot easily identify precision per se as a transhistorical virtue. For Restoration natural philosophers, 'precision' was associated with unwonted privacy and obsession, and for phlogistic chemists a century later it signified pedantry rather more than veracity. Indeed, Norton Wise argues that precision came to be associated with virtue only with the rise of *industrial* cultures of quantification in the nineteenth century.[64] In Chapter 2 I explore the meanings of precision and accuracy in this period as forms of intellectual virtue. These terms pertained only to moral virtue in measurement practice by means of their relation to the normative features of trustworthiness. Trust in a measurement claim depended on the extent to which the measurer was judged to have followed morally virtuous conduct – with honour and truthfulness – in relation to fellow practitioners. This meant excluding misleading sources of bias and error, reporting all results with all due openness and honesty, and making due reference to reliance on the work of fellow practitioners. Enacting such virtuous conduct was a major hallmark of trustworthiness and lent credibility to claims about accuracy and precision.[65]

Whilst the collective enterprise of electrical measurement hinged on the shared umbrella virtue of 'trustworthiness', it is important to bear in mind the politically fraught question of *whose* trust and trustworthiness actually mattered. It is all too easy for the historian to focus on members of a

---

land-owning elites to fulfil their acknowledged obligation to keep food prices at a 'fair' level. Edward P. Thompson, 'The moral economy of the English crowd', [1970], in *Customs in common: Studies in traditional popular culture*, London: Merlin Press, 1991, pp. 185–258.

[63] Daston borrows from Olesko's account of Neumann's physics teaching at Königsberg during the mid-1830s, in which initiates distrusted unobserved interpolations on graphs and sifted colleagues' results according to how trustworthy their work was deemed to be, thus creating a relatively unsociable and even idiosyncratic mode of quantification. See Kathy Olesko, *Physics as a calling*, London/Ithaca: Cornell University Press, 1991, pp. 61–127; Daston, 'The moral economy of science', pp. 3, 6, 11.

[64] Shapin, *A social history of truth*, pp. 310–54; Jan Golinski, '"The nicety of experiment": Precision of measurement and precision of reasoning in late eighteenth century chemistry', in M. Norton Wise (ed.), *The values of precision*, pp. 72–100; M. Norton Wise, 'Precision: Agent of unity and product of agreement – Part II, The age of steam and telegraphy', in ibid., pp. 72–100.

[65] In terms of moral virtue, accuracy could denote the qualitative 'care' expended in a measurement as well as the quantitative 'closeness to truth' that such care was presumed to guarantee. The numerical detail of this outcome was cited to within a 'degree of accuracy', only later designated as 'precision', appropriate to context. See Chapter 2 for more discussion.

voluble small elite and overlook the role of collaborators, technicians, assistants, and spouses who contributed to the rest of this complex division of labour.[66] The contribution of such less visible characters *did* matter. If any one of them were suspected of a significant slip or indiscretion, the trustworthiness of the measurement outcome would become moot. A relevant irony here emerges from Hollis' observation (see epigraph at the beginning of this section) that the ubiquity of communal trust is hard to reconcile with the highly *individualistic* accounts of human accomplishment that preoccupy Western culture. When attributing trustworthiness in the accomplishment of certain complex and important tasks, why do we so often attribute the relevant trust to prominent individuals when the labour undertaken in such tasks is so clearly collective in nature? Historians of the book have noted, for example, that it is customary to identify a unique author for a text, notwithstanding the complex cumulative and often collaborative nature of the processes involved in the long process of writing and publishing books.[67]

Such a deconstruction of claims for authorial uniqueness has already been extended beyond the literary sphere into the complex process of creative research. Shapin and Galison have undermined the notion that the title 'experimenter' can generally be applied to a single individual, just as Brannigan and Schaffer before them showed the arbitrariness of applying the epithet 'discoverer' to a unique fetishized heroic figure.[68] A directly analogous point can be made regarding the authorship of measurement: The specific ironic force of Hollis's comment is clearest in claims by *end-users* of a measuring instrument to be the unique 'author' of a measurement undertaken with it. Such a claim paradoxically gained plausibility only if the labours of those that had earlier designed, made, and calibrated such instruments were *so* utterly trustworthy that their contribution to the measurement process could easily be forgotten. Put another way, the performance of mundane and remote tasks with great trustworthiness could lead to the erasure of the bearer of such trustworthiness from the historical record of accomplishment. It was

---

[66] For discussion of the 'invisible technician' see Steve Shapin, 'The invisible technician', *American Scientist*, 77 (1989), pp. 554–62, and Shapin, *Social history of truth*. For recovery of 'wives' as 'invisible technicians', see the complexities and varieties of collaborative marital relations discussed in Helena M. Pycior, Nancy G. Slack, and Phina G. Abir-Am (eds.), *Creative couples in the sciences*, New Brunswick, NJ: Rutgers University Press. 1996. For an excellent case study in the instrument-making business, see Alison Morrison-Low, 'Women in the nineteenth-century scientific instrument trade', in Marina Benjamin (ed.), *Science and sensibility: Gender and scientific enquiry, 1780–1945*, Oxford: Blackwell, 1991, pp. 89–117.
[67] See Adrian Johns, *The nature of the book: Print and knowledge in the making*, Chicago: University of Chicago Press, 1998.
[68] Shapin, 'The invisible technician'; Peter Galison, *Image and logic: A material culture of microphysics*, Chicago/London: University of Chicago Press, 1997; Augustine Brannigan, *The social basis of scientific discoveries*; Schaffer 'Scientific discoveries and the end of natural philosophy', *Social Studies of Science*, 16 (1986), pp. 387–420.

such a self-aggrandizing amnesia among many end-users of measuring instruments that Norman Campbell took great pains to challenge (Chapter 2).

Given this problematic issue of inclusivity – of who should or should not be granted some shared responsibility for the outcome of measurement experiments – we should not assume any simple unity in a 'moral economy' of quantification. This is all the more so given that, as previously noted, different means of establishing trust were maintained by practitioners from different backgrounds – effectively creating fragmented subcommunities.[69] We shall see in the latter part of Chapter 4 that some were so sceptical of the performance of instruments made and used by other companies, that they preferred to undertake as many measurements as possible within their own institutional premises. This brings me to the most conspicuously heterogeneous component of the quantifying enterprise: the material culture of experimental measurement. In the final section we see how some interesting asymmetries arise in the attribution of trustworthiness and untrustworthiness to humans or non-human agents in quantification processes. The historical contingency in such matters can be seen clearly by comparing Shapin's account of the Restoration experimental philosophy with Steve Woolgar's account of the twentieth-century laboratory.

1.5. TRUST AND THE MATERIAL CULTURE OF MEASUREMENT

The Thomson and the Schuckert meters are perfectly applicable to alternate currents ... This [sic] is one of the very few meters that follows a simple law, and may therefore be considered trustworthy for alternate currents.

James Swinburne, 'Electrical Measuring Instruments', Lecture to Institution of Civil Engineers, 1892[70]

How are we to recover the respective statuses of human and material instruments in the generation of past forms of trustworthy quantification? The two mentioned in the preceding epigraph receive an interestingly asymmetrical role in Shapin's account of the social epistemology of trust in seventeenth-century gentlemanly philosophers. If a Restoration practitioner failed to produce results that could be construed as being in 'reasonable agreement' with others, a highly warrantable explanation lay in blaming the imperfections of instruments or the incompetence of artisan assistants. These agents entered into the moral economy of trust as entities in their own right only when experimental results were in doubt, an instrument or a technician being

---

[69] See Galison, *Image and logic*, for an account of science as 'an intercalated set of sub-cultures bound together through a complex of hard-won locally shared meanings', p. 841.

[70] James Swinburne, 'Electrical measuring instruments', *Proceedings of the Institution of Civil Engineers*, 110 (1892), pp. 1–32, quote on p. 28.

Moralizing Measurement 31

represented as the strategic bearer of the burden of distrust. However, when the 'right' answer was achieved, the responsibility for the 'rightness' lay unambiguously with the gentleman experimenter, no positive virtue being imputed to the role of the experimenters' tools in accomplishing this outcome. The implication here is that when Boyle's instruments worked appropriately they were either a morally neutral feature of laboratory hardware or a trivial and epistemologically 'transparent' extension of the experimenter's agency.[71] If we follow Shapin's reading of Boyle, there was thus no notion in Boyle's 'moral economy' of an instrument that bore, of itself, the hallmark of 'trustworthiness' – still less that it could meaningfully be labelled a 'precision' instrument. So in Shapin's account there were no trustworthy instruments *per se* in the moral economy of Restoration natural philosophy.

By contrast, according to Steve Woolgar, late twentieth-century instruments certainly are accorded a place in the 'moral order' of the laboratory. As a laboratory ethnographer, Woolgar examines how scientists construct 'representations' for the outcomes of experiments, and particularly the ambiguous role of instruments in such constructions. Nobel prize-winners, for example, are often intriguingly ambivalent about whether to represent themselves as responsible for extracting truths from the world or whether the relevant agency lies with trustworthy instruments, alongside which they stand as impartial operatives. Woolgar observes that many such scientists talk about the culture of the laboratory as if it comprised an ordered moral universe of entitlements, obligations, and capabilities differentially assigned to various agents.[72] Before a new instrument can be accorded the trust necessary for participation in the laboratory work of representation, it is put through a series of trials to become an 'adequately socialized' member of the community. By the reputation that it builds up from such results, an instrument may displace older techniques and become definitively trustworthy for its purpose.[73] Woolgar himself, however, noticeably keeps an ironic distance from this account, refusing to acquiesce in scientists' recurrent anthropomorphization of instruments as having a capacity for 'honesty' or 'treachery' in representing the world. Moreover, he adopts Shapin's scepticism about the distinction between the moral and the social, reducing the normative component of an instrument's trustworthiness to its obedience to socialized norms.

Scientists' attribution of trustworthiness to instruments is not a uniquely twentieth-century phenomenon. This we can see from the epigraph that opens this chapter in which James Swinburne described two particular alternate-current (ac) meters as being considerably more trustworthy than

---

[71] Shapin, *A social history of truth*, pp. 310–11.
[72] Woolgar, *Science: The very idea*, p. 102. Although Woolgar does not present any direct case studies, this is indeed a plausible and familiar sort of talk from Woolgar and Latour, *Laboratory life*.
[73] Woolgar, *Science: The very idea*, pp. 102–3.

others. However, the process of making instruments trustworthy was not as simple as contriving their initial socialization once and for all into the laboratory order. This was in part because instruments of the sort discussed by Swinburne would spend much of their working lives far beyond the orderly world of the laboratory. And even for those instruments which were destined to spend more time in the laboratory, workers observed that even highly familiar and finely tuned instruments could show deeply recalcitrant behaviour: They did not always behave as they 'ought'. J. J. Thomson made this statement in his presidential address to Section A of the BAAS in Liverpool in 1896:

Any investigation in experimental physics requires a large expenditure of both time and patience; the apparatus seldom, if ever, begins by behaving as it ought; there are times when all the forces of nature, all the properties of matter, seem to be fighting against us; the instruments behave in the most capricious way, and we appreciate Coutts Trotter's saying that the doctrine of the constancy of nature could never have been discovered in a laboratory.[74]

Thomson's mildly facetious use of normative 'ought' language shows how the everyday practice of dealing with instruments could involve more than just extracting regularized performances from known reliable instruments. In new investigations or unfamiliar contexts, hitherto 'trusted' devices could suddenly behave in the most perplexing and subversive fashion. Only with much varied experience of handling instruments over considerable periods of time could enable judgements to be made of their trustworthiness, especially of the conditions in which they were most likely to behave in a trustworthy fashion. It was only after extensive experience of many kinds of domestic meter that James Swinburne came to his conclusion about the only two forms that could be deemed trustworthy for monitoring ac supply. And that experience supported the view that only instruments with the *simplest* construction would have an intrinsic character that would not be misleading. Here then we have a valuable clue on how to recover the meaning and significance of the trustworthiness of instruments. Whereas Woolgar reduced the trustworthiness of instruments to their degree of socialization, I argue that this trustworthiness was judged rather more contingently by late Victorian instrument users according to three factors: the stability of materials used, the integrity of the theory employed, and the calibre of the construction. All of these factors would make measurements and their associated hardware far from simple to handle.

[74] Coutts Trotter was Fellow of Trinity College, Cambridge. For further discussion of this quote see Graeme Gooday, 'Instrumentation and interpretation: Managing and representing the working environments of Victorian experimental science', in B. Lightman (ed.), *Victorian science in context*, Chicago: University of Chicago Press, 1997, pp. 409–37.

1.5.1. Trust in Material Substances

Karen Knorr-Cetina has incisively noted that the everyday practice of science is an unavoidably decision-laden business.[75] Early practitioners of electrical measurement indeed had to make a tricky decision about which *particular* conducting medium or 'substrate'[76] they should use to mediate the phenomenon they sought to quantify. Although the human body could serve as a conductor for specific physiological, therapeutic, or punitive purposes, it proved too hard to calibrate in a reliable non-solipsistic manner. For the sorts of measurement discussed in this book, the conducting medium used was generally inorganic and, depending on the purpose and context of the measurement, could be a solid (metal or carbon), a fluid (mercury or electrolyte), or a gas (highly rarified). Given the particularity of the medium required and the history of the material sample employed, questions arose about the extent to which the outcome of any measurement was affected by the idiosyncrasies of the conducting material chosen – especially in the methods chosen for its preparation. The problem was not a trivial one. The recurrent experience of electricians and chemists was that the most versatile and malleable conducting materials were also the most susceptible to extrinsic influence. Thus a skilful trade-off had to be made between stability and convenience in choosing a conducting material.

Those familiar with the mercury, copper, and iron in electrical conduction in a range of different situations soon found that, although these were inexpensive, readily procured, and easily shaped, their performance was prone to the contingencies of handling and environment.[77] The performance of the (anomalously) fluid metal mercury was particularly sensitive to impurities and easily amalgamated with juxtaposed metals; the behaviour of solid-metal wires was dependent on their previous history of handling, their age, and impurities (Chapter 3). Magnetically susceptible iron was prone to complex time-dependent effects after carrying currents, and the performance of all materials to some degree varied with temperature – sometimes quite dramatically. The challenge for an experimenter using such materials was to

---

[75] Karin K. Knorr-Cetina, *The manufacture of knowledge: An essay on the constructivist and contextual nature of science*, Oxford: Pergamon, 1981.

[76] I use the term 'substrate' in the philosophical sense of a substance capable of supporting attributes and accidental (non-essential) properties, as opposed to the rather different uses of the term in the fields of geology, biochemistry, biology, and linguistics.

[77] See my preceding account of Shapin for a discussion of this in the case of Robert Boyle. For nineteenth-century problems of using the body to generate consistent electrical results, see J. Senior, *Rationalizing electrotherapy in neurology, 1860–1920*, Oxford: University of Oxford, 1994, unpublished PhD thesis. My work here follows the model of Bruce Hunt in studying the development of techniques to measure the electrical properties of insulators, notably in relation to submarine cables. Bruce Hunt, 'Insulation for an empire: Gutta Percha and the development of electrical measurement in Victorian Britain', in F. A. J. L. James (ed.), *From semaphore to shortwaves*, London: Royal Society of Arts, 1998, pp. 85–104.

minimize the effects of these chaotic material-dependent phenomena in the numerical outcome of a measurement. To borrow from Schaffer's analysis of Newton's uses of optical instruments,[78] the challenge was to render these materials 'transparent' so that any trait of their performance could be trustworthily attributed simply and only to their electrical or electromagnetic condition to be measurement. Rendering a material transparent in this way by suitable human intervention and manipulation, electrical-measuring devices could be relied on to produce results that unambiguously registered only what was claimed for it – an important criterion of trustworthiness.

Elsewhere I have shown that some late Victorian experimenters took great pains to acknowledge problems of environmental control in the laboratory, painstakingly detailing the artefactual results encountered and their probable origins.[79] Unless experimenters took such pains to show they had established the well-controlled integrity of the user–instrument environment, they risked scepticism about the possibly artefactual nature of their results or accusations of self-delusion in their claims. The crucial matter in establishing material stability in busy metropolitan environments and crowded workspaces was the exercise of 'care': The implementation of cautious preparations, fastidious attention to detail, and a preemption of all possible sources of interference or error. An important distinction here was drawn between the type and the degree of care exercised: 'ordinary', 'great', or 'absolute'. A report of measurement could be most readily trusted if the experiementer(s) involved had taken absolute care: All necessary time and resources were made available without any constraint. If great care was used, little consideration was given to time and repeat experiments were conducted at reasonable intervals so that initially undetectable errors became apparent. If ordinary care was exercised, however, this meant that the time and resources devoted to it were only such as would fit into the routine of the experimenters' working-day responsibilities. The amount of care taken in making measurements was directly linked to (related) claims of the trustworthiness and thus accuracy of the measurements. The more care that was taken, the fewer the errors or cases of interference or corruption and the closer the result was thus expected to be to its true value.[80] As we shall see in Chapter 3 in the controversies

---

[78] Simon Schaffer, 'Glass works: Newton's prisms and the uses of experiment', in David Gooding, T. Pinch, and S. Schaffer (eds.), *The uses of experiment: Studies in the natural sciences*, Cambridge: Cambridge University Press, 1989, pp. 67–104. See Simon Schaffer, 'Self-evidence' *Critical Enquiry*, 18(1992), pp. 327–62, for a historical discussion of the use of the body to gauge electrical discharge.

[79] Gooday, 'Instrumentation and interpretation', pp. 413–20.

[80] Augustus Matthiessen and Charles Hockin, 'Appendix C – On the reproduction of electrical standards by chemical means', *BAAS Report*, 1864, Part 1, pp. 352–67, quotes from Part 1, pp. 352–3. For a radically different interpretation of the moral ladenness of care as applied to a human subject, see the growing feminist literature discussed in Joan Tronto, *Moral boundaries: A political argument for an ethic of care*, London: Routledge, 1993.

concerning the use of mercury versus metal alloy for resistance standards, both Werner von Siemens and Augustus Matthiessen made telling attacks on each other's claims to have taken due care in preparing their metal samples in making standard resistance measurements.

A further question arose, however, about whether problems of trustworthiness in materials could ever be solved by any amount of care. Did the very versatility that made mercury and iron so useful in electrical equipment also render them too protean and complex to manage with sufficient control for refined electrical measurement? In Chapter 6 we find that, although domestic meter-makers found the fluidity of mercury convenient for furnishing a conducting path and the requisite mechanical friction, the metal was also blamed for oxidation and amalgamation effects that brought unreliable performance. Although Sebastian Ferranti devoted huge efforts to solving such problems in the 1880s (Chapter 6), even in 1891 James Swinburne considered it 'questionable' whether mercury should ever be used in meters, although he admitted his doubts could be overruled by further experience.[81] At the same lecture, Swinburne conceded that over the preceding ten years comparable problems had been solved in regard to iron, especially the widespread use of soft iron cores in current-measuring devices (Chapter 4). The care exercised by designers in overcoming errors caused by hysteresis – sometimes known as magnetic 'memory' – had been reduced by 'careful' designers to the point at which they were mostly negligible.[82] Yet for the ac-generating technologies covered in Chapter 5, the use of iron was nevertheless still problematic for reasons discussed in subsequent chapters.

1.5.2. Trust in Theories

The important electrical phenomena of resistance, potential difference, current, and self-induction were highly theoretical in nature and generally only immediately accessible indirectly by means of their manifestation in a technological embodiment.[83] The question for the designer was this: What kind of theory could be used to evince and control a quantifiable effect that represented what was to be measured? As Swinburne noted in his analysis of domestic ac meters, the trustworthiness of a measuring instrument or standard hinged importantly on the theory or theories and auxiliary assumptions deployed in its design and usage. The trustworthiest devices, from his point of view, were those which relied upon a 'simple' law: These were the least likely to incorporate potentially misleading approximations[84] or

---

[81] Swinburne, 'Electrical measuring instruments', p. 21.
[82] Swinburne, 'Electrical measuring instruments', p. 3.
[83] For the exceptions see Chapters 2 of this book and Schaffer, 'Self-evidence'.
[84] For an important discussion of how an approximation can introduce radical changes in the veracity of a mathematical law, see Nancy Cartwright, *How the laws of physics lie*, Oxford: Clarendon, 1983, pp. 104–27.

extrapolations. Trust in such theories – whether qualitative or quantitative – was obviously important for users of such hardware. For example, Werner von Siemens claimed that mercury was more trustworthy for constructing standards of resistance than any solid metal because according to his theory, its loose molecular structure would not vary over time or differ from one sample to another. The same could not be said, however, of solid metals prone to contingent structural flaws. Given the highly disputed speculations about the material ontology of electricity, magnetism, and electrical conduction, it was not easy to develop compelling qualitative theories of electrical matters. Much weight was placed instead on the phenomenologically derived laws of Oersted, Ampere, Faraday, Henry, and Lenz about how macroscopic conductors interacted with each other and other sources of magnetic action.[85]

Ampere's action-at-a-distance law of electromagnetism was widely used by designers of current-measuring devices and domestic meters: It gave an inverse relationship relating the distance between conductors and the force between them. Yet even for the special geometrically simple cases of the tangent or sine galvanometer and the Thomson mirror galvanometer, some approximations had to be introduced to arrive at an analytical formula that enabled a measurement of current to be calculated from the instrument's reading. For all other instruments, the complexities that their geometrical form generally entailed made it hard to derive even an approximate equation for the quantitative relation between the magnitude of current and the scale deflection it caused. Approximations or trial and errors were thus often required instead, especially for devices that used an electromechanical surrogate (see Chapters 2 and 4) to replace theoretical calculation to generate a direct reading of current or electrical consumption. The trustworthiness of these devices was dependent on how trustworthily the scale-reading technology could replace the need for the user to theorize the relation of deflection to current. A further theoretical problem arose in the construction of domestic meters concerning what it was that these devices should measure vis-à-vis what it was theoretically easiest for a designer to arrange for them to measure: Should householders be represented as consuming electric quantity, current, energy, or lighting? Because that last of these possibilities was theoretically very difficult to build into the operation of a domestic meter, customers were forced to trust a system of metering by methods more convenient for the designer, manufacturer, and supplier.

Equally troublesome was the question about whether it was possible to use any quantitative electromagnetic theory to ground an instrument for measuring self-induction, especially if this required a rather simplified version of theory to be used. The sometimes dramatic effects of self-induction

---

[85] See Peter Harman, *Energy, force and matter*, Cambridge: Cambridge University Press, 1983, pp. 30–3.

in creating sparks and voltage surges (when a simple current-carrying circuit was suddenly opened) were easy to see; but actually trying to quantify the value of self-induction for a fast-moving piece of complex electromagnetic machinery was a vastly more difficult task. In Chapter 5, I show how, in the 1880s, John Hopkinson trusted to the power of Cambridge mathematics and simple dynamical analogies to express self-induction simply as an analogue of mechanical inertia. He thus argued that the behaviour of ac machinery was easily amenable to the equations of simple mechanical oscillation, and implicitly measurable too. The mechanical-turned-electrical-engineer James Swinburne distrusted such applications of mathematical theories, having experienced only their tendency to misrepresent the complexities of engineering performance by the unwarranted use of *false* simplifying assumptions. From his own experience of both steam and electromagnetic machinery, he found that self-induction could *not* reliably be characterized as a form of 'inertia'. Indeed he argued from standard electromagnetic theory that the persistent relative motion of iron in an operative alternator made its self-induction a complex circumstantial variable, thus undermining Hopkinson's account. Given this conflicting field of interpretations, attempts by William Ayrton and John Perry to measure a presumptively near-constant self-induction on *static* alternators was theoretically problematic and did not proceed beyond a few suggestive results.

We shall see that Swinburne never trusted the results of the secohmmeter that Ayrton and Perry developed for this purpose, and its fate was tied to the persistent theoretical problems that arose in *any* attempt to measure self-induction in all but trivial cases. Thus we will see that trust in particular selected theories was important for judgements about those particular practical situations for which an instrument could be trusted to produce meaningful measurements and those for which it could not.

1.5.3. Trusting the Instrument's Construction

Finally, there is an important point to be made about how the trustworthiness of an instrument relates to the particular way that it had been constructed. As Schaffer has noted of James Clerk Maxwell's attempts to secure reliable apparatus for the early Cavendish Laboratory, this is partly a question of the quality of understanding between instrument-makers and those who design or commission instruments.[86] *Pace* Latour, Schaffer shows that so much tacit knowledge can be required for converting a paper blueprint into a piece of reliable hardware that the labour involved can entirely defeat the efforts of remote makers with whom there is little face-to-face communication.[87]

---

[86] I am grateful to William Ginn and Alison Morrison Low for discussing these matters in relation to their PhD researches.

[87] Schaffer, 'Late Victorian metrology', p. 34.

The contrasting case of Sir William Thomson/Lord Kelvin, who was a partner in White's instrument-makers of Glasgow, shows that this relationship could work much better when an instrument-maker was local to the client. Moreover, given such proximity, a tacit understanding could develop over a long time period about how an instrument should be made to fulfil detailed specifications.[88] In Chapter 2 we shall see, however, that sometimes users of instruments could just simply label makers as incompetent or dishonest: The physicist Frederick Guthrie complained in 1886 that many instruments sold for elementary instruction in physics were mostly overvarnished 'trash'.[89]

Speaking at the IEE in 1904, Kenelm Edgcumbe and Franklin Punga observed that there had indeed long been 'certain mistrust' felt by engineers for instrument-makers and all their works. Unlike them, few such makers had previously been engineers, and accordingly such makers did not know that engineers set little store by instruments finished merely by a 'lavish' use of lacquer and shellac varnish.[90] The two main electrical engineers I discuss in this book had in fact experiences of both instrument manufacture and electrical engineering. As I show in Chapter 4, they had significantly different views on the kind of 'reading technology' that an instrument should have, in ways that reflect their different early training and practical experiences. Having previously worked in telegraphy in India, Britain, and Japan, William Ayrton argued that the dials of heavier-duty electrical engineering instruments should be designed with equally spaced markings to maximize ease of reading – as many telegraph instruments were. This strategy required that the instrument mechanisms furnish a trustworthily linear relation between current and deflections over a significant range. Having been apprenticed into mechanical engineering in the 1870s, James Swinburne challenged the credibility of Ayrton's claims to have designed an effective form of such a mechanism. From his experience of using non-linear instruments in mechanical engineering, Swinburne argued that it was far better to have a trustworthy relationship between the mechanism's operation and overtly unequal scale markings and expect users to learn to read such calibrations. This devolved to users the responsibility for undertaking reliable interpolations and developing trustworthy visual faculties.

Thus there were two kinds of potentially problematic nexus in the linked issues of trust in measurement: the spatial/bodily relation between instrument and measurer, and the historical relation between the user of an

---

[88] See the numerous references to Thomson's intimate association with James White's company in S. P. Thompson, *The life of Lord Kelvin*, 2 Vols., London: 1910, Vol. 1, pp. 348, 419, 489, 493, 507, 575, 579; Vol. 2, pp. 680, 717, 741, 754, 769, 796, 994; 1155, and Smith and Wise, *Energy and empire*, p. 714.

[89] Frederick Guthrie, 'Teaching physics', *Journal of the Society of Arts*, 34 (1886), pp. 659–63, quote on pp. 660–61.

[90] Kenelm Edgcumbe and Franklin Punga, 'Direct-reading measuring instruments for switchboard use', *Journal of the IEE* (hereafter *JIEE*), 33 (1904), pp. 620–55, 655–93, 620.

instrument and its makers and calibrators. These themes recur in various different contexts in the chapters that follow.

1.6. CONCLUSION

Measurement certainly made it much easier for expert late Victorians to manage or comprehend electrical matters. Yet – *pace* William Thomson – measurement itself was not an unproblematic route to knowledge of electrical machines, standards, or instruments. This was because its practices were loaded with both questions of value and assumptions about what constituted trustworthy conduct and trustworthy material culture. I have shown that the ubiquity of issues pertaining to trust shows how the complex socio-technical character of measurement cannot be reduced to matters of well-chosen metrology or shrewdly exercised power. To that extent, this chapter has answered this question: What was the role of trust in electrical measurement, and what was the role of instruments in establishing – or not establishing – that trust? Subsequent chapters examine this role further in the measurement of several of the major electrical parameters: Resistance (Chapter 3), current (Chapter 4), self-induction (Chapter 5), and electrical energy (Chapter 6). Before to those discussions in the next chapter the basic questions are these: What did it mean to *measure* electrical quantities in the late nineteenth century and how did 'accuracy' or 'precision' figure in accounts of such measurement? In my Conclusion to this book I consider more generally how far historical explanations of measurement practices based on trust can legitimately appeal to 'moral' considerations. Finally I also relate my arguments about 'trustworthiness' in measuring instruments to debates on the 'social construction' of technology.[91]

---

[91] Trevor Pinch and Wiebe Bijker, 'The social construction of facts and artefacts: Or how the sociology of science and the sociology of technology might benefit each other', in Wiebe Bijker, Trevor Pinch, and Thomas P. Hughes (eds.), *The social construction of technological systems*, Cambridge, MA: MIT Press, 1987, pp. 17–50.

2

Meanings of Measurement and Accounts of Accuracy

Strictly speaking, to measure a thing of any kind is to ascertain the numerical relation between it and some magnitude of its own kind taken as a standard for comparison... Before methods of measurement can be devised, it is evident that clear conceptions must be formed of the things to be measured.

George Carey Foster, presidential address to STEE, 1881[1]

What is accuracy? The authors in one or two places speak [of] about 1/10 per cent, and in other places they hazard a guess that the ordinary switchboard instruments which they speak of might have an accuracy or a little more than that – five times, they suggest... I speak with some feeling in this respect because of the experiences we have at the Board of Trade [testing] laboratory. We have instruments sent us there that sometimes induce remarks which I am afraid the Chairman would not care to hear.

J. Rennie, discussion of paper on 'Direct Reading Instruments for Switchboard Use', at the IEE, 1904.[2]

Like William Thomson, George Carey Foster and his colleagues were sure that measurement furnished a reliable grasp of how the world worked. They did not need to ask searching questions about what measurement actually was nor about how measurement produced knowledge. In regard to electrical measurement, such matters were so self-evident for them that they devoted little effort to theorizing their certainties. And indeed, the development of measurement techniques in the nineteenth century cannot easily be told as the explicit application of a systematic theory of measurement to the practical problems of quantification. Most of the important innovations in electrical-measurement technique discussed in this book had taken place by 1887, the year pinpointed by José Diez as the point at which the first detailed and explicit theory of measurement was produced by Hermann von Helmholtz.[3]

---

[1] George Carey Foster, 'Inaugural address' [to the STEE], *JSTEE*, 10 (1881), pp. 4–19, quotes on pp. 9–10.
[2] Kenelm Edgcumbe and Franklin Punga, 'Direct reading instruments for switchboard use', *JIEE* 33 (1904), pp. 620–68, discussion on pp. 655–93, quote on pp. 658–9.
[3] José Diez, 'A hundred years of numbers. An historical introduction of measurement theory 1887–1990', *Studies in the History and Philosophy of Science*, 28 (1997), Part 1, pp. 167–85.

Meanings of Measurement and Accounts of Accuracy 41

Ironically, this process of post hoc theorization of measurement has subsequently revealed at least as many problems as it has solved. The instrument-designer Phillip Sydenham and the historian Philip Mirowski have argued that the systematic theorization of measurement construed the ontologies of length and mass – to which Victorian physicists were deeply committed – into little more than anthropocentric extrapolations of contingent bodily form.[4] Moreover, unlike Thomson and Foster, Sydenham sees no uncontentious way of linking measurements to the knowledge built from them; in fact, he argues somewhat radically that the meaning ascribed to measurement data is entirely a matter of 'codification' by the user.[5] His reply to William Thomson's claim that measurement was a prerequisite for knowledge[6] would presumably be this: Measurement does not tell you everything there is to know about something, and we have only a meagre and unsatisfactory understanding of how knowledge is based on measurement anyway.

Given our distance from the world of Thomson, Foster et al., it is difficult to accept Ian Hacking's suggestion that 'our' current conception of measurement was clearly established by the end of the nineteenth century.[7] By contrast, my aim in this chapter is to recover something of the credo and practice of Victorian measurement and their uneasy interrelations. Hacking does, however, raise useful questions that enable us to historicize our understanding of measurement work in the past. What was the actual *point* for Victorians of making measurements, especially those of high precision? Did their measurements capture anything but artefacts of how measurers theorize the world? Although my account indirectly addresses such concerns, the first question that I want to ask in this chapter appears much more mundane: What were late Victorian physicists and electrical engineers actually doing when they said they were making measurements? The public rhetoric of Foster and others often proclaimed that measurement was fundamentally a comparison of a number with a standard magnitude. Yet I show in Section 2.1 that measurements generally involved some process of reduction before comparison could take place and calculation afterwards: *Directly* comparative measurement was strictly possible only for length or mass. Many electrical measurements were similarly reduced either literally or metaphorically to the venerable paradigms of measurement: the mensuration of length

---

[4] Phillip Sydenham, *Measuring instruments: Tools of knowledge and control*, Stevenage, England: Peregrinus/IEE, 1979, pp. 10–14; Philip Mirowski, 'Looking for those natural numbers: Dimensions, constants and the idea of natural measurement', *Science in Context*, 5 (1992), pp. 165–88.

[5] Sydenham, *Measuring instruments*, pp. 15–22, and C. Finkelstein, 'Fundamental concepts of measurement: definition and scales', *Measurement & Control*, 8 (1975), pp. 105–11.

[6] William Thomson, 'Electrical units of measurement' 1883, in *Popular lectures*, London: 1891, Vol. 1, pp. 73–76, quote on p. 73.

[7] Ian Hacking, *Representing and intervening: Introductory topics in the philosophy of science*, Cambridge: Cambridge University Press, 1983, pp. 233–4.

or balancing with weighing scales. I explain therefore why it was a radical move for electrical engineers in the 1880s to introduce 'automated' apparatus that *allegedly* enabled the 'direct' measurement of electrical quantities at a glance.

In Section 2.2 I explore Foster's claim that a prerequisite of measurement work was a 'clear conception' of what was to be measured. I suggest conversely that the purpose of electrical measurement for Foster's teaching contemporaries was to familiarize novices with strange and ontologically problematic parameters such as resistance, current, and self-induction. Then I consider how the sharpening division of labour in designing, manufacturing, and calibrating instruments created problems in attributing 'authorship' to a measurement. Following an examination of the quantitative and qualitative meanings of the 'accuracy' of measurements, I consider in Section 2.4 how attributions were made regarding the agency and responsibility of accomplishing 'accuracy'. The issue of responsibility has a particularly close bearing on the role of 'care' in accomplishing accuracy and the notion of who was the bearer or 'trustworthiness' in a measurement. Finally, I consider how the notion of error was treated in measurement in the fields of physics and electrical engineering, noting some international differences in practice about how error was treated and concerns about it were enacted. Considerations of how the notion of 'degree of accuracy' functioned as a criterion of the elimination of error will help us to answer the 'notorious' difficulty identified by Norton Wise in efforts to identify how 'societal context' can in any important way affect how someone makes a measurement.[8]

2.1. COMPETING RHETORICS OF MEASUREMENT: COMPARISON VERSUS REDUCTION

Notwithstanding the very different characters of these quantities [Time, Temperature, Elasticity, Viscosity etc.] they are all measured by reducing them to the same kind of quantity... Every quantity is measured by finding a *length* proportional to the quantity, and then measuring the length.
        W. K. Clifford, 'Instruments Used in Measurement', Lecture on Scientific Apparatus, 1876[9]

Now a physical measurement – a measurement, that is to say of a physical quantity – consists essentially in the comparison of the quantity to be measured with a unit quantity of the same kind.
        R. T. Glazebrook and W. N. Shaw, *Practical Physics*, 1885[10]

---

[8] M. Norton Wise, 'Mediating Machines', *Science in Context*, 2 (1988), pp. 77–114, quote on p. 78.

[9] William K. Clifford, 'Instruments used in measurement', in *Handbook to the scientific loan collection of scientific apparatus*, London: 1876, pp. 55–59, quote on pp. 55–6.

[10] Richard T. Glazebrook and William N. Shaw, *Practical physics*, London, 1885, p. 1. Originally produced from notes used in teaching elementary Natural Sciences Tripos students

Meanings of Measurement and Accounts of Accuracy   43

Students at the turn of the twentieth century might have been somewhat puzzled to find two distinct answers to this question: What is measurement? Depending on which source they read, they might find that the measurement of a quantity was characterized either as a *reduction* of the task to the measurement of lengths or that it was a numerical *comparison* between any kind of quantity and the relevant standard. Thirty years after he was introduced to Glazebrook's and Shaw's teaching of measurement at Cambridge in the 1890s, Norman Campbell wrote of his long-held anxieties about the latter definition in his *Account of the Principles of Measurement and Calculation*.[11] Most particularly he objected to the assertion by his erstwhile Cavendish Laboratory teachers that measurement activities essentially involved a comparison of a quantities of the *same* kind: Surely, wrote Campbell in his Preface, there were grounds for contesting this opinion. A principal target of this attack was the discussion by Glazebrook and Shaw of Atwood's machine, a device consisting of a pendulum, a set of suspended weights, and stops located along a vertical rule for terminating the timed fall of these weights. By the late nineteenth century, this had become the canonical means of measuring the local gravitational constant g, and citation of it in this context was thus a recurrent feature of every edition of Glazebrook's and Shaw's classic textbook.[12] Somewhat scathingly, however, Campbell remarked that a student set to work with Atwood's machine 'may well wonder' at what stage gravitational acceleration was compared with a unit of the 'same kind'. It was all too obvious to him that the entire process of measuring g centred, as Clifford suggested, on making measurements of length. According to Campbell, even accomplished physicists on international standards committees had been 'apt to flounder' when discussing such features of the fundamental nature of measurement. After working as a research physicist at the Universities of Cambridge and Leeds, it was as an employee of the General Electric Company in London that Campbell set

---

at the Cavendish Laboratory in Cambridge, the preface explains that this book was aimed at a broad national audience of teachers and students. The first (1885) edition was twice reprinted, followed by a second edition in 1893, a third in 1905, and a fourth in 1918. For details of Glazebrook's and Shaw's teaching at Cambridge see *A history of the cavendish laboratory*, London, 1910, pp. 46–7, 110–11, 252, 264–5.

[11] Norman R. Campbell, *Account of the principles of measurement and calculation*, London: 1928. For more on Campbell, see Andrew Warwick, 'Cambridge mathematics and Cavendish physics: Cunningham, Campbell and Einstein's relativity, 1905–11, part II', *Studies in History and Philosophy of Science*, 24 (1993), pp. 1–25; Diez, 'A hundred years of numbers'; Silvio Funtowicz and Jerome Ravetz, *Uncertainty and quality in science for policy*, Dordrecht, The Netherlands/London: Kluwer, 1990, pp. 73–77, 81.

[12] For illustrations and explanation of the original purposes of this device first developed in the 1770s, see Simon Schaffer, 'Machine philosophy: Demonstration devices in Georgian mechanics', *Osiris*, 9 (1994), pp. 157–82, and Simon Schaffer, 'Atwood's machine' in Robert Bud and Deborah J. Warner (eds.), *Instruments of science: An historical encyclopedia*, London/New York: Garland, 1998, pp. 36–9.

out in 1928 to offer a more rigorous articulation of the conventions tacitly employed in measurement.[13]

The concise 'comparative' characterization of measurement attacked by Campbell had considerable currency in the late nineteenth century. It was by no means idiosyncratic to Foster's STEE address or the many editions of Glazebrook's and Shaw's widely used textbook *Practical Physics*. It was part of an introductory rhetoric common to physics, telegraphy, or electrical engineering. In his preliminary discussion, 'On the Measurement of Quantities', at the start of his *Treatise on Electricity and Magnetism*, James Clerk Maxwell analyses every expression of quantitative measurement to consist of a standard of reference and the 'number of times' the standard is needed to make up a given quantity – although he does allow that all measurement standards units *could* be reductively defined in terms of length, time, and mass.[14] Harry Kempe, in his *Handbook of Electrical Testing*, noted that in order to make measurements it was necessary to have standard units with which to make 'comparisons'.[15] And when the instrument maker Major Philip Cardew explained the matter to the 1891 Board of Trade Committee on Electrical Measurement, he said that the 'first idea' of measurement was to 'refer' all quantities to a calibrated unit standard.[16] Then again, in other textbooks, this comparative notion of measurement is so taken for granted that it is not explicitly mentioned at all. Readers of Thomson's and Tait's *Treatise on Natural Philosophy* (1867), Latimer Clark's and Robert Sabine's *Electrical Tables and Formulæ* (1871),[17] Stewart's and Gee's *Lessons in Elementary Practical Physics* (1885),[18] and James Swinburne's *Practical Electrical Measurement* (1888)[19] would search in vain for any such explanation of what constituted electrical measurement.

Although Campbell was specifically concerned to attack the simplistic comparative notion of measurement, it was the general incoherence of late Victorian and Edwardian accounts of physical measurement that attracted much of his derision. Indeed, when we turn to such texts to see what is said

---

[13] Campbell, *An account of the principles*, pp. iii–iv.
[14] James Clerk Maxwell, *Treatise on electricity and magnetism*, Oxford: 1873; 3rd edition, 1891, p. 1.
[15] Harry R. Kempe, *Handbook of electrical testing*, London: 1876, 2nd ed. 1881 (from which quote was taken), p. 1. Subsequent editions were 3rd (1884), 4th (1887), 5th (1892), 6th (1900), and 7th (1908). The origin of this book was in a series of articles published by Kempe in the *Telegraphic Journal and Electrical Review* 1874–5.
[16] Cardew's evidence cited in *Report of Board of Trade Committee on Electrical Standards*, 31 January 1891, p. 7.
[17] Latimer Clark and Robert Sabine, *Electrical tables and formulæ: For the use of telegraph inspectors and operators*, London: 1871.
[18] Balfour Stewart and William W. H. Gee, *Lessons in elementary practical physics*, London, Vol. 1, 1885, Vol. 2, 1887. Compare B. Stewart, *Lessons in elementary physics*, London, 1870, revised edition, 1874.
[19] James Swinburne, *Practical electrical measurement*, London: 1888.

about the *process* of measurement we find several levels of unresolved tension and complexity. Most important is that the very same texts that represent measurement as direct comparison with a standard *also* represent it as the reduction of all more complex quantities to those of length or mass.[20] Indeed, despite their initial subscription to the comparative notion of measurement in *Practical Physics*, Glazebrook and Shaw contend that direct comparative measurement was in fact possible *only* for length and mass. The comparative processes of true measurement were 'far from direct' in the great majority of electromagnetic, thermal, and acoustic processes because for them too the *actual* process of measurement 'really and truly' involved the determination of lengths or masses. As we shall see, though, the account of Glazebrook and Shaw sat rather uneasily alongside their account of the measurement of electrical resistance as direct comparison against a standard.

Although measurements of mass were often important, Glazebrook and Shaw emphasized that the measurement of length was for most issues *the* canonical form. Elapse of time was gauged by angular distance traversed on a clock face. Force was usually measured by the longitudinal extension of an elastic body, and both pressure and temperature were measured by their effect on the height of a column of fluid. Electric currents were determined by the deflection of a galvanometer needle or light-spot over a scale, and the transient angular 'throw' of a galvanometer needle served to ascertain coefficients of electromagnetic self-induction.[21] In this reductive mode of measurement as length determination, the comparative mensuration reduced to 'reading' an angular or linear displacement by means of a needle, column of mercury, or a spot of light against a calibrated scale. An important corollary of this reductive process was, of course, that some kind of explicit and theory-laden *calculation* was required for converting the lengths determined back to the original quantity to be measured. This was most obviously the case in the treatment of Atwood's machine of which Campbell complained: Students had to use an equation to calculate g from measurements of several distances traversed.[22] More relevant for our purposes, current measurement with a tangent galvanometer effectively began with the measurement of three distances. These were the size of the angular deflection of the compass needle by the current-carrying coil, the radius of this coil, and the local

---

[20] In his lecture to the South Kensington Loan Exhibition in 1876, Clifford strained somewhat to describe the balance measurement of mass as exclusively a determination of length; Clifford, 'Instruments used in measurement', pp. 57–8. William Thomson notably diverged from Clifford on this matter, contending in his lecture on 'Electrical Measurement' for the same Exhibition that measurement in physical science generally consisted of two methods: the null method of 'adjustment to zero' or a method of measuring some 'continuously varying quantity.' William Thomson, 'Electrical measurement', *Popular lectures and addresses*, Vol. 1, pp. 430–62, quote on p. 431.
[21] Glazebrook and Shaw, *Practical physics*, pp. 3–5. See Chapters 4–5 for further discussion.
[22] Ibid., pp. 133–8.

horizontal component of the Earth's magnetic field registered by the deflection of a magnetometer. To arrive at a value for the current, a calculation had to be performed with a notably approximate formula derived from Ampere's theory of magnet–current interactions.[23]

Electrical-light engineers who had not the leisure to perform such time-consuming calculations did not generally employ this particular reductive-calculational approach to current measurement. W. E. Ayrton and John Perry instead automated the human calculation process with a (fallible) electro-mechanical surrogate[24] that enabled instantaneous readings to be taken directly of amperes rather than of millimetres of degrees of deflection. This automatic process followed on the model of engineering exemplars such as the steam gauge that gave indications of pressure as *instantaneous* 'length' readings on a scale directly calibrated in the relevant unit (e.g., foot-pounds per square inch). Importantly, different kinds of *trust* were required for bridging the epistemological gaps in the operation of these two types of current-measurement instruments. Use of the tangent galvanometer for current measurement required trust in the application of an electromagnetic theory and the strategic approximations made in so doing, but otherwise users had to trust their own labours, especially their competence in undertaking reliable length measurements and calculations. By contrast, although users of the ammeter took responsibility for fine-tuning the instrument's calibration, they had to trust in the reliability of many others. They had to trust in the outcome of the trial-and-error methods employed in designing its mechanism, in the basic consistency of the calibration system, and in the design and manufacture of the iron and magnet components to limit their tendencies to wayward and unpredictable performance.

The direct-reading ammeter and voltmeter were partly modelled on the steam pressure gauges used in mechanical engineering, and these gauges were often referred to as being merely 'indicators' rather than as measuring devices. Yet if the semi-automated mechanisms of direct-reading instruments provoked doubts among electrical engineers as to whether these devices did 'measure' in the rigorous sense, they kept these doubts to themselves. For example, at a meeting of the STEE in February 1883 in which James Shoolbred explained the principles of early ammeters and domestic supply meters in a paper entitled 'The Measurement of Electricity for Commercial Purposes', no one present contested the legitimacy of the word measurement in his title.[25] Sir William Thomson's involvement in the production of direct-reading electrical devices appears, moreover, to have sanctioned their status as

---

[23] Ibid., pp. 386–90.
[24] My usage of this term borrows indirectly from Harry Collins' use of this term in *Changing order: Replication and induction in scientific practice*, London: Sage, 1985, pp. 74, 125–6.
[25] See, for example, James Shoolbred, 'The measurement of electricity for commercial purposes', *JSTEE* 12 (1883), pp. 84–107.

measuring instruments. At a meeting of the STEE in May 1888, for example, Thomson presented some of this 'new standard and inspectional electrical measuring instruments' and received nothing but admiration for his new design of a direct-reading 'portable marine voltmeter' for use on electrically lighted Atlantic liners.[26] Although some physicists later objected to ammeters masquerading as measuring devices in the physics teaching laboratory (see subsequent discussion),[27] I have identified only one instance in which electrical engineers publicly doubted whether direct-reading instruments were really measuring devices. Even this was a question regarding a very context-specific application. In a debate on power-station switchboard instruments at the IEE in 1904, E. B. Vignoles took to task the speaker, Kenelm Edgcumbe, for failing to note that these were not strictly measurement devices. As far as he was concerned, these were simply 'indicating' instruments that enabled switchboard attendants to 'find out what was happening' on their station circuits – the same diagnostic role played by steam gauges in a boiler room.[28]

At the 1888 STEE meeting mentioned in the preceding paragraph, Thomson also presented another important instrument for measuring current. This new 'current-weighing' device also subverted the orthodox reductive practice of measuring current as it did not involve the determination of several distinct lengths. Notwithstanding its name, it borrowed only metaphorically from the alternative 'mass' paradigm of measurement. The first current weigher in 1866 was developed for 'absolute' measurement by James Joule in his researches for the BAAS Electrical Standards Committee (Chapter 3): He used pre-calibrated 'weights' to balance the mutual attractions between current carrying wires to arrive at a 'null' condition of zero deflection.[29] In 1887 Thomson adapted Joule's devices into a series of double-sided absolute current balances that balanced the attraction or repulsion of an unknown

---

[26] William Thomson, 'On his new standard and inspectional electrical measuring instruments', *Proceedings of the Society of Telegraph Engineers & Electricians*, 17 (1888), pp. 540–56 and discussion on pp. 556–67. Thomson received praise from William Preece as being one of the 'great fathers' in the art of exact measurement, the other 'great father' present at this meeting being Werner von Siemens, discussion, ibid., p. 557.

[27] Graeme Gooday, 'The morals of energy metering: Constructing and deconstructing the precision of the electrical engineer's ammeter and voltmeter', in M. Norton Wise (ed.), *The values of precision*, Princeton, NJ: Princeton University Press, 1995, pp. 239–82, especially pp. 262–78.

[28] Edgcumbe replied – perhaps somewhat archly – that, given the extraordinary lack of interest on measuring instruments shown by station engineers at IEE meetings, it would 'almost appear' that Vignoles' contention about the use of such instruments as mere 'indicators' in their work was correct. Edgcumbe and Punga, 'Direct-reading instruments', pp. 680 and 688.

[29] See discussion of Joule's current weigher in James Clerk Maxwell, *Treatise on electricity and magnetism* 3rd edition, 1891, Vol. 2, p. 371. On the history of the balance, see Matthias Dörries, 'Balances, spectroscopes, and the reflexive nature of experiment', *Studies in the History and Philosophy of Science*, 25 (1994), pp. 1–36.

current against a known, or used the combined attraction and repulsion effects of current on itself. The only direct material vestige of the mass balance was the use of a lateral scale rider to achieve a very 'fine' balance condition.[30] Thomson went on to design forms of these that could 'weigh' currents over a range of 1/000 of an ampere (A) to 2500 A, and, over the next two decades, these were soon *the* definitive devices for measuring currents and calibrating ammeters used in testing and standardizing laboratories.

Thomson's comparative current weigher adapted the 'mass' paradigm of measurement by using the force between currents as a surrogate for the counterweight. It soon replaced another more time-consuming technique that was espoused by Glazebrook and Shaw, namely the measurement of currents by a mass generated as the surrogate for current. Faraday had shown in the 1820s that a constant current passing through a suitable electrolyte would deposit a mass of metal on an electrode that was proportional to the intensity of the current.[31] Lord Rayleigh and Eleanor Sidgwick used this principle in 1884 to make the first absolute determination of the ampere.[32] A year before then, electrolytic meters based on the same principle had already been installed in the homes of householders connected to the Edison lighting system. Edison officials visited monthly to remove the electrodes and sent customers a bill for their consumption according to the increased mass of the electrodes determined by a sensitive balance in a company laboratory. As we shall see in Chapter 6, though, Edison's customers could be (even) more distrustful of such unseen weighings than they were of commercial weighings undertaken more visibly in local shops.

The measurement of mass by the null balance method[33] metaphorically furnished yet another comparative means of measuring electrical parameters. When the bridge balance method was used, electrical resistance could be compared directly with a precalibrated standard of the same unit. Importantly, this was undertaken *without* resolution to submeasurements of length or mass, thereby undercutting the reductionist claims of Glazebrook and Shaw on this score. The Wheatstone bridge was a fast and convenient null method of measuring resistance that was often used in the laboratory, telegraph station, or power installation. As both Kempe and Glazebrook and Shaw explain, one of the four bridge arms would hold the unknown resistance, and another on the 'opposite' side of the circuit would be adjusted until

---

[30] Thomson, 'On his new standard and inspectional electrical measuring instruments', pp. 545–53.
[31] Glazebrook and Shaw, *Practical physics*, pp. 406–16. For more details, see Frank A. J. L. James 'Michael Faraday's first law of electricity' in John T. Stock and Mary V. Orna (eds.) *Electrochemistry past and present*, Washington D.C.: American Chemical Society, 1989, pp. 32–49.
[32] Lord Rayleigh and Eleanor Sidgwick, 'On the electrochemical equivalent of silver', *Philosophical Transactions of the Royal Society of London* 175 (1884), pp. 411–60.
[33] Glazebrook and Shaw, *Practical physics*, pp. 83–105.

a galvanometer between the two arms of the bridge gave a zero reading. This zeroing was accomplished by the switching of precalibrated resistances in or out of the adjustable side – by analogy to the addition or removal of weights from the adjustable arm of a mass balance.[34] Once this null balancing condition had been obtained from the values of the three known resistances in the Wheatstone bridge, the unknown resistance could quickly be calculated from a simple ratio formula. Contemporaries regarded such bridge techniques as highly reliable because, the bridge techniques, unlike deflectional methods, required that the user only pinpoint the condition of zero reading, and thus the user was unaffected by uncertainties about calibration at other points along the galvanometer scale. The fate of the more complex hybrid secohmmeter discussed in Chapter 5 reveals the problems, however, of attempting to extend such null bridge methods to the measurement of self-induction.

Comparative balancing methods for measurement, nevertheless, still raised a problem of trust – specifically trust in the prior work of others. Users of the mass balance had to employ manufactured sets of weights from a trustworthy maker, weights carefully built to ensure that, when handled with due care, their calibration would not be impaired. Users of the Wheatstone bridge similarly relied on sets of resistances that could by relied on not to change significantly in value from the cumulative effects of everyday working or from fluctuating environmental conditions. From the 1870s, such a set of resistances could easily be bought commercially (e.g., from the Post Office) in the form of a large solid box with internally protected resistances of values that could be added up to any desired quantity.[35] Such boxes were common within the contexts of academic and commercial life and were generally accepted as part of the division of labour in measurement. So long as the manufacturer was one that could be trusted to produce reliable standards, there was no need for measurers to trouble themselves to determine resistances for themselves by lengthy absolute methods. As we shall see in the next section, however, this sort of concession to expediency in measurement could provoke challenges from those physicists who placed great weight on the importance of measurers' trusting only their *own* efforts in the process of authentic measurement. Textbooks that recommended use of the Wheatstone bridge characteristically passed over such complicated and fraught issues as these in complete silence.[36] Unsurprisingly, generations of physicists and electrical engineers learned to use the Wheatstone bridge to

---

[34] For illustrations of the Wheatstone bridge and discussion of its history, see C. Neil Brown, 'Wheatstone bridge', in Robert Bud and Deborah Warner (eds.), *Instruments of science*, pp. 663–5.

[35] Maxwell attributed this innovation to the Siemens company, the minimum number of resistances required for most purposes being 1, 2, 4, 8, 16, 32, 64, etc. Maxwell, *Treatise*, 3rd edition, Vol. 1, pp. 470–1. According to Glazebrook and Shaw, an alternative arrangement was available for 1, 2, 2, 5, 10, 10, 20, 50, 100, 100, 200, and 500, presumably to ease the demands of high-speed mental arithmetic in telegraphy; see *Practical physics*, p. 428.

[36] Glazebrook and Shaw, *Practical physics*, pp. 426–8.

measure resistance without worrying that its operation either rested on fallible relations of trust or was at odds with the much espoused principle that only length and mass could be measured by direct comparative methods.

2.2. UNCERTAINTIES OVER THE IDENTITIES OF THE MEASURER AND THE MEASURED

It is impossible for us to think of our ancestors when they were utterly unacquainted with notions of length, mass, and of time, as we should have to go back to the ancestors of Silurian trilobites, but it is in no way difficult to conceive of human beings unacquainted with notions of resistance, quantity of electricity, difference of potential and electromotive force.

'The Ohm', editorial in the *Electrician*, 1881[37]

Once a late Victorian had undertaken a putative measurement – by whichever of the currently available definitions – a residual ambiguity might remain about both the authorship of the measurement and its epistemic significance. When Foster espoused the comparative notion of measurement to the STEE in 1881, he did not specify whether putative measurers had to undertake to ascertain *every* single aspect of the quantitative process themselves in order to count as an author of the measurement. A major theme of Campbell's 1928 critique was that his contemporaries and predecessors, such as Foster, were confused or deluded on this issue because they had an insufficient grasp of the complete division of labour involved in constructing and calibrating measuring instruments. He challenged common sense notions of *who* precisely had undertaken a given measurement by questioning whether users of pre-calibrated instruments had ever really been taking measurements at all. Referring particularly to the case of the commercial voltmeter, he suggested that most practitioners left 'all true measurement' to instrument-makers and experts in standardizing laboratories because it was they, and not the users of their products, who had originally assigned the numerals to dials to represent properties. As an employee of the electrical manufacturer GEC, Campbell had doubtless considerable experience on this issue: Whilst he considered the relevant division of labour to be indispensable, he maintained that it was 'dangerous' for users to forget that most of their operations in laboratory practice were 'not complete measurement'.[38] This particular anxiety was one which had been avoided by Victorian physicists who solipsistically persuaded themselves that they were largely reliant on only their own calibrational efforts. There was, however, one incident in the 1890s which threw into sharp relief the commercialized division of labour in Victorian measurement practice.

---

[37] Editorial, 'The ohm', *Electrician*, 7 (1881), p. 24.
[38] Campbell, *An account of the principles of measurement and calculation*, pp. 11–12.

Although by the time Campbell was writing voltmeters had become commonplace in the physics laboratory, efforts by William Ayrton in 1894–5 to introduce ammeters and voltmeters into this domain from the practice of electrical engineering had proved to be highly controversial. Ayrton's paper to the Physical Society of London, written collaboratively with his student H. C. Haycraft, argued that such meters could speed up the 'determination' of Joule's mechanical equivalent of heat. Ayrton and Haycraft claimed that an ordinary student could use these industrial devices to measure this 'constant' in a matter of minutes – in contrast to the hours or days hitherto required for such an experiment.[39] Such claims raised major problems about the attribution of authorship of measurements to the student in question. Of the heated responses received, one of the most irritable came from C. V. Boys, Assistant Professor at the Royal College of Science just over the road from Ayrton's City & Guilds Central Technical College.[40] Boys considered that Ayrton's experiment would tend to give students reprehensible delusions of grandeur about the importance of their role in the putative measurement of Joule's constant:

The measurement attributed to the student is really made by the original investigators who made the absolute determinations, e.g., of the electro-chemical equivalent of silver, employed later for standardising a Kelvin [current] balance, employed later for calibrating an ammeter – who made the absolute determination of the electromotive force of a Clark cell, or who found the value of the resistance of a piece of wire called an ohm. These people did the work of making the determination. Various middle men passed this work along and copied or compared, and after a time the actual ammeter and voltmeter worshipped by the student became calibrated, and finally a dummy in the form of a student, who, according to the provision of the Paper, is carefully guarded from doing anything which involved either knowledge, pain, or skill, shakes out the magical figure possibly under the impression that he has created it.[41]

One of Ayrton's allies in the technical education movement, the irenic Quaker Silvanus P. Thompson, tried to resolve the matter by pointing out a kind of *reductio ad absurdum* in the physicists' position – one that was hinted at

---

[39] William E. Ayrton and E. Haycraft, 'Students' simple apparatus for determining the mechanical equivalent of heat', *Proceedings of the Physical Society*, 8 (1894–5), pp. 295–309. Note, however, that the textbook of Glazebrook and Shaw textbook included a version of this experiment that used an electrochemical instrument, viz., a voltmeter, for a 2-min period in 1885, Glazebrook and Shaw, *Practical physics*, pp. 416–20.

[40] Originally opened with sponsorship from the City & Guilds of London as the Central Institution in 1884–5, renamed the Central Technical College in 1893. During 1907 it became the City and Guilds College as a constituent of the newly formed Imperial College of the University of London.

[41] Charles V. Boys, '"Labour-saving apparatus" as an instrument of education', letter to *Electrican*, 34 (1894–5), p. 376. See discussion in Gooday, 'The morals of metering', 1995, pp. 267–75.

in the previous section. Surely, Thompson suggested, it was now acceptable in every laboratory to use some pre-calibrated apparatus, most obviously standard boxes of resistances, sets of pre-weighed weights for the balance, and indeed a mechanical clock for measuring time. In what sense did the use of such apparatus, calibrated so obviously by someone else, cohere with the narrow notion that it was necessary to do *all* the labour in the various stages of a measurement experiment oneself in order to count unequivocally as its author? If pre-calibrated resistance boxes were allowed in the laboratory for measurement, he asked, why not also the pre-calibrated ammeter and voltmeter? Where, Thompson asked rhetorically, 'should the line be drawn?'[42]

Elsewhere I have suggested that physicists drew the line by embracing the precalibrated resistance standards that had been partly gestated in their universe of academic life. They excluded apparatus made in and for the increasingly alien culture of commercial lighting engineering, moralistically condemning direct-reading ammeters and voltmeters as decadent labour-saving 'automatic' apparatus.[43] Here I can suggest another related view of the same topic pertaining to the theme of trust. Those who had for many years taught the measurement of current by means of the tangent galvanometer, such as Boys and Foster, considered that the risk of using ammeters and voltmeters was not only that students could fall victim to an untrustworthily capricious calibration. The more serious problem was that unreflective deployment of such equipment encouraged inexperienced users to develop an unhealthy self-delusion that they had undertaken authentic measurements when they had not in fact done so: They had merely leaned unwittingly on the trustworthy labours of others earlier in the production process. That is why Foster replied to Thompson that, if anyone attained an accurate value for Joule's equivalent by Ayrton and Haycraft's method, all they had done was to show that their voltmeters and ammeters were correctly calibrated. As far as Foster was concerned, users of the ammeter or voltmeter could not claim to have measured anything with such an instrument unless they had also been responsible for the labour involved in its calibration.[44]

Electrical engineers such as Thompson and Ayrton, on the other hand, treated readings from a pre-calibrated ammeter as bona fide measurements. As far as they were concerned, this was not significantly different from regarding results obtained with the Wheatstone bridge with pre-calibrated resistances as also a genuine and complete form of measurement. If in the latter case the labour of calibrating resistance boxes could be bracketed out of

---

[42] See analysis of the debate recorded in *Electrician* 34 (1894–5), pp. 220–1 in Gooday, 'The morals of metering', pp. 265–8.
[43] Gooday, 'The morals of metering', pp. 267–75.
[44] George C. Foster, 'Direct reading instruments and the teaching of physics', *Electrician*, 34 (1894–5), pp. 548–9 and Gooday, 'The morals of metering', p. 269.

consideration in identifying the resistance measurer, so too could the labour of the person who originally calibrated an ammeter be ignored in considering whether the ammeter user had made a real measurement. Using an ammeter to measure current was therefore just as comprehensive and self-contained a measurement technique, maintained Ayrton, as the use of a 'foot rule' to measure the dimensions of a room.[45] In ensuing chapters this problem of the identity of the measurer arises in a number of indirect ways, most generally in regard to the invisibility of the presumed trustworthy calibrator, leaving the responsibility for a trustworthy measurement to fall apparently only on the shoulders of the user.

In Chapter 3 we can see that an important asymmetry emerged in regard to the authorship of resistance measurements (for the mercury or absolute standard). Only when challenges were raised about the competence and trustworthiness of the calibration did the calibrator feature alongside the end-user as a possible 'co-author' of the measurement. In Chapter 4 we shall see questions arise about the best instrument for the simultaneous measurement of a current by two separate individuals, and in Chapter 5 the problem will emerge of how to devise a commercial instrument for measuring self-induction which is amenable for use by only one person. In Chapter 6 the authorship of a measurement becomes most problematic of all. The domestic meter appears prima facie to be a self-recording instrument that required no human intervention for a measurement to be made. However, customers and suppliers could and did argue about their respective prerogatives in access to meter readings and in being trusted to take readings most accurately. We shall see that Edison's attempt to install domestic meters that could be read only by company officials was controversial because it denied consumers the prerogative of being authors of measurements of their own consumption.

Whilst the issue of authorship of measurements is relatively opaque in Foster's 1881 STEE lecture, he did explicitly stress the importance of the measurer's holding a 'clear conception of the things to be measured. Such conceptions, he maintained, usually grew up by degrees among many workers from indistinct beginnings until in some one mind, they took definite shape and received sufficient 'precise expression' to be subjected to mathematical manipulation.[46] Only for the cases of electrical resistance and self-induction could this account conceivably be reconciled with historical evidence. As yet, however, no historian has definitively shown that either Ohm's theoretical development of the notion of electrical resistance[47] or Clerk Maxwell's claims concerning electromagnetic self-induction preceded any attempts to

---

[45] Gooday, 'The morals of metering', p. 267.   [46] Foster, 'Inaugural address', pp. 9–10.
[47] Christa Jungnickel and Russell McCormmach, *The intellectual mastery of nature: Theoretical physics from Ohm to Einstein*, 2 Vols., Chicago/London: University of Chicago Press, 1986, Vol. 1, 'The torch of mathematics 1800–1870'. Debate over whether Ohm's law was an analytical definition or empirical claim continued, however, into the 1890s, see *Electrician*, 34 (1895), pp. 136–7.

measure the parameters in question.[48] I suggest that few kinds of electrical measurement were premissed on the prior formulation of a 'precise' theoretical conception. There was, as I mentioned in the Preface to this book, much disagreement on what constituted electricity.[49] Latimer Clark warned student readers of his *Elementary Treatise on Electrical Measurement* of 1868 that natural philosophers were 'not in accord' as to the nature of electricity. He thus pragmatically advised students that, until their views were 'more matured', they should regard it as a single substance having a 'veritable existence' like gas or water, amenable to being 'pumped' out of the Earth at one point, and 'poured' into it at another.[50]

If it was difficult to accomplish a consensual articulation about what electricity was, then it was perhaps even more difficult to give a theoretical articulation of what constituted an electrical current. Thus, although in Maxwell's *Treatise on Electricity and Magnetism* of 1873 one can see electric charge formulated rather abstractly in terms of the 'displacement' where matter is placed in an electric field,[51] there is no account anywhere to be found of what a current actually was. What Maxwell offered instead was a phenomenology of what brought currents into being: Differences of potential and voltaic batteries, and on the detectable and measurable *effects* of currents, notably electrolytic and magnetic.[52] Notwithstanding this, Maxwell was as much an accomplished expert on the measurement of electrical current as he was in mathematizing its behaviour. In Chapter XV of the fourth section of the *Treatise*, Maxwell gave his readers a wide-ranging account of galvanometers, illustrative of the extent to which Maxwell used these instruments regularly in his own researches.[53] Evidently a 'clear' conception of current was not necessary to be able to measure it, nor was it even necessary to give 'precise expression' to the behaviour of current in order to subject it to mathematical manipulation.[54]

One can in fact invert the force of Foster's comments and observe that, for those who had little or no clear conception of electrical parameters, the process of measuring them could serve another important purpose: convincing the sceptical or uninformed of the supposed 'reality' of empirical phenomena. An editorial in the *Electrician* for 1881 jested that whilst only the ancestors of Silurian trilobites would be 'utterly unacquainted' with notions of

---

[48] See further discussion on this in Chapter 5.
[49] Edmund E. Fournier d'Albe, *Electron theory*, London: 1906; for further discussion see Graeme Gooday, 'The questionable matter of electricity: The reception of J. J. Thomson's "corpuscle" among electrical theorists and technologists', in Jed Z. Buchwald and Andrew Warwick (eds.), *Histories of the electron: The birth of microphysics*, Cambridge, MA/London: MIT Press, 2001, pp. 101–34.
[50] Latimer Clark, *Elementary treatise on electrical measurement*, London: 1868, p.vii.
[51] Maxwell, *Treatise*, 3rd ed., Vol. 1, pp. 65–9.   [52] Ibid., pp. 354–61.
[53] Maxwell, *Treatise*, 3rd ed., Vol. 2, pp. 351–73. See more on this topic in Chapter 4.
[54] Foster, 'Inaugural address', pp. 9–10.

Meanings of Measurement and Accounts of Accuracy 55

length, mass, and time, it was not difficult to conceive of human beings unacquainted with notions of resistance, quantity of electricity, or electromotive force.[55] Teachers of physics and engineering certainly found such humans in their classes. Fleeming Jenkin's experience of lecturing engineering students at Edinburgh University was that, whilst his best students apprehended the notion of resistance from theoretical exegesis, the 'second-best young man' only really grasped it through the process of 'numerical comparison' called measurement. To an International Health Exhibition in London in 1884 he made this declaration:

> You may define the absolute unit of electrical resistance as accurately as you will, and your definition shall affect the average brain to no perceptible extent; but a young man of very ordinary education and intelligence can learn to measure resistances in ohms, and having learnt this, an ohm becomes a reality to him. Not only does the knowledge he has acquired make him a more valuable assistant to the engineer and contractor, but having acquired a working faith in the existence of ohms, he is prepared to take some trouble to understand the scientific definition.[56]

If some such individuals apprehended the meaning of 'resistance' only by repeated performances of controlled laboratory measurement, we need not take at face value Jenkin's claim this was due simply to intellectual deficiency. An acquaintance with the daily workings of electrical technology might well have convinced a novice engineer that Ohm's law failed to capture fully the sometimes disorderly relation between the potential difference across a conductor and the current through it. Even so, by the mid-1880s, at least, many practising electrical engineers were familiar enough with the notion of resistance as a property of a metal – albeit one dependent on environmental conditions – to find the notion of self-induction a rather peculiar challenge. The effects of self-induction in ac circuits seemed to constitute a new form of resistance which appeared to contradict Ohm's law insofar as its magnitude was frequency dependent and contingent on the material configuration and juxtaposition of circuit elements. This point was addressed by Jenkin's close associates in electrical engineering, Ayrton and Perry, when attempting to introduce their standardized means of measuring self-induction at the STEE in April 1887:

> ... during the fifty years that have elapsed since Lenz employed the 1 foot of No. 11 copper wire as his unit of resistance, the electrical world has not only learnt to regard resistance as a definite property of a definite piece of matter in a definite state, but has become so fully imbued with the idea, that it positively resented [the] experimental

---

[55] Editorial, 'The ohm', *Electrician*, 7 (1881), p. 24.
[56] Fleeming Jenkin, 'On science teaching in laboratories', in Sidney Colvin and James Alfred Ewing (eds.) *Papers of Fleeming Jenkin*, 2 Vols., London: 1887, Vol. 2, p. 185. My thanks to Bruce Hunt for drawing my attention to this source.

proof, last year, that the resistance of a conductor for an intermittent current was a variable.[57]

This 'clear perception' about resistance had not been acquired by the perusal of theoretical treatises. Ayrton and Perry contended that it had been established by the measuring of 'hundreds of thousands' of resistances during the past twenty-two years in terms of a unit of resistance with a 'simple name' – namely the BA unit, first produced in 1865 (see Chapter 3). They considered that only when practical men had made 'many' measurements of the coefficients of self-induction of various coils, electromagnets, and dynamos that were part of their regular work, and had expressed in a unit of self-induction with a simple name, would they come to have the 'same instinctive feeling' about self-induction as they did about resistance.[58] In Chapter 5 I examine the extent to which Ayrton and Perry succeeded in their campaign of familiarization through a direct-reading device for measuring self-induction.

A related problem of familiarization arose for Edison when attempting to persuade potential British and American customers that electricity supply could be metered like gas or water. His official biographers reported that there was 'infinite scepticism' around him on the subject. The public took it for granted that anything so 'utterly intangible' as electricity, that could not be seen or weighed, and gave secondary evidence of itself only at the exact point of use, could not be 'brought to accurate registration'.[59] Thus *pace* the *Electrician* and Foster, Edison's problem was not 'ignorance' nor lack of clarity on the part of consumers, but rather the very extent of their familiarity with the immaterial nature of electricity. In Chapter 6 we shall see that Sebastian Ferranti effectively won customers' faith over to his meters by making them closely resemble the operations of the more familiar – if not entirely trusted – domestic gas meter.

Finally, it should be noted that it was not even just theoretical ignorance or scepticism that posed persistent problems for practitioners in establishing the nature of what a measurement device actually registered. There was also the specifically material spatial problem of *interference* to be addressed: The difficulty of preventing possible disturbing effects from undermining the credibility of measurements. Any uncontrolled or unaccounted source in the environment might indeed compromise – or even comprise – the identity

---

[57] William E. Ayrton and John Perry, 'Modes of measuring the coefficients of self and mutual induction', *JSTEE*, 16 (1887), 1887, pp. 292–3. See Chapter 5 for a discussion of Hughes' controversial experiments.

[58] Ibid., 292–3. Ayrton and Perry returned to this notion of 'instinct' at the IEE two years later, Ayrton and Perry, 'Laboratory notes on alternating current circuits', *Electrician*, 24 (1889), pp. 284–8.

[59] Frank Lewis Dyer and T. M. Morton, *Edison: His life and inventions*, 2 Vols., New York/London: 1910, Vol. 1, p. 406.

Meanings of Measurement and Accounts of Accuracy

of what is being registered in a measurement.[60] Thus the practice of measurement consisted in much more than registering a number on a scale. It consisted in organizing and isolating the target of the planned measurement so that there could be as little doubt as possible as to what was being registered by an instrument. This was an especially difficult matter within a building whose architecture and administration might not have been at all conducive to this process. As we shall see in Chapter 4, this was a tricky point not just for the laboratory in the busy urban setting, but also in the windswept telegraph hut or among the gyrating metal machinery of a power station. Subtler, but equally problematic, was the business of establishing reliable mechanisms for measuring at a distance: The domestic electrical meter was supposed to register energy consumption reliably, not least by being robust enough to pre-empt householders' interference with its readings (see Chapter 6). For all such contexts, the readings of an electrical instrument might not necessarily represent the electrical quantity putatively undergoing measurement – even if users thought they were certain of both who the measurer was and the theoretical nature of their task.

2.3. 'REASONABLE AGREEMENT' AND THE MULTIPLE MEANINGS OF ACCURACY

In Section 2.1 we looked at the allegedly comparative nature of late Victorian accounts of the measurement enterprise. We found there that reconciliation of such accounts with the actual practices of measurement was not straightforward. Insofar as measurement was some form of comparative practice, there are two remaining issues to apprehend regarding how such comparisons were conducted. First, there is the comparative relation between the number accomplished by measurement and the putatively 'true' number that was *expected* on theoretical grounds or by reference to a definitive empirical determination. Since the middle of the twentieth century, the closeness of this sort of relation has been labelled specifically as the accuracy of a measurement, although before that the terms precision and exactitude were often applied synonymously. Second, there is the comparison of *different* and invariably non-identical attempts to make the same measurement. Within the 'scattered' results obtained from many readings, the numerical refinement with which a measurement could be confidently specified has come in the past half century to be known specifically as 'precision'; before that it was known as the 'degree of accuracy' of a measurement. These two comparative notions are not unrelated, for a *quantitative* specification of the closeness of a measurement to its true value carries some (implicit) claim of

---

[60] See Sophie Forgan and Graeme Gooday, "'A fungoid assemblage of buildings:" Diversity and adversity in the development of college architecture and scientific education in nineteenth century South Kensington', *History of Universities*, 13 (1994), pp. 153–92.

its numerical refinement, if not necessarily vice versa. And as Kathy Olesko has shown, German observers in the late nineteenth century generally took greater trouble to differentiate between these notions than did most British practitioners.[61]

Thomas Kuhn's classic 1961 paper, 'The Function of Measurement in Modern Physical Science',[62] bears closely on the first kind of comparison, concerning accuracy. There he explains how scientists decide that measurements and theoretical predictions are in 'reasonable' agreement with each other so that measurement work – and controversies associated with them – can be brought to a close. As Kuhn noted, no honest practitioner *ever* observed a perfect digit-for-digit consonance between measurements and predictions; those who claimed to have done so were dismissed as fraudsters.[63] Given that perfect agreement between measurements and predictions is axiomatically impossible, the grounds for 'reasonableness' are necessarily contingent: It is simply not self-evident how *close* an imperfect numerical agreement has to be in order for it to be a *reasonable* agreement. This can be seen in relation to empirical tests of Maxwell's electromagnetic theory of light outlined in 1865 and reiterated in the latter part of his *Treatise* of 1873.

According to that theory, the velocity of light in air should be numerically equal to the inverse square root of the product of the electromagnetic and electrostatic constants.[64] Schaffer claims that a discrepancy of 4% between this and Foucault's value of the speed of light of 29.84×10^7 m/s was judged to be a 'disagreement' in 1865. Maxwell himself contended in 1873 only that the quantities were at least 'of the same order of magnitude' and hoped that having the two values 'more accurately determined' would resolve the matter. By the early 1880s the discrepancy had been reduced by various sympathetic experimenters to 0.8% so that, according to Schaffer, the two figures matched each other 'rather well'.[65] The problem raised by Kuhn is why an agreement to within 4% was so much less persuasive than one established to within 0.8%: Did manifest disagreement somehow transmute into reasonable agreement once a discrepancy fell below a specific threshold of, say, 1%? Noting that rules for such judgement are very hard to find explicitly

---

[61] Kathy Olesko, 'Precision, tolerance, and consensus: Local cultures in German and British resistance standards', in *Archimedes*, Dordrecht, The Netherland/Boston/London: Kluwer, 1996, pp. 117–156, esp. pp. 118 and 124–31.

[62] Thomas S. Kuhn, 'The function of measurement in modern physical science', originally published in 1961 reproduced in T. S. Kuhn, *The essential tension: Selected studies in scientific tradition and change*, Chicago: University of Chicago Press, 1977, pp. 178–224.

[63] Kuhn notes that this inevitable discrepancy has become so well institutionalized that scientists have granted it facetious recognition as the fifth law of thermodynamics: that 'no experiment gives quite the expected numerical result'. Ibid.

[64] Maxwell, 3rd ed., articles 784–7, pp. 434–36. See also Peter Harman, *The natural philosophy of James Clerk Maxwell*, Cambridge: Cambridge University Press, 1998, pp. 64–8.

[65] Schaffer, 'Accuracy is an English science', pp. 148–163.

articulated in canonical texts or textbooks, Kuhn observes that novices learn the tacit criteria for reasonable agreement from the example set by fellow practitioners in their field. These criteria are in fact highly field dependent: An order-of-magnitude similarity suffices in contemporary cosmology but not in optics, in which agreement is expected to be within at least a millionth part. Kuhn also pointed out that such thresholds changed over time, each generation of practitioners demanding closer numerical agreement than its predecessor, especially as new instruments, new techniques made possible ever more refined standards of work.[66] As W. H. Preece said of William Thomson's 1888 paper on current balances, 'some of us are rather anxious to know when Sir William is going to stop. All his instruments are marked by wonderful accuracy, but he will not cease.'[67]

Kuhn's account enables us, by analogy, to draw out an important point about how judgements of accuracy are made in electrotechnology. 'Reasonable agreement' between measurement and predictions is analogous to what counts as 'sufficient accuracy' in a measurement for technological purposes, particularly in the reproducibility of measurements to meet a given specification. What counts as *sufficient* accuracy for an electrical measurement depends on the actual purpose of the measurement. Consider, for example, the issue of electrical resistance in manufacturing testing. Whilst a late Victorian maker of electric light bulbs might reasonably allow a variation of 1% in the filament resistance, this would not be sufficiently accurate for quality control in the manufacture of standard resistance coils, for those customers might well have demanded an accuracy of the order of 0.1% or even 0.01% instead. The difference here is in an importance sense economic as well as epistemological. Whilst an electrician might have been interested to know the resistance of a particular light bulb to within a small fraction of a per cent, it would have made little difference to the financial feasibility of the lighting enterprise to know this; on the contrary, it would probably be an imprudent use of time and money to use equipment that was more refined – and thus more expensive – than was needed for the task. This financial issue also matters when one considers the diachronic aspect of the analogy. As transatlantic telegraphy developed after 1866, the average length of cables increased from hundreds to thousands of miles. Accordingly more accurately manufactured coils were needed for the speedy use of resistance-based methods to locating cable faults that impeded the lucrative flow of telegraph communication. Such faults needed to be identifiable to within one or two knots in order to be mended with due rapidity, a task that became ever more challenging as longer telegraph cables were laid and more pressing as the volume of telegraph traffic increased.

[66] For further analysis of the *learned* nature of similarity judgements, see Kuhn, 'Second thoughts on paradigms', *The essential tension*, pp. 293–319, esp. 305–18.
[67] William Preece in discussion to Thomson, '...inspectional instruments', 1888, p. 558.

I would like to go further than Kuhn in arguing for the contingency of what counts as reasonable agreement or sufficient accuracy. Kuhn's arguments for the field dependence of quantitative criteria are framed with the assumption of a consensus about 'normal science' among practitioners in a given 'paradigm'. Yet, unlike Kuhn, I emphasize that the level of quantitative (dis)agreement that was deemed reasonable was very much issue dependent. For example, in the mid-1880s it is striking that, whereas electrical researchers held 0.8% to be a sufficiently close agreement between Maxwell's electromagnetic theory and measurements of the speed of light, they expected agreement of 0.1% in the BAAS determination of its absolute resistance. Accordingly a discrepancy of just over 1.2% in 1881 was a serious problem for them. Yet, as we shall see in Chapter 6, the same practitioners later that decade would have judged an (in)accuracy of 1.2% in the readings of their domestic electricity meters to be extremely good as gas consumers were long reconciled to using gas meters that were guaranteed only to within 3%–4%. What was sufficient accuracy in this context was, to some extent, the same as the level of accuracy to which the relevant constituencies were accustomed. Rather more pragmatically, however, what was sufficient accuracy was also a matter of what was available and affordable: Even if significantly more accurate meters had existed, it is unlikely that many consumers could have afforded to hire such refined apparatus.

Kuhn's arguments can be radicalized ever further. Even within one disciplinary paradigm such as electrical research, the criterion of what was satisfactorily accurate measurement for a given purpose could be the subject of dispute, even to a point beyond reconciliation. In the controversy between Augustus Matthiessen and Werner von Siemens concerning the best metal to use in making resistance standards, each side demanded considerably higher standards of reasonable agreement of their opponents' results than of their own. And each mobilized a variety of ways of challenging their rival's claims to have accomplished agreement to within a small fraction of a per cent specified standard (Chapter 3). Even where there was something approaching a Kuhnian consensus about what standards of accuracy *ought* to be expected, these were not necessarily thereby attained. In his Presidential address to the STEE in 1881, Foster reported repeated complaints of the 'want of agreement' between resistance boxes issued by different makers and even between those issued by the same maker. Foster considered that this 'very serious evil' could be remedied only if an authoritative body were to authenticate the calibration of resistance coils to within sufficient accuracy for practical purposes.[68] Such widespread complaint that electrical-measuring

---

[68] Foster significantly concluded by proposing that this role should fall to the STEE; Foster, 'Inaugural address', pp. 15–19. Compare with Schaffer's account of the movement of the standards programme to Cambridge in Maxwell's period in Simon Schaffer, 'Late Victorian

technology was not of sufficient accuracy for its purposes is historiographically significant. Mere demand for a specific level of accuracy did not immediately bring electrical equipment of the relevant qualities into existence, as a simplistic creed of economic functionalism might have supposed. Desired standards of accuracy were attained – if at all – only after sustained long-term efforts in developing new techniques.

This is evident in the development of mechanical engineering in the century before the growth of electrical-lighting technology. This was a field in which accuracy was not prima facie a closeness of fit between numbers, but rather a closeness of fit between mechanical parts. According to Samuel Smiles' *Industrial Biography*, James Watt's first steam engines were made with such little accuracy that the lack of fit could be seen, heard, and felt. Although Watt was proud of an 18-in.-diameter cylinder that varied by no more than 3/18 in. (less than 1 part in 100), even such expedients as packing the piston with 'chewed paper and greased hat' were insufficient to prevent an alarming waste of steam – or often complete seizure. By the time Smiles was writing in 1863, a discrepancy of 1 part in 5000 generally warranted the outright rejection of a machined cylinder: evidence he used to argue for the progressive effects of automated manufacture.[69] Six years before that, however, Joseph Whitworth had told the Institution of Mechanical Engineers that he expected his Manchester workmen to work within twice this accuracy.[70] A certain pluralism of expectations about minimum accuracy persisted in engineering for decades thereafter. In 1896, *The Engineer* declared that less than a hundred of the thousands of engineering firms that operated over the globe insisted on a 'close approximation to accuracy of work'. Many others were content with 'less perfect' work, yet turned out technology that performed its assigned task, in some cases even with 'great efficiency'.[71] We can thus easily understand how Foster found that not all resistance coils were manufactured to within the same limits of accuracy.

---

metrology and its instrumentation: a manufactory of ohms', Robert Bud and Susan Cozzens (eds.), *Invisible connections: Instruments, institutions, and Science*, Vol. IS09 of the SPIE Institute Series, Bellingham, WA: SPIE, 1992, pp. 24–55. This centralized calibration work in Britain was undertaken by the Board of Trade Standardizing Laboratory ca. 1891.

[69] Samuel Smiles, *Industrial biography: Iron workers and tool makers*, London: 1863, p. 181. Smiles specified that a variation of 1/80 in. in a 5-ft cylinder would entail rejection.

[70] Joseph Whitworth, 'On a standard decimal measure of a length for mechanical engineering work', reprinted in Joseph Whitworth (ed.), *Miscellaneous papers on mechanical subjects*, Manchester: 1870 (each paper separately paginated).

[71] 'Accuracy', *The Engineer*, 81 (1896), p. 90. Donald Mackenzie argues that, during the Cold War, the US military attached an entirely contingent value to the necessity of ballistic 'accuracy' in the targeting of its intercontinental ballistic missiles, policies fluctuating with changing perceptions of the strategic value of blanket vis-à-vis pinpoint bombing. Donald Mackenzie, *Inventing accuracy: A historical sociology of nuclear missile guidance*, Cambridge, MA/London: MIT Press, 1990.

Thus far I have talked about the term accuracy in a strictly *quantitative* sense. In this sense it was often used in the late nineteenth century synonymously with precision and exactitude, despite efforts by Mansfield Merriman to promote a statistical redefinition of 'precision' as the reciprocal of 'probable error.'[72] This same synonymity was also deployed in an article titled 'Precision of Observation as a Branch of Instruction', published by the American journal *Science* in 1883, and in T. C. Mendenhall's address on 'Measurements of Precision' to students at Johns Hopkins University in 1894.[73] Both pieces contain clues, however, to further *qualitative* meanings of accuracy and precision. First, as outlined in the previous chapter, it was generally assumed that to accomplish a given amount or degree of accuracy, an appropriate amount of care had to be exercised. Thus the writer for *Science* complained that when students of chemistry, engineering, or electricity had to use an unfamiliar instrument they would too often develop the 'wildest notions' as to the degree of accuracy attainable by exercising merely 'ordinary' degrees of care. Better professional courses of systematic instruction were required that would enable students to judge when it would be appropriate to exercise great care in aiming for high accuracy, and when it would not:

A good routine observer is one who, being informed of the accuracy desired at each step, is able to take just care enough to attain it, without wasting time and energy in uselessly perfecting certain parts of the work. Our professional observer must add to this the good judgement which is able to discover the relative accuracy required in different parts of a complex observation, and to decide how accurately to aim to make a single performance of the whole.[74]

In the context of measurement, care in measurement seems to have referred to attention to detail – a form of fastidiousness or 'taking pains' in matters of experimental preparation that eliminated actual or potential sources of

---

[72] See subsequent discussion for Mansfield Merriman's examples of the application of the notion of precision in error theory in his *Elements of the method of least squares*, London, 1877. Charles Pierce, who, like Merriman, worked for the US Coast and Geodesic Survey, used the term precision interchangeably with accuracy; see Charles S. Pierce, 'The doctrine of necessity examined', *Monist*, 2 (1892), pp. 321–37, cited in Hacking, *Representing and intervening*, p. 240. In analysing the meaning of these terms Schaffer notes the etymological origins of accuracy as denoting *care* expended ('accurate' = past participle of 'to perform with care'), contrasting this with precision as 'abbreviation' (from *praecisio* = to cut off), and exactitude as 'evoking exacting demands' (from *exactus* = driven out). Schaffer does not overlook the importance, especially for a Wittgensteinian account of meaning, of the significant degree of interchangeability among such terms in everyday language. Schaffer, 'Accurate measurement is an English science', p. 139.

[73] Thomas C. Mendenhall, 'Measurements of precision', *American Journal of Science*, 50 (1894), pp. 584–87, esp. p. 584.

[74] Anonymous, 'Precision of observation as a branch of instruction', *Science* 2 (19 October 1883), pp. 519–20.

inaccuracy (see the case of Thomson and Preece discussed in the next section). The epistemological upshot of this was that, if sufficient care were taken to pre-empt error in experimental work, accuracy *qua* truthfulness would then be accomplished. This form of care can be seen as a typically masculine preoccupation with time-consuming technological management that depended on the virtues of perseverance and energy.[75] It can thus be distinguished from the notion of care to which feminist scholars have rightly drawn attention: The practice of sustaining family dependents traditionally associated with women's work in nineteenth-century family life.[76] Whilst care for apparatus was very important in the accurate work of physicists and chemists, we shall see in Chapters 4–6 that designers and manufactures exercised sufficient care in the production of electrical engineering measuring instruments that their users would not (ideally) need to exercise great care themselves.

Insofar as accuracy was the bearer of value-laden qualitative meanings concerning care, it is important to note that accuracy per se was generally not attributed to *humans* as an intrinsic quality. Rather, it was applied to describe human activities or accomplishments such as observations, reasoning, calculations, speech, writing, and mental processes.[77] Thus accuracy was a quality that practitioners of science and technology might aspire to in their work and for which they might be praised for accomplishing. It was certainly applied as a flattering 'virtuous' epithet to the career work of individuals on public occasions: In 1869 the President of BAAS Section A, J. J. Sylvester, praised the Association's President, Professor G. G. Stokes, as being 'distinguished for accuracy and extent of erudition and research'.[78] Other commentators represented accuracy not just as a virtue but rather more strongly

---

[75] Samuel Smiles, *Character*, London: 1873, see esp. p. 143 on 'energy and perseverance.' See Foster's comment on Adolf Weinhold, (trans., ed. Benjamin Loewy), in *Introduction to experimental physics*, London, 1875, that 'the most indispensable qualification for one who wants to make real use of this book is *perseverance*', *Telegraphic Journal*, 4 (1875), p. 59.

[76] For the feminist analysis on two distinct notions of care as 'disposition', and care as 'practice' see Joan C. Tronto, *Moral boundaries: A political argument for an ethics of care*, London/New York: Routledge, 1993; Elizabeth B. Silva, 'Transforming housewifery: Dispositions, practices, technologies', in E. B. Silva and C. Smart (eds.), *The 'new' family: Dispositions, practices and technologies*, London: Sage, 1998.

[77] I have shown elsewhere that the strategic importance of 'accuracy' for later Victorian experimental physicists lay in a variety of concerns. Through it they could identify their work as representing a virtuous practice and accomplishment; accuracy was, they argued, crucial to improving industrial efficiency, achieving progress in physics and technology, and or guaranteeing rigour in the process of (scientific) education. Rhetorical invocations were often made by academic physicists and electrical engineers, in the 1860s to 1880s, seeking to promote their newly founded laboratories as ideal venues for accomplishing these virtuous aspects of accuracy through specialized forms of measurement practice and training. Graeme Gooday, 'Precision measurement and the genesis of physics teaching laboratories', *BJHS*, 23 (1990), pp. 25–51.

[78] John Joseph Sylvester, Address, *Report of the BAAS*, 1869, Part 2, Section A, p. 2.

as a moral obligation. Augustus de Morgan, Professor of Mathematics at University College, London, who taught Frederick Guthrie, William Ayrton, and Alexander Muirhead, reportedly put great stress on accuracy as a 'duty'. An instrument manufacturer in later life, Muirhead had this sense of duty inculcated in him from attending de Morgan's lectures circa 1866–8. Some sixty years later, Elizabeth Muirhead reported what her late husband took to have been de Morgan's influence:

> To mislead another is a greater crime than to murder, [de Morgan] would say, for killing stops with its accomplishment, whereas to state what is not true is to spread errors that may lead generations astray. This doctrine, Alexander confessed to me, had always kept him from publishing his own work, for fear that he should mislead others, and yet when – late in life – I asked him what had, in his opinion, been the great feature of his work, his reply came promptly 'Absolute accuracy.'[79]

I previously mentioned that accuracy could be predicated on human accomplishments if not usually on persons themselves. Similarly the accomplishments of human technological labour were also treated as entities capable of bearing the qualities of accuracy and precision in both quantitative and qualitative senses. Machine tools could be described as accurate or precise if they had the capacity to produce parts that fitted 'true' together to within a numerically specific tolerance. And the artefacts made by such technology (with at least some kind of human input) were accurate or precise by virtue of the fine engineering techniques by which they were constructed. For instruments, however, there is another contemporary sense in which the meaning of precision was more flexible than accuracy: An instrument could be precise insofar as it showed *sensitivity* or *delicacy* in its behaviour, quite irrespective of whether it could be accurate *qua* truthful. Smith and Wise demonstrate that, in his Glasgow University natural philosophy lectures and laboratory work, William Thomson framed the value of precise instrumental measurement within an aesthetic discourse of 'delicacy', 'sensitivity', and 'beauty' rather than of 'truthfulness'.[80] The mirror galvanometer introduced by Thomson to the fraught transatlantic cable-laying expeditions of 1858 showed both types of quality, its sensitivity being apparent in its facility for detecting exceedingly faint transatlantic electrical signals.[81] In Chapter 4, however, we will see that sensitivity and accuracy could also be complementary instrumental qualities, the sensitivity to

---

[79] Elizabeth Muirhead, *Alexander Muirhead, 1848–1920*, Oxford: Blackwell, privately printed, 1926, p. 23.

[80] Crosbie W. Smith and M. Norton Wise, *Energy and empire: A biographical study of Lord Kelvin*, Cambridge: Cambridge University Press, 1989, pp. 127–8, 243–4, 282, 454–6, 685–6, 700.

[81] Ibid., pp. 669–78, 714, 711. From 1869 onwards, Thomson's patented instruments were the only ones used in signalling work on long-distance submarine telegraphy; ibid., p. 704.

extrinsic sources of interference making some instruments prone to 'inaccurate' measurements.

Then again, an instrument could be precise without any capacity for furnishing measurements, let alone quantitative accuracy. In February 1878, George Forbes, Professor of Natural Philosophy at the Andersonian College in Glasgow, wrote a short piece for *Nature* describing Alexander Graham Bell's newly invented telephone as 'an instrument of precision'. He claimed that it was capable of outperforming the mirror galvanometer in receiving transatlantic cable transmissions, the telephone responding to very small currents with an audible click even when the galvanometer spot hardly moved at all.[82] If the telephone could thus be precise in registering minute phenomena without ever being quantitatively accurate, it could nevertheless be accurate in representation of communicated spoken language. And in this communicational context we should note that linguistic precision did indeed have a quite distinct meaning from linguistic accuracy in late nineteenth-century accounts. Thus, for example, when *The Electrician* attacked the 'want of precision' in nomenclature for electrical units sanctioned by the relevant national committees in 1889, it was not criticizing the language of farads, ohms, and so forth as being technically misleading or untruthful in any sense. Rather, it was commenting on residual ambiguities resulting from a lack of sufficient *refinement* in the definitions of the relevant terminology.[83] Thus in qualitative descriptions of both instruments and language the term precision had a common connotation of 'refinement.'

In Section 2.5 we shall see how Merriman and others attempted to import this usage into measurement work by creating a distinctively new statistical meaning of precision. But before then, having explored some of the various possible meaning of accuracy and precision, I turn now to the morally loaded question of who or what was *responsible* for their accomplishment.

2.4. RESPONSIBILITY FOR ACCURACY AND ERROR: THE POLITICS OF AMBIVALENCE

We have found that a great many of the errors which in the author's paper are put down to electrical causes are really due to want of accuracy in mechanical workmanship.

R. E. B. Crompton, discussion of 'Direct Reading Instruments for Switchboard Use', IEE, 1904[84]

---

[82] Forbes described this instrument as perhaps the 'most delicate test' of a current then available and proposed that it be used instead of the galvanometer in telegraphic work; George Forbes, 'The telephone as an instrument of precision', *Nature* (London), 17 (1878), p. 343.

[83] 'Electrical nomenclature', *The Electrician*, 24 (1889), pp. 146–47, especially p. 147.

[84] Edgcumbe and Punga, 'Direct reading instruments for switchboard use', discussion, quote from pp. 658–9.

An interesting ambivalence recurs throughout later nineteenth-century discussions as to whom or to what the accuracy (or error) of a measurement or technological production could be attributed. This hinged on two sorts of questions: To what extent did accuracy depend on the qualities of the measuring technologies, and if so was it in virtue of their design or of the labour that created them? On the other hand, was the accomplishment of accuracy instead to be attributed more directly to the human labour involved, but if so, to whose role within the division of labour? In the nineteenth-century hagiographical literature on engineering we see these issues thrown into sharp relief. Samuel Smiles identified the 'love of accuracy' as a quality that marked out Henry Maudslay as the greatest of engineers. And under Smiles' heavy editorial hand, James Nasmyth declared that he cherished the 'class' of technical dexterity in workmen which enabled them to produce mechanical parts with perfectly true plane surfaces.[85] On other accounts, and for somewhat obviously political reasons, the accuracy of finished technologies was attributed to the introduction of automated machine tools rather than that of the skilled artisan. Deftly passing over the troubled industrial relations that precipitated the introduction of such machine tools by Manchester manufacturers Joseph Whitworth and Richard Roberts, William Fairbairn made much of the qualities of these technologies in his Presidential Address to the BA at Manchester in 1861:

> It is to the exactitude and accuracy of our machine tools that our machinery of the present time owes its smoothness of motion and certainty of action. When I first entered this city [in 1817], there were neither planing, slotting, nor shaping machines, and, with the exception of very imperfect lathes and a few drills, the preparatory operations of construction were effected entirely by the hands of workmen. Now everything is done by machine tools, with a degree of accuracy which the unaided hand could never accomplish.

Despite this attribution of meritorious agency to machine tools, Fairbairn then praised Whitworth for being their originator, contending that mechanical science was indebted to him for some of the 'most accurate and delicate' pieces of mechanism ever executed.[86] Ironically, Whitworth himself did not agree with Fairbairn that humans could not be disciplined to achieve the same results as his machines. Whitworth declared in 1870 that he would not 'rest satisfied' until the 'evils' of poor workmanship had been overcome by making his workmen learn to make machinery with an accuracy corresponding to that of his famous machine tools. In a move familiar to readers of Karl Marx's critique of contemporary society, *Das Kapital*, Whitworth represented the agency of insufficiently disciplined workers as a force

---

[85] Smiles, *Lives of the engineers*, p. 227. For a similar discussion see James Nasmyth (ed. Samuel Smiles), *James Nasmyth, engineeer: An autobiography*, London, 1883, p. 148.
[86] William Fairbairn, *BAAS Report*, 1861, Part 1, pp. lxiii–lxv.

tending to counteract the accuracy of his machine tools – and implicitly also the profits of his company.[87]

Smiles' judgements on such matters were also sometimes as nakedly political. In writing *Industrial Biography: Iron Workers and Tool Makers* in 1863, Smiles sympathized with the aims of Richard Roberts, Whitworth's contemporary, as a virtuous man who made automated machine tools that could produce parts with 'mathematical accuracy' to evade the obstructions caused by 'refractory' unionized workers.[88] A contrasting representation had appeared in his ever-popular *Self-Help* published four years earlier. In contexts in which machines were not predominant and in which (not coincidentally) political disputes with workers were not as prevalent, Smiles laid the responsibility for accuracy on the orderly *self-discipline* of the autonomous individual. In his evangelizing survey of the defining virtues of successful men of business, Smiles identified 'application', 'method', 'punctuality', and 'despatch', as important matters, but a distinctly personal responsibility. Accuracy was indeed a prerequisite for accomplishing trust:

> It is the results of everyday experience that steady attention to matters of detail lies at the root of human progress; and that diligence, above all, is the mother of good luck. Accuracy is also of much importance, and an invariable mark of good training in a man. Accuracy in observation, accuracy in speech, accuracy in the transaction of affairs. What is done in business must be done well; for it is better to accomplish perfectly a small amount of work than to half-do ten times as much... With virtue, capacity, and good conduct in other respects, the person who is habitually inaccurate cannot be trusted; his work has to be gone over again; and he thus causes an infinity of annoyance, vexation, and trouble.[89]

Although such concerns, Smiles conceded, might appear to be small matters at first sight, they were of essential importance to human happiness, well-being, and 'usefulness.'

In the world of Victorian physics and electrical engineering, we can find not only explicit references to Smiles' moralism, but also to similar persistent ambiguities about how to attribute responsibility for quantitative accuracy

---

[87] Whitworth's obituarist remarked that his workshop came to be known worldwide as 'a synonym for all that was best, most accurate and most trustworthy of its kind' as a result of the combined discipline of machinery and skilled workers. T. Healey and J. Harrison, *Sir Joseph Whitworth and the Whitworth scholarships*, London: 1887, pp. 1–2. For political studies of machinery, see Maxine Berg, *The age of manufactures, 1700–1820: Industry, innovation and work in Britain*, 2nd edition, London: Routledge, 1994; Judy Wajcman and Donald Mackenzie, *The social shaping of technology*, 2nd edition, London: Routledge, 1999.

[88] Smiles, *Industrial biography*, pp. 271–4.

[89] Samuel Smiles, *Self-help: With illustration of character and conduct*, London: 1858. Quotation from pp. 233–34, 1921 revised edition. See similar comments by Smiles on Richard Roberts' 1848 innovation of machine tools to punch metal plates: These performed with 'a despatch, accuracy and excellence' that he alleged to be impossible with 'refractory men'; Smiles, *Industrial biography*, pp. 271–2.

and inaccuracy in measurement. Just as with questions of the trustworthiness and authorship of measurements, the issue was complicated by the division of labour among instrument designers, manufacturers, users, and the instruments themselves. During the discussion at the STEE in 1888 on Sir William Thomson's new marine voltmeter and current 'weighers' as made by Whites of Glasgow, much was made of Thomson's role as designer in the accomplishment of accuracy. In his elaborate explanation of the shaping and orientation of iron and magnets in his instruments, Thomson flattered the members of the Society present as being all 'scientific' enough to know that consideration of details was the 'very essence' of success in the design of measuring instruments. The working out of such details was, he averred, the 'life-work' of anyone who was engaged in practical science.[90] To this, chief electrician W. H. Preece, of the Post Office responded with this comment:

It was, I think, Mr. Samuel Smiles who stated that attention to details was at the root of human progress. We have learned from Sir William Thomson tonight that attention to detail has been his guiding spirit throughout all his scientific career, and it is attention to detail that we know is the very root of the accuracy of these instruments that he has brought before us.[91]

On Preece's account, an instrument could be the bearer of accuracy in virtue of the quality of human labour invested in it – and *before* it was used in a measurement. There is an ambiguity, of course, as to whether Preece meant an 'accurate instrument' to signify one that was 'accurately constructed' or that was 'able to facilitate an accurate measurement'. The two were interrelated as the removal of errors and uncertainties in the constitution of an instrument was considered a prerequisite of attaining accuracy in measurement work. The electrical manufacturer R. E. B Crompton made just this sort of connection for the case of *inaccuracy* in accounting for the errors of switchboard instruments (see the epigraph that opens this section). Following the paper by Edgcumbe and Punga on this subject at the IEE in 1904, Crompton challenged their explanation that it was the design of the electrical fittings that were to blame for errors in readings. He suggested from his own company's experiences in such matters that the source of inaccuracy lay in the refractory factory workers:

... we have already, by sufficient precision in mechanical manufacture, got over many of the difficulties which have for many years past stood in the way of making instruments sufficiently correct to be interchangeable and so to agree with standards. These difficulties only existed so long as we depended on the personal skill of the workman who made the instruments, and not on the perfection of the tools with which they

[90] Thomson, 'On his new standard and inspectional electrical measuring instruments', discussion, p. 543.
[91] Preece in discussion of Thomson, ibid., p. 557.

are now made. This is an old story as regards the production of repetition work by modern machinery, but in no case are the advantages of it so marked as in the case of turning out moving-coil instruments for switchboard use.[92]

Thus Crompton followed the 'old stories' told by Fairbairn and Smiles in presenting accuracy of instrumentation as matter of bypassing the subversively non-uniform 'skill' of workmen and allowing a fully mechanized Whitworthian system of precision production. This perspective on the origins of error in instruments was specific to those within the managerial stratum of electrical-manufacturing industry. It hinged on the politically and morally loaded assumption that automated machinery was generally more trustworthy than an artisan in producing accurate work. Because many late nineteenth-century instruments were still in part handmade, it was nevertheless impossible for users to distrust *all* skilled human workers as incapable of exercising due care in creating accurate measurement equipment. Indeed, some inexpert end-users might have shared Charles Darwin's difficulty in distrusting the work of remote instrument-makers.[93]

This troubled question about whom in the division of labour of measurement should take the responsibility for accuracy or error – or be distrusted as incapable of acting effectively on such a responsibility – is clearly illustrated by the case of H. G. Wells. The youthfully subversive Wells became an unwilling instrument-maker as a trainee teacher under Frederick Guthrie at the Normal College of Science (later the Royal College of Science) in South Kensington, London. Guthrie was Professor of Physics in South Kensington, where he was employed alongside T. H. Huxley and Edward Frankland in training the burgeoning numbers of Department of Science and Art teachers. Guthrie's greatest challenge was to equip such teachers with classroom demonstration apparatus when instruments sold for elementary instruction were mostly overvarnished 'trash', and the 'instruments of precision' supplied by the best makers were prohibitively expensive.[94] So from 1871 to 1872, Guthrie collaborated with his London colleagues George Carey Foster and William Fletcher Barrett in devising a scheme that enabled students to make and calibrate their own thermometers, barometers, and galvanometers. This is what Foster rather ironically described as acquiring knowledge by 'self-tuition and self-help', despite the fact that laboratory demonstrators,

---

[92] Edgcumbe and Punga, 'Direct reading instruments for switchboard use', discussion, pp. 658–9.

[93] Francis Darwin (born 1848) reports that the whole instrument-making trade was a complete 'mystery' to his father, Charles Darwin: He had such great 'faith' in the micrometers he used in research at Down House that he was 'astonished' when one produced a reading that differed from another. Francis Darwin, *Life and letters of Charles Darwin*, 3 Vols., London: 1888, Vol. 1, pp. 147–8. I thank John F. M. Clark for giving me this reference.

[94] Frederick Guthrie, 'Teaching physics', *Journal of the Society of Arts*, 34 (1886), pp. 659–61, quotes on pp. 660–1.

notably Charles Vernon Boys, were often on hand to assist students with difficulties.[95] Guthrie reported that, once students learned that their final results depended on the 'trustworthiness' of their apparatus, they learned to be 'exact' in their manipulations; they could thus make instruments 'far more accurate' than they could conceivably buy, such as a galvanometer accurate to one part in a thousand. Their handmade instruments were 'sufficiently' exact, Guthrie averred, to convince students of the 'faithfulness of nature' and the 'trustworthiness' of the statements of science.

For most trainee teachers Guthrie's scheme was moderately effective, if rather lenient – it was not often that a student failed.[96] Although Wells passed the course graded in the second-class category in 1886, shortly before Guthrie died, he nevertheless vigorously objected to being forced to rely on his own resources to make his instruments accurate. Having spent one year in T. H. Huxley's biology laboratory studying in a well-ordered regime of microscopy, Wells disliked the 'confused workshop' methods adopted by Guthrie to teach him physics:

After breaking a fair amount of glass and burning my fingers severely several times, I succeeded in sealing a yard's length tube, bending it, opening out the other ends, tacking it on to the plank, filling it with mercury, attaching a scale to it and producing the most inelegant and untruthful barometer the world has ever seen...I was then given a slip of glass on which to etch a millimetre scale with fluorine. Never had millimetre intervals greater individuality than I gave to mine.[97]

Even if he had seriously attempted to satisfy the course requirements, Wells contended that the inattentive clumsiness that had troubled his short career as a shop-assistant would have introduced an element of 'absurdity' into his instruments anyway. As it was, the 'distinguished badness' of his apparatus won the admiration of fellow pupils and was preserved in a cupboard for several years thereafter. Yet it was specifically as a comment on Guthrie's conception of education – one of 'colossal ineptitude'[98] – that Wells agreed this apparatus was worth preserving. He personally blamed Guthrie for making the study of experimental physics so preposterous that it hardly merited serious effort in calibrating his instruments. He thereby attributing the

[95] See Foster's review of Weinhold, in *Introduction to experimental physics*, cited in footnote 75.
[96] Guthrie, 'Teaching physics' p. 661. See G. C. Foster, [Obituary of Frederick Guthrie], *Proceedings of the Royal Society*, 8 (1887), pp. 9–13.
[97] Herbert G. Wells, *Experiment in autobiography* 2 Vols., London: 1934, Vol. 1, pp. 212–13.
[98] Ibid. Wells commented that 'the time of the class was frittered away in the most irrelevant and stupid "practical work" a dull imagination had contrived for the vexation of eager spirits.' For an alternative view of Guthrie's work, see Gooday 'Frederick Guthrie' in the *New Dictionary of National Biography* (forthcoming, Oxford University Press), and Gooday '*Precision measurement and the genesis of physics teaching laboratories in late Victorian Britain*, Canterbury, England: University of Kent at Canterbury, unpublished PhD thesis, Chapter 8.

responsibility for the inaccuracy of his constructions beyond his own anarchic labours to the allegedly deficient teachers who had failed to provide him with suitably precalibrated apparatus.[99]

There is a certain irony, then, in the complaints by Wells' other *bête noire*, C. V. Boys, in 1894 that some modern students enjoyed the use of instruments that were unduly precalibrated and had too easy a time in making measurements (see Section 2.2). Clearly Boys the teacher and Wells the student had radically different expectations about practices of accuracy – each considered the *other* to be predominantly responsible for its accomplishment. The further irony of giving unskilled novice students responsibility for the trustworthiness of their own measurements is clearest in G. C. Foster's contemporary letter about his students' work in the University College physics laboratory:

> Our results have not always been good – indeed, they are often very bad – but . . . even a rough result obtained in this way is, I believe, of greater educational value than a far more accurate one got by help of a direct reading contrivance, for the accuracy of whose indications the experimenter is not personally responsible. My contention is that we ought to try to teach students of physics to rely as much as possible on themselves and to depend as little as they can upon the accuracy of the instruments.[100]

As I argue in Chapter 4, however, such extreme self-reliance in guaranteeing the accuracy of measurements was simply impracticable for practitioners of commercial telegraphy and particularly electrical-light engineering. The temporal constraints of their work and the diversity of other technological responsibilities made it unfeasible for them to devote time to making themselves personally responsible for every aspect of an instrument's accuracy. They were effectively forced to work within a division of labour that devolved much of the responsibility for accuracy of measurements to the designers and manufacturers of their instruments. It became the job of the designer to develop instrument mechanisms that performed sufficiently consistently – whatever users later did with them – that they would not undermine any calibration then assigned to them by the manufacturer. The remaining responsibilities for the *users* of an ammeter in securing accuracy were to procure the most trustworthy design of instrument from the most trustworthy manufacturer and check the calibration of their instrument at regular intervals.

---

[99] Wells was unaware that the scheme was not Guthrie's own and later admitted that had he known Guthrie was dying of throat cancer at the time he might have conducted himself differently. Wells, *Experiment in autobiography*, p. 217.

[100] George C. Foster, letter to editor, 'Direct reading instruments and the teaching of physics' *Electrician*, 34 (1894–5), pp. 548–9. See Gooday 'The morals of meeting', pp. 268–72 for further discussion.

I have now considered the complex and contested question of who was or should be responsible for the accuracy of a measurement and indicated how this matter is closely related to the overall authorship of a measurement. The contingent answers to this question in particular contexts will be explored in later chapters. In the next section I explore some further issues that arise in the practical management, reporting, and evaluation of errors in measurement, by whomsoever undertaken.

2.5. REPORTING ACCURACY: THE PROTOCOLS AND LANGUAGES OF ERROR

If your experiment needs statistics, you should have done a better experiment.
                  Remark to Cambridge physics students, attributed to
                  Ernest Rutherford, attributed.[101]

Notwithstanding the anxieties of Boys, Foster, and later Campbell, it was increasingly commonplace by the late nineteenth century to credit end-users of precalibrated instruments as the authors of 'their' measurement, occluding the roles of all others earlier involved in the work of design, making, and calibration.[102] This account of authorship gave a privileged status to the labour of producing the final numerical result for a *particular* measurement. This involved making adjustments to eliminate errors, taking the visual readings from the scale, dial, or calibration chart, and the act of reporting the outcome with analysis of results. In such 'real-time' technological enterprises as the monitoring of supply levels in electrical power-stations oral or 'mental' reportage of measurements was the norm, the crucial matter was to enact adjustments in response to measurements rather than to record the numbers concerned. The situation was rather different in contexts in which a text was to be generated to accompany a measurement – such as in an instrument-maker's or cable manufacturer's testing room or physics laboratory. The individual end-users who performed the calculations of results and wrote the final report generally did so as if they were somehow the *de facto* authors of the measurement.

How then were interested observers or readers to judge the reliability or accuracy of a measurement? For telegraphic electricians it could be the

---

[101] Cited in Wise (ed.), *The values of precision*, p. 11.
[102] Some relatively invisible wives acted as unpaid laboratory assistants for Victorian physicists and electrical engineers. See Sophie Forgan and Graeme Gooday, 'The married lives of the electrical engineers' (forthcoming) and Steve Shapin, 'The invisible technician', *American Scientist*, 77 (1989), pp. 554–63. I owe much to Adrian Johns' analysis of the problematic notion of authorship in Restoration publishing. Adrian Johns, *The nature of the book: Print and knowledge in the making*, Chicago: University of Chicago Press, 1998. The best recent study is Mario Biagioli and Peter Galison (editors), *Scientific authorship. Credit and intellectual property in science*, London: Routledge, 2003.

Meanings of Measurement and Accounts of Accuracy 73

efficiency with which they could use a measurement to identify the location of a distant line or cable fault; for power-station engineers, the best guide was the stability and economy of the supply system. Matters were somewhat different for those who read textual accounts of measurement produced in the testing room or laboratory. For them, much could be gleaned from detailed reports of the precautions – 'care' – taken to eliminate errors and sources of interference, as well as from explicit claims concerning the accuracy of an experiment, and estimates of error if any were given. I have argued elsewhere that late Victorian British physicists presented claims for the accuracy of measurement in a qualitative manner that hinged on the credibility of this reportage about experimental precautions. Rather than judging a measurement simply on quantitative claims concerning its accuracy, an informed reader would assess the credibility of the claimed accuracy by reference to the internal evidence of this reportage and any experience of their own they had had of the problem of undertaking measurements of that sort.[103] It is this sort of approach that Augustus Matthiessen and the BAAS Electrical Standards Committee used in the 1860s to condemn as absurd the claims to accuracy made by Werner von Siemens and his assistants in using mercury to establish a standard of resistance (Chapter 3).

Contemporary German physicists (and engineers) tended rather to offer their readers comprehensive quantitative analyses of experimental error, deploying protocols developed from the corpus of Legendre/Gauss theory concerning the expected distribution of observational errors according to the 'normal' curve. From the distribution of a set of several attempts to make the same measurement, the protocols of least-squares analysis drawn from error theory enabled them to calculate the 'probable error'[104] of a result above or below the mean value. This probable error was reciprocally related to the 'sharpness' of the data set around the mean value: A smaller error was likely to be associated with a greater narrowness of data spread. As Olesko emphasizes, the term *Schärfe*, or sometimes *Präzision*, was thus used literally to refer to the sharpness of the data set, and had a further significance in relation to the number of significant figures for a measurement that could be specified with any reasonable certainty.[105] According to the Gaussian theory, the sharpness or *Präzision* increased, and the probable error decreased, as the square root of the number of observations minus one. Thus whilst a

---

[103] Gooday, 'Experimentation and interpretation'. In regard to the BAAS Committee's approach to citing accuracies, see Schaffer, 'Late Victorian metrology'.
[104] 'Probable error' here meant the value of error for which there was an equal probability (i.e., probability of 0.5) that the actual error was above or below this value. For a set of n attempts at an observation, the probable error was given by the formula $+ 0.6745 \times \sqrt{S} \div \sqrt{n(n-1)}$, where S are the sums of the squares of the errors from the arithmetical mean.
[105] Olesko, 'Precision, tolerance and consensus', p. 119.

single measurement might perhaps have a very high *Genauigkeit* [truthfulness or accuracy] it would have zero *Präzision*. Accordingly, adherents of error theory, such as Friedrich Kohlrausch, considered it essential to undertake a large number of measurements to decrease the error to a certain minimum threshold. Their British-based counterparts, by contrast, were more likely to have undertaken fewer measurements and devoted more effort to maximizing *accuracy* by enhancing the number of antecedent precautionary measures against experimental error. That view accounts for Rutherford's reported derision towards the superfluity of statistics in a properly designed and conducted experiment.

Olesko has shown that the Germanic practice of error analysis and citation was nurtured in Franz Neumann's Königsburg physical seminars from 1834 and was promulgated throughout German physics education and research through the use of such textbooks as Kohlrausch's *Leitfaden der praktischen Physik*.[106] Her evidence on the Germanic case certainly undermines the generality of Ian Hacking's claim that physicists outside of astronomy did not report estimates of error until the 1890s.[107] The international difference of practice in this matter might be explicable on cultural grounds: Olesko suggests that for Germanic readers the reduction of error shared a particular social significance as the elimination of 'deviation.'[108] By contrast, the first British proponent of continental error theory, Astronomer Royal George Biddell Airy, introduced his readers to the notion of error to stand for the less politically loaded 'uncertainty' of data.[109] Certainly in Britain, Hacking is right to make an exception for astronomy. Although 'observers in natural philosophy' were the audience that Airy explicitly anticipated for his *Theory of the Error of Observations* in 1861, over the next two decades only mathematicians and astronomers used his distillation of error theory into a form suitable for everyday practical use.[110]

---

[106] Friedrich Kohlrausch, *Leitfaden der praktischen Physik*, Leipzig, 1870. The simplified summary of error analysis which occupies the very first chapter explains to readers how to quantify experimental error, and succeeding chapters explain how to conduct experiments so that this error should be *minimized*. See Olesko, *Physics as a calling discipline and practice in the Königsberg seminar for physics*, London/Ithaca, NY: Cornell University Press, 1991, pp. 403, 411-2.

[107] Hacking, *Representing and intervening*, p. 234. This somewhat undermines his earlier claim that measurement practices in the late nineteenth century were essentially those subsequently adopted in the twentieth century.

[108] Olesko, 'Precision, tolerance and consensus', esp. pp. 131-5.

[109] See introduction to George Airy, *On the algebraical and numerical theory of errors of observations*, Cambridge/London, 1861. This is not discussed in Schaffer's analysis of Airy's astronomical regime at the Greenwich Observatory; see Schaffer, 'Astronomers mark time: Discipline and the personal equation', *Science in Context*, 2 (1988), pp. 115-45.

[110] Augustus de Morgan, 'On the theory of errors of observation', *Transactions of the Cambridge Philosophical Society*, 10 (1864), pp. 409-27; Isaac Todhunter. 'On the method of least squares', *Transactions of the Cambridge Philosophical Society*, 11 (1869), pp. 219-38;

Meanings of Measurement and Accounts of Accuracy   75

One British reinterpretation of error theory, a translation of Kohlrausch's revised text, published as *Introduction to Physical Measurements* in 1874, tellingly presented it as a treatise on *accuracy*. It offered readers an explanation not only of how to relate the overall 'probable error' of their measurements to the 'inaccuracy' of their individual observations, but also suggested that error analysis might enable them to judge the accuracy of their overall measurements too.[111] This was in contrast to the approach taken by US civil engineer and geodetic survey expert Mansfield Merriman. He translated the German statistical notion of *Präzision* by proposing a new quantitative meaning for the English term precision as the reciprocal of probable error. By implication this was distinct from any already existing meaning for the term accuracy.[112] Merriman's *Elements of the Method of Least Squares* produced in 1877 (revised 1885) became the standard elementary work on the application of error theory to measurements, and it was cited thereafter in textbooks by physicists at Owens College, Manchester, and by the Leeds-based chemist, Sydney Lupton.[113] However, only in the second and third decades of the twentieth century was Merriman's approach widely adopted in Britain. And from the derision of Rutherford noted in the epigraph that

---

James Glaisher, 'On the law of facility of errors of observations, and on the method of least squares', *Memoirs of the Astronomical Society of London*, 39 (1872), Part II, pp. 75–124.

[111] Friedrich Kohlrausch (trans. T. H. Waller and H. R. Procter from 2nd German edition 1873), *An introduction to physical measurements*, London/New York: 1874. pp. 1, 4. Henry Richardson Procter (1848–1927) studied at the Royal College of Chemistry, leaving in 1871, and was a Fellow of the Chemical Society by the time he translated this book with Waller.

[112] Manfield Merriman (1848–1925), *Elements of the method of least squares*, London: 1877. Merriman was an instructor in civil engineering at the Sheffield Scientific School, Yale University, 1875–8, Professor of Civil Engineering at Lehigh University 1878–1907, and an assistant on the US Coastal & Geodetic Survey 1880–5. For biographical details, see *Who was who in America*, Chicago, 1943, 5 Vols. (1897–1947), Vol. 1. Merriman originally expressed the error formula as $y = c \exp(-h^2 x^2)$, in which the empirically determined parameter h was labelled the 'measure of precision' of a set of measurements. Merriman, *Elements*, pp. 10, 18. Compare Merriman, *A textbook of the method of least squares*, London: 1885, pp. 1, 27; Sidney Lupton, *Notes on observations*, London: 1898, p. 75; Arthur Schuster and Charles Lees, *Advanced exercises in practical physics*, Cambridge: 1901, pp. 2–3.

[113] Stewart and Gee, *Lessons in elementary practical physics*, 1885, Vol. 1, pp. 266–75. Stewart's earliest use of error theory can be traced to his work on meteorology at Kew Observatory 1861–70; Stewart cited examples of the application of error methods in General Sabine's 'Report of a magnetic survey of Scotland', *BAAS Report*, 1836 (Stewart and Gee, p. 275). Sidney Lupton, *Chemical notes*, Harrow/Leicester: 1881; idem, *Elementary chemical arithmetic*, London: 1882; idem, *Numerical tables and constants in elementary science*, London: Macmillan, 1884; idem, *Notes on observations*, London: 1898; Schuster and Lees, *Advanced exercises in practical physics*, pp. 1–8. In a significant textual homology with Kohlrausch's volume, the first 8 of the 367 pages in the book by and Schuster and Lees are devoted to 'Errors of observation'.

opens this section, enthusiasm for this innovation was apparently not great at the Cavendish Laboratory in Cambridge.[114]

Rutherford's refusal to countenance statistical reporting of errors of measurement was in fact long precedented in Cambridge physics. Even before his first arrival at the Cavendish Laboratory in the mid-1890s, students had been encouraged to focus not on quantifying errors after a measurement had been made, but on *pre-empting* the occurrence of error by careful arrangement of the apparatus before a measurement was made. The sections on galvanometric technique in James Clerk Maxwell's 1873 *Treatise on Electricity and Magnetism* advised readers, for example, on how to determine instrumental constants with least error and how to use a particular type of galvanometer so that its deflections fell in the region of minimum error. But Maxwell's readers were not told how to establish the probable error of a series of measurement *after* they had been made.[115]

Following in this Cambridge tradition, under the heading 'Errors and Corrections', readers of *Practical Physics* by Glazebrook and Shaw were told about the 'very great increase in care and labour' required in the *preparation* of an experiment to secure a further decimal place in a measurement process. Only later on, however, would readers have found a footnote to Airy's treatise on error theory to support the claim that the 'so-called' probable error of a measurement decreased as the square root of the number of observations. Yet instead of then using Airy's version of least-squares analysis to explicate how a probable error for a measurement outcome might be calculated after the experiment, Glazebrook and Shaw interpreted this to as a clue to yet another *pre-emptive* means of minimizing error. They argued simply that it was 'advisable' to take at least several observations for a given measurement as measurers would thereby get a 'more accurate' result. In a further divergence from continental error theory, the authors recommended that a mere arithmetical analysis of differences between observations would enable them to determine the 'degree of accuracy' of which their method was capable. For example, in analysing attempts to determine a length with obtained results of 3.333, 3.332, 3.334, and 3.334 in. they determined the 'true' length was 'accurately represented' by 3.333 in. to the fourth significant figure, there being a variation of 'only' two parts in that last place in the data set.[116] Readers of Glazebrook and Shaw were evidently meant to infer that a measurement cited as 3.333 in. *implied* that the method employed furnished a 'degree of accuracy' to the fourth figure, and that the

---

[114] David Brunt, '*The combination of observations*, Cambridge, 1917; Norman Campbell, 'The adjustment of observations', *Philosophical Magazine*, 6th series, 39 (1920), pp. 177–94; H. L. P. Jolly, 'Observations, the combination of', in Richard T. Glazebrook (ed.), *Dictionary of applied physics*, 5 Vols., London: 1923, Vol. 3, pp. 644–65.

[115] Maxwell, *Treatise on electricity and magnetism*, 3rd edition, 1891, pp. 382–3.

[116] Glazebrook and Shaw, *Practical physics*, 1885, pp. 31, 34.

error would be no more than one part above or below the figure cited at that place.[117]

So whilst this tradition of reporting measurements did not involve reporting explicitly or theoretically quantified probable errors, practitioners writing such reports could (vaguely) *imply* what they considered to be the upper and the lower limits of uncertainty or error by the placing of the last digit of a number. The assumption was evidently that, if all the important possible errors had been eliminated in the design of a measurement set-up, the sole remaining determinant of the accuracy of a measurement lay in the degree of accuracy associated with the *method* of measurement. The didactic analysis of Glazebrook and Shaw thus continued with a summary of the technology dependence of the possible accuracy of measurements in different fields of physics. Whilst the mass balance could give results to within 1 part in a 1,000,000,[118] angular measures with a vernier could be made to 1 part in 40,000, some length measurements (e.g., with micrometers) could be made to 1 part in 10,000, but a thermometer could not 'without great care' be made to measure temperature to within more than 1 part in 100. Most physical parameters, like length, were susceptible to determinations of only around 1 part in 1000 (0.1%).[119] In October 1888, Sydney Lupton wrote a piece for *Nature*, in which he too argued that the great majority of measurements in physics, chemistry, geodesy, navigation, and crystallography were 'not to be trusted' beyond the fourth or fifth figure. He complained, moreover, that too many convinced themselves that their everyday work could match Airy's measurements to 1/100 of a second in a day, (1/864,000) and Whitworth's mechanized measurements of length to within 1/1,000,000 of an inch.[120]

This problem of overestimating the possible accuracy of methods was a theme that had previously bothered Lord Rayleigh, Glazebrook's and Shaw's former director at the Cavendish Laboratory. A year before he gave his Presidential Address to Section A of the BAAS in 1882 he had demonstrated that the BAAS Electrical Standards Committee of the 1860s had misdetermined the absolute determination of resistance by 1.2%. The committee had unfortunately neglected to measure the self-induction of their whirling-coil apparatus to within a sufficient degree of accuracy. Perhaps alluding back to that episode in his BAAS address, he encouraged his audience

---

[117] Had Merriman's simplest error protocol been followed, this would probably have been expressed in the form 3.333 + 0.001 in.; see Merriman, 1877, p. 30.

[118] Two decades before, Thomson and Tait concluded that few measurements of any kind were 'correct to more than *six* significant figures', the limit that might be achieved is that of a 'fine' weighing balance registering a difference of 1 part in 500,000; William Thomson and Peter Guthrie Tait, *Treatise on natural philosophy*, Oxford: 1867, p. 333. Italics in the original.

[119] Glazebrook and Shaw, *Practical physics*, 1885, pp. 35–6.

[120] Sydney Lupton, 'The art of computation for the purposes of science', *Nature* (London) 37 (1888), pp. 237–9, 262–3, quote from pp. 237–8.

to consider quantifying the error of their experiments *after* they had taken place:

> I hope I shall not be misunderstood as underrating the importance of great accuracy in its proper place if I express the opinion that the desire for it has sometimes had a prejudicial effect. In cases where a rough result would have sufficed for all immediate purposes, no measurement at all has been attempted, because the circumstances rendered it unlikely that a high standard of precision could be obtained. Whether our aim be more or less ambitious, it is important to recognise the limitations to which our methods are necessarily subject, and as far as possible to estimate the extent to which our results are uncertain. The comparison of estimates of uncertainty made before and after the execution of a set of measurements may sometimes be humiliating, but it is always instructive.[121]

Notwithstanding Rayleigh's advice, few physicists in Britain bothered to estimate uncertainties in experimental error for a few decades thereafter.

Like the textbook instructions to physicists previously discussed, successive editions of the *Handbook of Electrical Testing* (1876) by the Post Office engineer Harry Kempe advised novice telegraphists to adopt a pre-emptive approach to the error of measurement and gave them no advice on post hoc estimation of errors. In Chapter 4, I give an account of how Kempe showed his readers how to treat a galvanometer and its working environment so as to remove all normal sources of error from current measurement. Kempe then showed them how to quantify the maximum degree of accuracy attainable in different kinds of tests on telegraph equipment – a theme picked up in the subsequent text by Glazebrook and Shaw. Kempe also emphasised that this was not just a technology-dependent, but also a method-dependent, matter: The degree of accuracy that could be secured with one method of testing the internal resistance of a galvanometer or battery could be twice that of an alternative method. Given an ultrasensitive Thomson mirror galvanometer which could register a 1/1,000,000,000 part of an ampere in a scale deflection of one unit, Kempe reported that Latimer Clark had been able to measure the potential difference of a battery cell to within 1 part in 1,000,000. Such an extreme degree of accuracy mattered for electricians involved in checking for tiny flaws in the gutta-percha insulation of submarine cables. Yet for pinpointing faults in remote lines or cables, they appear to have been satisfied with measurements of resistance taken to within ten thousandths of an ohm or less – such were the orders of magnitude by which 'sufficient accuracy' might range.[122]

---

[121] Lord Rayleigh, *BAAS Report*, 1882, Address to Section A, p. 438.

[122] Kempe, *A handbook of electrical* testing, 2nd edition, London: 1881, pp. 38, 53–70, 114, 154–5. For a discussion of gutta-percha insulation see Bruce Hunt 'Insulation for an empire: Gutta Percha and the development of electrical measurement in Victorian Britain', in Frank A. J. L. James (ed.), *From semaphore to shortwaves*, London: Royal Society of Arts, 1998, pp. 85–104.

Electrical-lighting engineers were also primarily concerned with precluding error before measurement took place. They were comparatively less concerned, however, with assessing the accuracy of *methods* of measurement or with the minutiae of setting up instruments. When the first major wave of centralized power stations was launched in 1888, most engineers were satisfied with measurements accurate to the order of around 1%, but in their busy daily practice they needed instruments that could be read trustworthily at a glance without much preparatory work. So they needed inexpensive and portable instruments that could relieve them of the need to exercise great care in the act of measurement. The resulting division of labour devolved much of the task of error avoidance to the designer and manufacturer. Even so, users were still presumed to have responsibility for maintaining the instruments in good working order and occasional recalibration. Thus manuals on electrical engineering instruments often focused on the optimization of the internal design and shielding so that the user could achieve sufficiently accurate results in any circumstances, even next to a high-speed dynamo. For example, James Swinburne in his *Practical Electrical Measurement*, was emphatic that his work was devoted to instruments that gave accuracy in 'ordinary' everyday engineering use, not those which gave 'absolute' accuracy but only in the conditions of 'mere laboratory practice.' I show in Chapter 4 how Swinburne was also highly critical of the error proneness of certain ammeters' mechanisms and was especially critical of William Ayrton and John Perry for the design compromises they made to accomplish a supposedly 'proportional' calibration in their direct-reading ammeters.[123]

Over the next 16 years the expected accuracy of switchboard instruments increased by a whole order of magnitude, but the experience of working with instruments in power-stations taught engineers that they could not always enjoy the accuracies claimed by designers and manufacturers. This was evident in a paper delivered to the IEE in 1904 by Kenelm Edgcumbe and Franklin Punga. They reported that instruments reading to within 0.1% in a test-room might nevertheless be 5%–10% in error on a power-station switchboard. Another station device, allegedly reliable to a quarter of an ampere in test conditions, would then swing wildly over a range of 10–20 amperes when used in a commercial power circuit. Attempts to overcome such problems by improved shielding or ventilation nevertheless brought into focus the still-unresolved problems of 'intrinsic' error: mechanical friction, magnetic hysteresis, inaccurately engraved calibrations, and parallax. To quantify the relative importance of these problems, electrical engineers began to cite maximum limits of error for each of these parameters in the performance of their instruments. They did not, however, follow the new practice of some physicists in citing errors for individual discrete measurements because switchboard instruments were engaged in a continuous *process* of measurement.

[123] James Swinburne, *Practical electrical measurement*, London: 1888, p. iii.

Yet citing a single overall error for an instrument was tricky. Whilst mechanical errors were effectively constant, the magnitude of likely electrical errors usually varied in proportion to the magnitude of the reading. Some manufacturers compromised by citing the accuracy of their devices at maximum deflection when the highest errors were likely, with the obvious implication that the actual errors would generally be less than this. Edgcumbe and Punga, however, were not satisfied with this. Following an unnamed company of consulting engineers they went further than this to demand not only a certain minimum accuracy throughout an instrument's range expressed as a fraction of the maximum reading; they also demanded that, above two thirds of its full load, there should be a *higher* minimum guaranteed percentage of accuracy, that is, lower error. Performance at high loads merited special consideration in view of the deleterious effects of a fluctuating or ill-quantified supply on large numbers of customers (Chapter 6).[124] Yet again we see how conventions on the setting of standards of sufficient accuracy and the reporting of accuracy and error were specifically related to the social–commercial purpose of the measurement and did not arise from purely epistemological considerations regarding the nature of electrical knowledge.

Clearly, the operative standards of a sufficient degree of accuracy in power engineering was as context dependent as in physics and telegraphy and were prone to perpetual redefinition as ever more was learned about the errors of such instruments in everyday use. As Matthias Dörries has shown, nineteenth-century research on instruments was importantly reflexive in nature. He argues that instruments generated new ideas, ironically, not through their utility for measuring, but through their 'increasingly problematical limitations'.[125] We can argue the same with regard to the everyday usage of measurement instruments that generated new problems and new questions concerning the nature of accuracy, how much accuracy was required for a given task, how it could be achieved, and how that accuracy could be quantified.

2.6. CONCLUSION

In this chapter, I have explored the complexity of late nineteenth-century notions of what constituted authentic and accurate measurement, mapping the ambiguities of what was entailed in accomplishing a measurement and also the multiple qualitative and quantitative meanings of the term accuracy. Looking at the history of electrical measurement in this way enables us to resolve the 'notorious' difficulty that Norton Wise has seen in attempts to show how 'societal context' can affect in an important way how someone makes a measurement. From the preceding account it should be clear that there were unavoidable *decisions* to be faced in considering how to conduct

---

[124] Edgcumbe and Punga, 'Direct reading instruments for switchboard use', pp. 621–2.
[125] Dörries, 'Balances, spectroscopes, and the reflexive nature of experiment', pp. 1–36.

and represent a measurement. Should it be made by methods of comparison, reduction–calculation, or surrogate techniques? Was it necessary to have a clear prior conception of what was being measured or should that emerge only as the result of a measurement? To whom should be attributed authorship of the measurement – the end-user alone or the makers and calibrators of the instruments too? What degree of care was required, what sort of errors should be pre-empted, and by whom, and to what extent to achieve what 'degree of accuracy' for a given particular purpose? As I previously showed, and will further explore in ensuing chapters, practitioners in different disciplines answered these questions in different ways at different times between the 1860s and the 1890s. This is one of the important morals to be drawn from studying the history of measurement.

3

Mercurial Trust and Resistive Measures: Rethinking the 'Metals Controversy', 1860–1894

[A]nd as the true value for the resistance of the mercury unit, as defined by Messrs. Siemens, we may take 0.961 B.A. Units, a value differing from their 1864 issue by about 0.5 per cent, and when corrected for specific gravity, by about 0.8 per cent... Now why do these differences exist? Are we not led to think from the papers written by these gentlemen, and others working in their laboratory, that the reproduction of the mercury unit is the most simple thing possible?

        Augustus Matthiessen, 'Some Remarks on the So-Called Mercury Unit', *Philosophical Magazine*, 1865[1]

Professor Matthiessen and Mr Fleeming Jenkin ... have attacked my proposition ... in a way not hitherto customary, I think, in scientific critiques. The plan followed by these gentlemen in common does not consist in opposing the principle of the system by any reasonable grounds, but in attacking the trustworthiness of my labours.

        Werner Siemens, 'On the Question of the Unit of Electrical Resistance', *Philosophical Magazine*, 1866[2]

The development of the BA's 'absolute' units and standards of electrical resistance from the 1860s is well documented.[3] William Thomson and his allies eventually persuaded many others to express their electrical measurements in interrelated units of length, mass, and time through the principle of energy

---

[1] Augustus Matthiessen, 'On the specific resistance of the metals in terms of the B.A. unit (1864) of electrical resistance, together with some remarks on the so-called mercury unit', *Philosophical Magazine*, 4th series, 29 (1865), pp. 361–70, quote on p. 366–7.

[2] Werner Siemens, 'On the question of the unit of electrical resistance', *Philosophical Magazine*, 4th series, 31 (1866), pp. 325–6, quote on p. 330.

[3] Arnold C. Lynch, 'History of the electrical units and early standards', *Proceedings of the Institution of Electrical Engineers*, 132A (1985), 564–73; Crosbie W. Smith and M. Norton Wise, *Energy and empire: A biographical study of Lord Kelvin*, Cambridge: Cambridge University Press, 1989, pp. 684–98; Simon Schaffer, 'Late Victorian metrology and its instrumentation: A manufactory of ohms', in Robert Bud and Susan Cozzens (eds.), *Invisible connections: Instruments, institutions and science*, Vol. IS09 of the SPIE Institute Series, Bellingham, WA: SPIE, 1992, pp. 24–55; Bruce Hunt, 'The ohm is where the art is: British Telegraph Engineers and the development of electrical standards', *Osiris*, 2nd series, 9 (1993), pp. 48–64; Kathy Olesko, 'Precision, tolerance, and consensus: Local cultures in German and British resistance standards', in *Archimedes*, Dordrecht, The Netherlands/London: Kluwer, 1996, pp. 117–156. My reading of the story is closest to Olesko's, and I am especially grateful to her for sharing early versions of this paper with me.

conservation.⁴ Nevertheless, as Schaffer rightly observes, there was nothing *inevitable* about the way that the BA absolute unit of resistance (later the 'ohm') displaced its chief rival, the Siemens' mercury unit. The largest constituency of resistance users in the telegraph industry certainly did not need to use absolute units⁵; such units became commercially important only when domestic energy metering for electrical lighting became widespread in the late 1880s (see Chapter 6). This chapter takes instead as its starting point the multilateral international agreement at the Paris Electrical Congress of 1881, which was incorporated with revisions into British law in 1894 that the material embodiment of resistance standard should be a column of pure mercury. From this perspective I re-examine the debate in the 1860s, aptly dubbed by Olesko as 'the metals controversy', challenging previous interpretations that this was essentially a dispute over the relative merits of absolute versus arbitrary units of measurement. My claim is that the argument focussed on the choice of *trustworthy metal* to embody the material standard, with Werner von Siemens advocating a mercury column and Augustus Matthiessen promoting a wire of solid alloy.⁶

The first Act of Parliament defining legally enforceable British standards for absolute electrical measurement and their statutory 'limits of accuracy' was approved by Queen Victoria on 23 August 1894, a deed followed in Germany four years later by her grandson, Kaiser Wilhelm II.⁷ The unit of resistance, the ohm, was defined as 10^7 cm s^{-1} in absolute 'electromagnetic' units, and was required to be reproducible to within 0.01%, the ampere and volt only to within 0.1%. To accomplish this reproducibility, the practical quantitative definition of this absolute ohm was a 106.3-cm column of regular cross section containing 14.4521 g of mercury at the melting point of water.⁸ Representatives from ten nations unanimously agreed to this protocol at a special congress in Chicago in 1893, albeit only after twelve years of wrangling over the third and later fourth significant figures of the length

---

⁴ Smith and Wise, *Energy and empire*, pp. 687–93; Graeme Gooday, 'Precision measurement and the genesis of physics teaching laboratories', *BJHS*, 23 (1990), pp. 23–51.
⁵ Schaffer, 'Late Victorian metrology', pp. 26–9.
⁶ Olesko, 'Precision, tolerance, and consensus', p. 126.   ⁷ Ibid., pp. 146–8.
⁸ Fleeming Jenkin and William Thomson both took trouble to explain why, in the 'electromagnetic' system of units, absolute resistance turned out not to be analogous to the lengths of wire used for resistance, but rather to have the dimensions of *velocity*. See Fleeming Jenkin, 'On the New Unit of Electrical Resistance', *Philosophical Magazine*, 4th series, 29 (1865), pp. 477–86, esp. pp. 484–6, and William Thomson, 'Electrical measurement', given to the Mechanics Section of the conferences on the Special Loan collection on scientific apparatus at the South Kensington Museum in May 1876, reproduced in William Thomson, *Popular lectures and addresses*, 3 Vols., London: 1891, Vol. 1, pp. 430–62; the resistance as velocity metaphor is explained on pp. 440–9. For a dimensional analysis of the 'electromagnetic' and 'electrostatic' systems of units, and a comparison with the telegraph engineers' 'practical system' of electric units based on lengths of conductors, see James Clerk Maxwell, *Treatise on electricity and magnetism*, 3rd edition, Oxford: 1891, Vol. 2, pp. 263–9.

of the mercury column.⁹ This prolonged difficulty in articulating the mercury standard constituted evidence for the British contingent that the Paris Congress had been mistaken in rejecting the BAAS proposal for a metal-alloy standard. Whilst historians have stressed that the specification of the absolute unit was a major victory for the British contingent at Paris Congress,[10] they have not hitherto explained the rejection of the alloy proposal nor acknowledged its wider significance. In what follows I offer an account of both these issues.

Long after the Paris 'defeat', the British contingent refused to trust resistance measurements involving mercury, many doubting whether they were reproducible to even three significant figures. This view is palpable in the discussions of the Board of Trade Committee, formed in 1891 to frame the new electrical standards legislation. Instrument manufacturer Alexander Muirhead contended, for example, that different people at different sites could not independently set up mercury columns and agree on their electrical resistance to within 0.3%. He thus followed the recommendation of his late chemistry tutor, Augustus Matthiessen, that a solid metal or 'definite' alloy was much more reliable in this regard.[11] By contrast, in his self-aggrandizing *Lebenserinnerungen* of 1892, the elderly Werner von Siemens expressed lingering resentment at the way that Matthiessen had 'violently opposed' his proposals for a mercury resistance standard three decades previously. Siemens claimed that he and his anglicized brother Charles had invested 'much trouble and labour' in using a 100-cm column of mercury to make possible the world's first universally comparable electrical measurements. His self-fashioning as a man of high virtue continued with an account of how it had been 'somewhat hard' for him to accept the decision of the 1881 Paris Congress to adopt a mercury column 6 cm longer in order to accommodate the slightly larger absolute ohm. Yet he did so as a voluntary 'sacrifice' offered up to 'science and the public interest'.[12]

---

[9] The 'nations' were the USA, Britain, France, Italy, Germany, Mexico, Austria, Switzerland, Sweden, and 'British North America' (later Canada). See reproduction of the Act's text in William Ayrton and Thomas Mather, *Practical electricity*, 3rd edition, London: 1911, pp. 144, 488–93. For further analysis see Lynch, 'History of the electrical units and early standards', p. 568; Larry Lagerstrom, 'Universalizing units: The rise and fall of the international electrical congress, 1881–1904', unpublished manuscript, Personal Communication, November 1994.

[10] Schaffer, 'Late Victorian metrology,' p. 26; Olesko, 'Precision, tolerance, and consensus', p. 141; Lagerstrom, 'Universalizing units', pp. 5–7.

[11] Alexander Muirhead, *Report of Board of Trade Committee on standards for the measurement of electricity*, London, 1891, questions, pp. 123–4.

[12] Werner von Siemens, *Inventor and entrepreneur: Recollections of Werner von Siemens*, Berlin, 1892, London, 1893; new edition, London: Lund Humphries, 1966, pp. 166. and 264. In 1888 Siemens was raised to the peerage by Kaiser Friedrich III, and was thereafter entitled to insert 'von' before his family name.

Siemens' morally loaded account reveals that the metals controversy was about both the evaluation of the stability of metals and a highly personalized conflict about personal integrity. Yet although several historians have noted that there were many factors involved in this complex conflict, none has satisfactorily accounted for the *origin* of the controversy in 1860–1. Neither have they explained why it reached a heated climax in the period 1864–6, nor why resentments arising from the controversy lingered into the early 1890s – long after even Siemens himself had acquiesced in the absolute unit. After explaining the difference between my approach and previous scholarship on this subject, in Sections 3.2 and 3.3, I outline Siemens' and Matthiessen's respective proposals for a metal standard of resistance in 1860–1. In Sections 3.4 and 3.5, I show how the debate soon broadened in 1861–3 with the explicit incorporation of other allies, as Matthiessen enrolled the sympathies of the BAAS Electrical Standards Committee and Siemens leaned heavily on the assistance of the Siemens' company staff. In Section 3.6 I explain how the peak of the disputes in 1864–6 was linked to the Siemens mercury unit, gaining wide currency in telegraph engineering well before the BAAS was able to validate any commercial coils calibrated to its own absolute units. I show in Section 3.7 that British distrust, after the disputes faded away in 1866, shifted away from Siemens himself to be directed exclusively at the peculiar qualities of mercury.

3.1. RETHINKING THE 'METALS CONTROVERSY': SIEMENS VERSUS MATTHIESSEN

'In the establishment of a general unit of resistance, its practical advantages, rather than the scientific harmony of system of measurement, should be taken most prominently into consideration.
    Werner Siemens 'On the Question of the Unit of Electrical Resistance', 1866[13]

No doubt most people think they know what a metre of mercury of 1 millimetre section means, and comparatively few understand the definition adopted by the [BAAS] Committee. But who in practical life, or in the use of standards, refers to their definition? What Frenchman, measuring the contents of a brick wall, thinks of the earth's diameter? What Englishman, using a foot, thinks of pendulums? For practical use the material standard, not the definition, is the important point.
    Fleeming Jenkin, 'Reply to Dr Werner Siemens' Paper "On the Question of the Unit of Electrical Resistance,"' *Philosophical Magazine*, 1866[14]

Historians have understandably attached a great deal of importance to the creation of the absolute electrical standards that have since become the

---

[13] Werner Siemens, 'On the question of the unit of electrical resistance', *Philosophical Magazine*, 4th series, 31 (1866), pp. 325–36, on p. 326.
[14] Fleeming Jenkin, 'Reply to Dr Werner Siemens' paper "On the question of the unit of electrical resistance"', *Philosophical Magazine*, 4th series, 32 (1866), pp. 161–77, quote on pp. 163–4.

central underpinnings of contemporary science. In historiographical terms this story had been treated as the ultimate vindication of Wilhelm Weber's prognostication in 1851 that electrical measurement need not be an arbitrary matter of localized standards, but could be unified by being subjected to a universal definition based on length, mass, and time. Historians have also rightly linked the 'rise' of resistance standards to the contemporaneous evolution of international networks of submarine telegraphy. These standards and commercial copies of them were produced in response to a demand from telegraphic workers, and the active take-up of such standards by telegraph practitioners explains much about their widespread geographical 'spread'.[15] But the symbiosis of submarine telegraphy and resistance was not specifically tied to the development of absolute standards. The absoluteness of such standards did not guarantee the technical and commercial success of a telegraphic enterprise; and conversely, Siemens non-absolute standards were used in many successful missions to lay, operate, and test both submarine cables and landlines. As Bruce Hunt has shown, much more critical problems concerned matters of mechanical engineering, signalling equipment, and quality control over the copper cable cores and gutta-percha insulation.[16] Even Muirhead, who manufactured absolute electrical standards for the BAAS,[17] told the Board of Trade Committee in 1891 that submarine cable engineers seldom needed to make absolute measurements of current and therefore 'need not be concerned' in the absolute value of the BA unit of resistance.[18] What mattered was that telegraphic electricians had ready access to a set of robust resistance coils that were mutually calibrated with sufficient consistency to test cable resistance to required tolerances of error and speedily locate remote faults on lines or cables.

Until the first coils calibrated to the BA absolute ohm appeared in 1865, the only resistances widely available to telegraphists were copies of non-absolute standards. Commonly used were miles of iron wire, Jacobi copper étalons, and coils of a copper–zinc–nickel alloy known as 'German silver', calibrated to the Siemens mercury standards of 1860 or later. When the

---

[15] Hunt, 'The ohm is where the art is', idem., 'Doing science in global empire: Cable telegraph and electrical physics in Victorian Britain', in Bernard Lightman (ed.), *Victorian science in context*, Chicago/London: University of Chicago Press, 1997, pp. 312–33.

[16] See Bruce Hunt, 'Scientists, engineers and Wildman Whitehouse: Measurement and credibility in early cable telegraphy', *BJHS*, 29 (1996), pp. 155–69. For a critical discussion of technical and commercial issues in historical evaluations of 'success' and 'failure', see Graeme Gooday, 'Re-writing the "book of blots": Critical reflections on histories of technical failure', *History and Technology*, 14 (1998), pp. 265–91.

[17] Lynch, 'History of the electrical units and early standards'; Elizabeth Muirhead, *Alexander Muirhead: 1848–1920*, Oxford: Blackwells, (privately published), 1926.

[18] Muirhead's evidence cited in *Report of Board of Trade Committee on Standards for the Measurement of Electricity*. p. 126.

British government appointed the newly anglicized C. W. Siemens and his brother Werner as special advisors to assess the failure of many state-owned submarine cables in 1859,[19] they provided the Siemens brothers opportunities to try out prototypes of their new mercury resistance unit. The brothers reported results of deploying this unit in gauging the underwater erosion of gutta-percha insulation to Section A of the BA meeting at Oxford in September 1860.[20] The practical effectiveness of successive forms of this mercury unit was well known to members of the BAAS Electrical Standards Committee. William Thomson publicly admitted in 1876 that many of the 'most important' measurements in connection with submarine cables were stated in terms of the Siemens unit.[21] And, in his Presidential address to the STEE in 1881, G. C. Foster remarked that it would simply be 'tedious' to have to list the many ways in which electrical practice had been promoted by the 'multiplication' of resistance coils adjusted to either the BA ohm or to the Siemens mercurial standard.[22]

Werner Siemens was himself not consistently opposed to absolute *units* of resistance, but pointed out that it was only 'rarely' necessary to use them. They were indeed *required* only for what Siemens called 'strictly scientific' studies of dynamic values pertaining to energy,[23] as in James Joule's use of the BA 1865 absolute unit for an electrical redetermination of the mechanical equivalent of heat in 1867 (see Section 3.7). My contention is that, the extent to which the metals controversy was ever about the relative merits of absolute versus arbitrary units, this feature emerged as a secondary issue only in late 1861 *after* Matthiessen joined the BAAS Electrical Standards Committee. Before that Matthiessen even *agreed* with Siemens in 1860–1 that it was very difficult, perhaps impossible, to produce an absolute form of resistance standard to any useful degree of accuracy. Ironically, in 1862 Siemens himself indicated a willingness to compromise and accommodate his mercury standard into the absolute system of units if only the BAAS Committee would adopt a mercury construction for their physical standard. And it is worth noting, moreover, that the debate about whether mercury

---

[19] Wilfried Feldenkirchen, (trans. B. Steinbrunner), *Werner von Siemens: Inventor and international entrepreneur*, Columbus, OH: Ohio State University Press, 1994, p. 73. Not insignificantly, this government contract was won after Wilhelm had become a naturalized British citizen in 1859, changed his name to William and married Anne Gordon, the sister of Newall's consultant Lewis Gordon.

[20] Werner and C. W. von Siemens, 'Outline of the principles and practice involved in dealing with the electrical conditions of submarine electric telegraphs [abstract]', *BAAS Report*, 1860, Part 2, pp. 32–4.

[21] Thomson, 'Electrical measurement', p. 452.

[22] G. C. Foster, 'Presidential address', *Journal of the Society of Telegraph Engineers and Electricians*, 10 (1881), quote pp. 14–15.

[23] Siemens, 'On the question of the unit of electrical resistance', p. 326.

could serve *practically* as the basis for a reproducible unit ran from 1860 to 1893, much longer than the debate on the *theoretical* merits of the absolute unit that lasted only from 1861 to 1881.

The contexts of British and Germanic nationalistic rivalry, legal metrology, and conceptual traditions are unquestionably important in understanding the conduct of the metals controversy. Olesko points out that Germanic participants differentiated among three different kinds of material standard: *Urmaas* [foundational standard], *Normalmaase* [reproduction of the standard], and *Étalon* [calibrated resistance scale for daily use]. Moreover, Olesko plausibly suggests that Siemens' decision to base these three forms of standard on a one metre column of mercury was made with a view to their ready assimilation into the extant Prussian metric system of weights and measures. By contrast, Matthiessen and others in the UK showed little sympathy or understanding for this tripartite system and set no store by the congruence of resistance standards into *any* national conventions of metrology.[24] Apart from that no scholar has suggested that issues of national identity or competition were part of the original *cause* of the disagreement about resistance standards. In that regard it is worth remembering that Werner Siemens was primarily a Prussian until 'Deutschland' was created in 1870; it was not until the international congresses of the 1880s and after that anyone adopted a stance on electrical standards for reasons pertaining simply to national loyalty.[25]

In other respects, a neat British–German dichotomy does not fully capture the complexity of the metals controversy. After all, the branch of the Siemens company engaged in international cable-laying enterprises in the 1850s was based in London and managed by Werner's brother C. W. Siemens, by then a naturalized Briton. And the young Fleeming Jenkin assigned to be Werner Siemens' assistant in 1858–9 would have observed Siemens' testing techniques using 'miles' of copper wire[26] and used the prototype mercury-calibrated coils tried out in both the Red Sea cable laying and investigations of the British Government's inoperable submarine cables.[27] Moreover, one of Siemens' chief assistants in Berlin from 1861 to 1867 was the Briton Robert Sabine, whilst Werner's anglicized brother had joined the BAAS

---

[24] Olesko, 'Precision, tolerance and consensus', p. 125.
[25] Lagerstrom, 'Universalizing units'.
[26] Siemens, *Inventor and entrepreneur*, p. 165. Siemens had apparently produced units (1–100) of 1 geographical mile of copper wire 1 line diameter at 20 °C at the Berlin branch since 1848; Siemens, 'On the question of the unit of electrical resistance', p. 335. According to Siemens, this unit came from the suggestion of the Russian physicist Jacobi in 1848 that the standard of electrical resistance should be a length of copper wire, and indeed the Siemens Berlin company manufactured resistances that were calibrated in 'miles' of such a unit and used across the Germanic states. For Fleeming Jenkin's involvement, see Gill Cookson and Colin Hempstead, *A Victorian scientist and engineer: Fleeming Jenkin and the birth of electrical engineering*, Aldershot, England: Ashgate, 2000, pp. 40–48.
[27] Siemens, *Inventor and entrepreneur*, pp. 128–31.

Electrical Standards Committee. More plausibly it has been suggested that there were important divergences in German and British practice in electrical measurement which derived from very different forms of training and values in reportage, especially in the reporting of errors.[28] Differences in the reporting of error are certainly apparently in the publications of Matthiessen and Siemens. And, important as this issue is, an appeal to national styles of pedagogy could not explain the origin or the ferocity of the dispute as both Matthiessen and Siemens received their training in the Germanic states. Also, in the very first paper that Matthiessen published in opposition to Siemens in 1861, the British chemist himself used a 'least-squares analysis' characteristic of a Germanic commitment to error theory (see Chapter 2) and cited numbers claimed to be 'probable errors' at least as explicitly as Siemens did.

In mapping how the dispute first emerged in Section 3.2 below, I discuss how the arguments between Matthiessen and Siemens were about the interrelations of trust, rivalry, and reputations. Both to some extent staked their 'personal' reputation on specialized expertise in metals, in particular their ability to judge the stability of mercury or solid alloys as reliable 'substrates' for context-independent electrical measurement and the expert 'care' with which they handled metal conductors in measurement experiments. This will be in part a social history of metals:[29] of the apparently stolid character of 'artificial' alloys widely used in engineering, and of the famously 'ticklish' behaviour of mercury – a substance so common in thermometers, barometers, and electrical circuit contacts.[30] I show how Mattheissen's and Siemens' accounts of the behaviour of both metals were linked to the amount of *care* exercised in handling it.[31] When Siemens and Matthiessen (or their later allies) disagreed about the value of each other's proposal and of measurements accomplished within it, each represented the criticisms of the other in

---

[28] Kathy Olesko, 'The meaning of precision: The exact sensibility in early nineteenth century Germany', in M. Norton Wise, *The values of precision*, Princeton, NJ: Princeton University Press, 1995, pp. 103–34; Olesko, 'Precision, tolerance, and consensus', pp. 131–35.

[29] Insofar as social constructivist histories have dealt with the history of materials, they have been with manifestly artificial products of human ingenuity rather than with 'natural' elements; e.g. T. Misa, 'Controversy and closure in technological change: Constructing steel', in W. Bijker and J. Law (eds.), *Shaping technology/ Building society*, Cambridge, MA/London: MIT Press, 1992, pp. 109–139. An exception is Hunt's study of gutta percha; Bruce Hunt, 'Insulation for an Empire: Gutta Percha and the development of electrical measurement in Victorian Britain', in Frank A. J. L. James (ed.), *From semaphore to shortwaves*, London: Royal Society of Arts, 1998, pp. 85–104. Historians of a psychoanalytic persuasion may note that the domineering Werner von Siemens often treated his manufacturing company as if it were as protean as the mercury substrate of his resistance standard, whereas the nervous Matthiessen preferred the solidity of metal alloys, excoriating the tendency of mercury to be corrupted by contact with foreign impurities.

[30] I thank John Christie for pointing out the use of this phrase in eighteenth-century chemistry.

[31] On the economy of time in relation to efficiency of research see, Ben Marsden, 'Engineering science in Glasgow: W. J. M. Rankine and the motive power of air', *BJHS*, 25 (1992), pp. 319–416.

several different ways. Each employed a discursive slippage among five different types of expressions of distrust in his opponent's work: distrust in the competence of the measurer; in honesty and openness in reporting results; in the reliability of measurement practices; in the (in)vulnerability of the metal to unpredictable behaviour; and sometimes in the reliance on the research of third parties. The irony of the story that I tell is that, whilst all these modes of expressing distrust were in play at various stages of the metals controversy, the longer-term outcome was hardly predictable. The reputations of Matthiessen and Siemens and their work survived remarkably resiliently, but the characteristics of mercury itself became the sole bearer of suspicion for British observers.

3.2. THE MERCURIAL SOLUTION: SIEMENS' 1860 PROPOSAL

> There can be no doubt that in this way standard measures of resistances can be reproduced of any degree of accuracy.
>
> Werner Siemens, 'Proposal for a New Reproducible Standard Measure of Resistance', 1860[32]

> People ordered coils from the most celebrated firm in Europe and took what was given them – the miles of copper wire before 1860, and the mercury units afterwards.
>
> Fleeming Jenkin, 'Reply to Dr Werner Siemens', 1866[33]

In 1860–1 Werner Siemens published in five German-, French-, and English-language journals his new scheme for calibrating resistance coils by reference to a chemically and geometrically defined column of mercury.[34] To understand this move and its wider significance, it is necessary to consider his previous career and the commercial nature of Siemens' company enterprises in Berlin and London that he had developed in an often fraught collaboration with brother Carl Wilhelm (anglicized to Charles William from 1859 onward). Werner attended the Artillery and Engineering Academy in Berlin from 1835 to 1838, studying mathematics with G. S. Ohm, physics with H. G. Magnus, and chemistry with Otto Erdmann. Thereafter he had ten years of practical field engineering experience as a Prussian army officer,

---

[32] Werner von Siemens, 'Proposal for a new reproducible standard measure of resistance to galvanic currents', *Philosophical Magazine*, 21 (1861), pp. 25–38, quote on p. 34. As Kathy Olesko has pointed out to me, the use of 'standard' here does not fully capture the Germanic distinction between *Urmaas* and *Normalmaas*.

[33] F. Jenkin, 'Reply to Dr Werner Siemens' paper', quote on p. 163.

[34] Werner Siemens, 'Vorschlag eines reproducirbaren Widerstandmasses', *Poggendorff's Annalen*, 110 (1860), pp. 1–20, *Telegraph Vereins Zeitschrift*, 7 (1860), pp. 55–68, *Annales de Chimie*, 60 (1860), pp. 250–56, published as 'Proposal for a new reproducible standard measure of resistance to galvanic currents', *Philosophical Magazine*, 4th series, 21 (1861), pp. 25–38, and *The Electrician*, 4 (1861), pp. 63–4. Citations here refer to Frederick Guthrie's English translation prepared for the *Philosophical Magazine*.

taking responsibility for his orphaned siblings in 1840. In 1847 Werner went into the telegraph business with Berlin mechanic Johann Georg Halske,[35] buying himself out of the Prussian army after having won promising contracts for the Siemens–Halske company to build lines linking the Germanic states. By 1850, Siemens had developed and published articles on some of the earliest effective techniques of using resistance coils (Jacobi's 1848 copper units) in locating faults on long-distance telegraph lines.[36] Nevertheless, as Hunt notes, problems with the gutta-percha insulation Siemens used on the Rhineland lines during 1849–51 prompted cancellation of all his Prussian contracts.[37]

Fortunately for Siemens, Russian Tsar Nicholas I was then importing new imperial technologies from Western Europe, and through personal contacts, Siemens secured a deal to supply seventy-five pointer telegraphs to accompany the new St Petersburg to Moscow railway in 1852. Adapting rapidly to Russian customs for dealing with administrative officials, the Siemens & Halske company soon won spectacularly lucrative contracts for laying thousands of miles of telegraph lines across Russia. And shrewd bargaining for long-term maintenance and surveillance contracts helped sustain the Siemens–Halske fortunes until at least 1867.[38] Meanwhile, the London branch set up in 1850 succeeded through Carl Wilhelm's personal contacts in setting up a working relationship with the wire rope and submarine cable makers R. S. Newall & Co. from 1853 to 1860. With the Siemens brothers' sophisticated expertise in electrical testing of gutta-percha insulation and of cable conductivity, this collaboration ventured into cable laying, notably the Sardinia–Algeria line (1857) that the hitherto predominant Brett brothers had twice failed to complete. By the time it had laid cables among Constantinople, Crete, and Alexandria and across stretches of the Red Sea and the Indian Ocean in 1858–9, the Siemens company had over 400 employees and ranked among the largest telegraph concerns in the world.[39]

For some time, however, Siemens had been dissatisfied with the 10% variation in resistance coils based on the Jacobi copper wire unit, a problem not helped by Jacobi's deposition of the only existing reference standard (*Urmass*) some distance from Berlin in Leipzig. And it cannot have escaped

---

[35] Feldenkirchen, *Wener von Siemens*, pp. 24–49.
[36] Ayrton and Mather, *Practical electricity*, p. 470, probably a reference to Werner von Siemens' paper, 'Sur la télégraphie électrique', *Comptes Rendus de l'Academie des Sciences*, 30 (1850), pp. 434–87, *Annales de Chimie*, 29 (1850), pp. 385–430, and 'Ueber telegraphische Leitungen und Apparate', *Poggendorf's Annalen*, 79 (1850), pp. 481–500. Charles Bright followed two years later with his patented method of using resistance coils to the same end in 1852.
[37] Bruce Hunt, 'Michael Faraday, cable telegraphy and the rise of field theory', *History of Technology*, 13 (1991), pp. 1–19.
[38] Feldenkirchen, *Werner von Siemens*, pp. 55–67. Siemens, *Inventor and entrepreneur*, pp. 96–118.
[39] Siemens, *Inventor and entrepreneur*, p. 161.

Siemens attention that there was a potentially large market of other telegraph companies for more reliable coils.[40] Werner's plan was thus to define a new resistance calibration standard (*Normalmaase*) that could in principle be set up *de novo* anywhere – and thus be manufactured by the Siemens company at any location. Claiming to draw on an earlier suggestion made by the chemists Marié-Davy and De la Rue,[41] Werner adopted a specification for a resistance based on a uniform metre column of pure mercury. After about two years of trials with prototypes, Werner widely published his 'Proposal for a New Reproducible Standard Measure of Resistance to Galvanic Currents' in 1860. Siemens presented this as if it were a standard that any competent experimenter could duplicate with basic laboratory resources. Showing the rigorous *Wissenschaft* that underpinned the trustworthiness of the Siemens' standard, it was more probably published with a view to promoting his company's future sales of commercial copies. Certainly later critics of Siemens coils attacked this paper (and his later publications) as if it were the principal evidence for the integrity of the mercury standard and the commercially produced coils copied from it.

In the 'Proposal' Siemens introduced what became his characteristic representational strategy to undermine the credibility of alternatives to mercury. This strategy was to attribute the inconsistencies with the old copper wire unit to problems generic to the use of *solid* metals in resistance standards. Siemens argued that the elapse of time, shaking in transport, and the passage of electrical currents lead inevitably to alterations in their resistance. The resistance of a wire would be unavoidably contingent on its 'molecular state' – the haphazard internal structure determined by the history of its handling – and would anyway be compromised by the 'almost unavoidable' chemical impurity of such metals. Siemens considered a more reliable technique for guaranteeing the location-independent reproducibility of a resistance standard was a canonical *spatial* form quantitatively specified by a 'geometrical' definition – in contrast to the dynamical definition of the absolute unit. Mercury was the ideal substance to fill this space because, as a liquid, different samples of it would not have 'different molecular states'. Moreover, its resistance was less temperature dependent than any other simple metal, and it could easily be sufficiently purified by chemical means. Thus he proposed a resistance standard of a metre column of mercury of 1-mm^2 cross section at 0 °C: this was – Siemens claimed – amenable to being reproduced to

---

[40] Ibid., p. 165.
[41] Ayrton and Mather cite Marié Davy and De la Rue as having shown that mercury could be cleaned and purified by chemical methods, making it especially suitable for the construction of electrical standards; Ayrton and Mather, *Practical electricity*, p. 470. See, for example, Edme Hippolyte Marié Davy, 'Sur la transmision des courants au travers des liquides conducteurs', Archives d'Electricité, 5 (1845), pp. 35–67. Siemens does not cite either Marié Davy or De la Rue in his publication, but significantly makes no attempt to claim priority for using chemically cleansed mercury as a reliable medium for resistance standards.

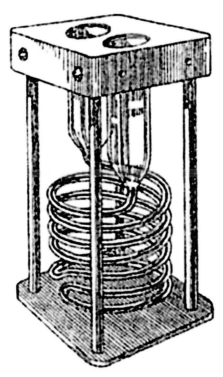

Figure 1. Werner von Siemens' design of apparatus to hold mercury column for resistance standard. Originally designed for a 1-m column of mercury in 1860, this later variant, ca. 1890, is for a 1.063-m column of mercury representing an absolute ohm. (*Source*: Werner von Siemens, *Inventor and Entrepreneur: Recollections of Werner von Siemens*, London, 1893, p. 133.)

'sufficient accuracy' with 'ease' by anyone using mundane laboratory techniques. Given data on the specific gravity of mercury, a sensitive weighing balance could be used to prepare the correct mass of mercury required to fill the glass spiral tubes, and Siemens supplied a formula to compensate for non-uniformity in the dimensions of the glass tubes[42] (see Figure 1).

Siemens significantly acknowledged no problems that might arise from the idiosyncratic qualities of mercury, although he did emphasize that 'great care' was required for preventing changes in its temperature and the incursion of air bubbles.[43] Some ironies are apparent, though, in Siemen's specification

---

[42] Siemens, 'Proposal', pp. 25–6.
[43] Siemens ended this piece with very specific instructions for those who wished to perform this reproduction for themselves: 'It is necessary to warm the mercury to be employed for several hours under a covering of concentrated sulphuric acid mixed with a few drops of nitric acid, to get rid of all metallic impurities, as well as the free oxygen, which greatly *increase* its conducting power.' Siemens, 'Proposal', p. 38.

for care in such matters when reporting the reproducibility of his methods. He reported comparisons by his Berlin assistant, Ernst Esselbach,[44] of independently prepared six mercury standards, citing the following ratios of calculated resistance to that measured against a Jacobi copper étalon: 1.008, 1.00 [sic], 1.0008, 0.992, 0.994, 1.005. Siemens and Esselbach avoided citing statistical detail concerning error, presumably to avoid drawing attention to the substantial (1.6%) range in the results. Siemens merely noted that each individual result differed from unity by 'under 1 per cent', and declared these to be 'fully accounted for' by temperature variations of 2–3 °C in the environment (0–2 °C for the ice cooling the mercury, and 19–22 °C for the Jacobi standard).

Notwithstanding the indeterminacy produced by an apparent lapse in care in temperature management, Siemens claimed that this method could reproduce standard measures of resistances to any 'degree of accuracy' that their instruments permitted or required.[45] Yet, as Olesko notes, although Siemens had supplied Esselbach with the finest available chemical balance, resistance bridge, and mirror–galvanometer, it would not have been obvious to all readers of Siemens' paper that these had secured the high accuracy he claimed possible for his standard.[46]

Evidently Siemens did not consider these features of his paper to offer potential critics any significant hostages to fortune. Indeed, he published it to attain the widest publicity in the telegraphic community: After it appeared in *Poggendorf's Annalen*, *Telegraph Verein's Zeitschrift*, and *Annales de Chimie* in 1860, Siemens also commissioned an English translation from Frederick Guthrie, which was published in 1861 in both the early (short-lived) British telegraphic journal *The Electrician*, and in the January issue of the *Philosophical Magazine*. Notwithstanding the democratic rhetoric of Siemens' paper – that effectively *anyone* could reproduce his standard – only a young disabled chemist in London published anything approaching a critique of Siemens' proposal for a mercury standard.

3.3. MATTHIESSEN'S CASE FOR THE ALLOY: TRUST IN SOLIDITY

The only reference in Siemens' paper to Wilhelm Weber's scheme of absolute resistance measurement was that it was as hopelessly 'ill-adapted' to general use as it could be reproduced only by means of 'very perfect instruments', in places 'specially arranged' for the purpose, and by those with 'unusual'

---

[44] Esselbach was a former student of Weber and Hemlholtz who had studied for his PhD in physics under Gustav Karsten at Kiel. See Olesko, 'Precision, tolerance, and consensus', pp. 123–4.
[45] Siemens, 'Proposal', p. 34–6.
[46] Olesko, 'Precision, tolerance, and consensus', pp. 124–5.

manual dexterity.⁴⁷ And Augustus Matthiessen uttered a similar view in a not-dissimilar paper published in the February 1861 issue of the *Philosophical Magazine*, 'On an Alloy Which May Be Used as a Standard of Electrical Resistance'. He maintained that, although it was the 'best' theoretical means of expressing a resistance standard, the practical fulfilment of Weber's absolute system was 'beyond the means' of most experimenters.⁴⁸ Interestingly, though, the criticisms Siemens made against the technical feasibility of the absolute definition were just like those Matthiessen used against Siemens' supposedly easily reproducible standard. Indeed, Mathiessen's alternative proposal for a metal-based standard had nothing to say in favour of the recent Siemens' proposal in the same journal, merely dismissing it in a rather arch footnote.⁴⁹ Historians have not previously discussed Matthiessen's alloy proposal, yet it is crucial for understanding the position he adopted in the latter stages of the metals controversy. To understand the origins of this, we need to look at his previous specialist research on the electrical conductivity of metals.

Born in 1831, Matthiessen suffered a 'paralytic seizure' (probably a stroke) at the age of 2 or 3 years, and this left him with a permanent and uncontrollable twitching in his right hand. He developed a passion for chemistry and learned to manage his disability sufficiently to secure a PhD under Justus von Liebig at Giessen in 1853. As his interests in agricultural chemistry waned, Matthiessen studied metallic chemistry at the University of Heidelburg for four years, working with Bunsen on the electrolytic production of alkali metals, and then with Kirchhoff on the electrical conductivity of metals in general. The results of these appeared in the *Philosophical Magazine* of 1857–8 and *Poggendorf's Annalen* for 1858, from which it is clear that Matthiessen gained important comparative experience in the manipulability and measurability of mercury vis-à-vis solid metals.⁵⁰

By this time, Matthiessen had returned to London and continued at his own private laboratory with extensive measurements on the electrical conductivity of metals and alloys,⁵¹ publishing the results of the perturbing effects of temperature change and impurities in both British and German

---

⁴⁷ Siemens, 'Proposal', p. 25.
⁴⁸ Augustus Matthiessen, 'On an alloy which may be used as a standard of electrical resistance', *Philosophical Magazine*, 4th series, 21 (1861), pp. 107–15, quote on p. 115.
⁴⁹ Augustus Matthiessen, 'On an alloy'; this was probably not a direct response to Siemens' paper published in *Poggendorf's Annalen*, as Matthiessen's footnote criticizing Siemens' paper appears to have been added rather as an afterthought – Siemens' name is not mentioned in the main text at all.
⁵⁰ Augustus Matthiessen, 'On the electric conducting powers of the metals', *Philosophical Magazine*, 16 (1858), pp. 219–23, *Poggendorff's Annalen*, 103 (1858), pp. 428–43. Also published in *Philosophical Transactions of the Royal Society*, 148 (1858), pp. 383–88, *Annales de Chimie*, and *Nuovo Cimento*.
⁵¹ See obituaries subsequently cited for fuller details of Matthiessen's career.

journals.[52] His examination of the effects of impurities on the conductivity of copper was considered important enough to be drawn on by the Committee of Enquiry into the failure of the 1858 Atlantic cable. Through that connection Matthiessen came into close contact with William Thomson and was elected a Fellow of the Royal Society in 1861.[53] He was appointed a lecturer in chemistry at St Mary's Hospital, London, in 1862, and then at St Bartholomew's Hospital from 1868. In obituaries published after his suicide two years later, it is evident that contemporaries were impressed both by Matthiessen's management of his disability and his accomplishments in metals research. *Nature* judged Matthiessen's achievements to be remarkable, given his 'physical disadvantages', and praised his qualities of 'perseverance'.[54] Although the obituarist for the *American Journal of Science and Arts* passed over his bodily affliction in silence,[55] it attributed to him instead special mental powers: an 'acuteness' which led to the perception of truth, and a great facility for detecting and eliminating 'sources of error'.[56] Nevertheless, none alluded to the way in which Matthiessen spent the years 1861–6 in sustained polemics against the 'sources of error' he detected in Werner Siemens' work on the mercury standard.

Matthiessen's proposal for a resistance standard echoed Siemens' in arguing that it should consist of a geometrically defined body of a metal with a stable physical constitution and that it should be reproducible in any location without excessive care. Where he sharply diverged from Siemens, however, was in his choice of metal. Because the resistance of pure metals, such as silver, copper, and mercury, was greatly affected by the presence of even small amounts of chemical impurities, Matthiessen dismissed these as ineligible. He noted in passing a 'grave objection' to Siemens' proposal that contact with any other metal in an electric circuit would render the mercury 'impure'.[57] He argued that the ideal metal for building a material standard of

---

[52] Augustus Matthiessen and M. Holzmann, 'On the effect of the presence of metals and metalloids upon the electric conducting power of pure copper', *Philosophical Transactions of the Royal Society*, 150 (1860), pp. 85–92; A. Matthiessen, 'On the electric conducting power of alloys', *Philosophical Transactions of the Royal Society*, 150 (1860), pp. 161–76. This was published in German as 'Ueber die electrische Leitungsfähigkeit der Legirungen', *Poggendorf's Annalen*, 110 (1860), pp. 190–221; *Zeitschrift Berlin Telegraph Vereins*, 8 (1861), pp. 9–14; idem, 'On the specific gravity of alloys', *Philosophical Transactions of the Royal Society*, 150 (1860), pp. 177–84.

[53] Hunt, 'The ohm is where the art is', p. 53.

[54] Anonymous, 'Augustus Matthiessen', *Nature* (London), 2 (1870), pp. 518–19.

[55] For a discussion of the bodily dimension to scientific practice see Chris Lawrence and Steven Shapin (editors), *Knowledge incarnate: Human embodiments of natural knowledge*, Chicago/London: University of Chicago Press, 1998.

[56] Anonymous, 'Obituary: Dr. Matthiessen', *American Journal of Science and Arts*, 3rd series, 1 (1871), pp. 73–74. The circumstances of his death are discussed in Section 3.7.

[57] As Olesko has pointed out, Matthiessen's contention here might have rested on a misapprehension that the Siemens mercury standard was to be used constantly in measurements as a

resistance was an alloy of equal volumes of pure gold and pure silver (by weight, two parts gold to one part silver); at this particular constitution, 1%–2% gold could be added without altering the 'conducting power' of this alloy to any 'appreciable' extent. If all due precautions were followed, Matthiessen reckoned that a hard-drawn wire of this gold–silver alloy of 1-mm length and 1-mm diameter calibrated to 100 resistance units would be the best possible reference for comparisons of resistance.[58]

To show that this alloy's resistive properties could be reproduced by different practitioners at many sites, he showed the results of eight attempts to follow his protocol for constructing a wire standard out of 6-g gold and 3-g silver of 0.5-mm diameter. In addition to two sets produced by London metallurgists, Matthiessen drew on his international network of contacts in Frankfurt, Paris, Brussels, and New York to supply the others. Each compared the electrical-conducting power of his pair of alloy samples with a locally made comparative standard of hard-drawn silver. Having calibrated the results and calculated their respective errors from the mean, Matthiessen reported one null result and the others as having errors of +0.2%, +0.2%, −1.2%, +0.4%, −0.3%, +0.3%, and +0.1%. By ignoring the anomalously low third result, he impressionistically concluded that these results proved the alloy could be reproduced at different sites by 'anybody' with a mere 0.3% variation in resistance. This he compared favourably with the consistency of similar tests on 'pure' forms of silver, copper, gold, and mercury by Becquerel, Lenz, Siemens, and himself: These showed variations of as much as 5%–10% between sites, and considerably more for mercury.[59]

Matthiessen further dismissed the feasibility of using mercury in a standard as contact with any other metal compromised the purity of both; use of mercury would thus require impractically frequent replacement with fresh supplies of the pure fluid metal. Much care certainly had to be taken to prevent it from compromising his alloy's electrical properties by chemical contamination. In a peremptory footnote, Matthiessen attacked Siemens' chemical expertise, contending that the Prussian was incompetent to evaluate the corrupting effect of impurities on mercury: These, he alleged, *decreased* the 'conducting power'[60] of mercury, rather than increased it, as Siemens had supposed.[61] Matthiessen also criticized Siemens for assuming the increase of metallic resistance with temperature was a linear relation rather than a

---

workaday étalon rather than as a *Normalmaass* set up only occasionally for the sole purpose of calibrating the former; Olesko, 'Precision, tolerance, and consensus', pp. 125–7.

[58] Matthiessen, 'On an alloy', p. 117.

[59] Ibid., pp. 109–10. The comparative conducting powers cited by Matthiessen for mercury with reference to a silver wire standard were 1.86 at 0 °C (Becquerel), 3.42 at 18.7 °C (Lenz), 1.72 at 0 °C (Siemens), and 1.63 at 22.8 °C (Matthiessen).

[60] This term was used by many scientists, including William Thomson, as the algebraical reciprocal of resistance (see Chapter 1).

[61] Matthiessen, 'On an alloy', p. 108.

quadratic relation, as Lenz had shown it to be. The English chemist then used a formal least-squares analysis to determine the most probable values of the relevant coefficients in the equation for his alloy,[62] Matthiessen having learned this error protocol from his chemical training with Kirchhoff.[63] Making a further implicit attack on Siemens and his mercury standard proposal, Matthiessen presented data which purportedly showed that the resistance of his gold–silver alloy was less temperature dependent than mercury, varying 6.5% for a hard-drawn alloy as compared with 8.7% for mercury between 0 and 100 °C.[64]

Although subsequently Matthiessen changed his position on the practical feasibility of the absolute system, he never deviated from the conviction that an *alloy* was a more reliable material than mercury could ever be for constructing resistance standards of any sort. Matthiessen thus staked his reputation as a chemist on his authority on the useful properties of metal alloys and on the problems in handling mercury. To ensure that Siemens and other German and British telegraphic practitioners saw his paper, Matthiessen had it published in the *Philosophical Magazine* for February 1861 and shortly afterwards in *Poggendorff's Annalen* and the *Telegraph Verein's Zeitschrift*.[65]

3.4. CONTROVERSY BEGINS: CHALLENGING ACCURACY AND METALLIC UTILITY

Soon after Matthiessen's paper was published, Siemens' angry response appeared in German in *Poggendorf's Annalen* and in the Berlin *Telegraph Zeitschrift*.[66] Siemens contended that, although at first sight Matthiessen's proposal had much merit, on closer consideration there were 'very overwhelming arguments against it' [*sehr überwiegende Gründe dagegen*].[67] Throughout he vigorously attacked both Matthiessen and his alloy proposal in notably personalized terms, employing a range of polemical strategies. Siemens particularly scorned Matthiessen's claim that the absolute system was theoretically the best since Weber and Thomson had shown that an absolute resistance standard could not be determined to within even several

---

[62] Ibid., pp. 111–14.
[63] I thank Kathy Olesko for confirming the origins of Matthiessen's training in error theory.
[64] Matthiessen, 'On an alloy', p. 115.
[65] Matthiessen's, 'On an alloy' was also published in *Poggendorf's Annalen*, 112 (1861), pp. 352–64, and *Telegraph Vereins Zeitschrift*, 8 (1861), pp. 73–5.
[66] Werner von Siemens, 'Über Widerstandmaasse und die Abhängigkeit des Leitungswiderstandes der Metalle von der Wärme', *Poggendorf's Annalen*, 113 (1861), pp. 91–105. In his later critique of this paper, Matthiessen translated the title as 'On standards of electrical resistance, and on the influence of temperature on the resistance of metals'.
[67] Siemens, 'Über Widerstandmaasse', p. 91.

per cent agreement.[68] More damningly, Siemens used Matthiessen's own evidence to challenge his claim that the gold–silver alloy standard could be reproduced anywhere to within 0.3%. If one included the anomalous New York data which the chemist ignored in arriving at the error figure of 0.3%, the total variation amounted to 1.5%, that is, ±0.75% from the mean. Even if one omitted the numbers least in mutual accordance, Siemens contended that the agreement between Matthiessen's figures was only 1 part in 100 – a conclusion drawn from the maximum variation in Matthiessen's results, not from an explicit error analysis. Siemens further contended that Matthiessen's proposal had thoroughly failed [*durchaus verfehlt ist*] because, for good-quality standards, agreement was needed to within 1 part in 10,000 (0.01%).[69] Siemens did not, however, explain why he adopted this particular threshold of 'reasonable agreement' (see Chapter 2) and, unsurprisingly, did not acknowledge that his own mercury standard had not yet met this requirement either.

Siemens did not stop there, for evidently more was at stake for him than showing Matthiessen's proposal to be unworkable. Having been lambasted for his misjudgements on chemical matters, Siemens mocked Matthiessen's competence as a chemist in handling mercury and interpreting its behaviour. The regular replacement of mercury to prevent contamination with copper connectors was a trivial problem because removing, cleansing, and replacing mercury was not difficult in practice. By contrast, Siemens declared, the difficulties of maintaining the integrity of a resistance standard based on a solid wire were much greater, especially in guaranteeing the truly cylindrical form and homogeneity of the wire when handled by humans.[70] And in regard to the effects of impurities on conductivity, Siemens replied that Matthiessen was wrong for there was ample evidence that, unlike the conductivity of other metals, the conductivity of mercury *did* increase with the addition of impurities.[71] Although this was not a matter that directly impinged on the practical merits of the metals in question, this was clearly an attempt to

---

[68] The original wording: 'Es kann wohl mit Bestimmtheit behauptet werden, daß [sic] auch die geübtesten und mit den vollkommensten Instrumenten un Localitäten ausgerüsteten Physiker nicht im Stande seyn [sic] werden, absolute Widerstandsbestimmungen zu machen, die nicht um einige Procente von einander verschieden wären! Ein maaß [sic], welches so wenig genau ist, würde aber nicht einmal den anforderungen der Technik genugen.' Siemens, 'Über Widerstandmaasse', p. 92. For Matthiessen's translation of this passage, see 'Some remarks on Dr Siemens' paper', *Philosophical Magazine*, 22 (1861), p. 195. Throughout this chapter I use Siemens' original spelling instead of anachronistically converting into modern German.

[69] Siemens, 'Über Widerstandmaasse', p. 93. [70] Ibid., pp. 93–4.

[71] Ibid., pp. 93–4. See Augustus Matthiessen, 'On the electric conducting power of alloys', *Philosophical Transactions of the Royal Society*, 150 (1860), pp. 161–76, quote on p. 162. Responses to Siemens' criticisms came in Augustus Matthiessen and Carl Vogt, 'On the influence of traces of foreign metals on the electric conducting power of mercury', *Philosophical Magazine*, 4th series, 23 (1862), pp. 171–179; see subsequent discussion.

show that Matthiessen knew far too little about the behaviour of mercury to be a reliable spokesman for its allegedly corrupting behaviour. If there were indeed any problems with the performance of mercury, Siemens contended that sufficient effort and ingenuity on the part of the experimenter readily overcame these.

Matthiessen responded angrily in 'Some Remarks on Dr Siemens' Paper' that appeared soon afterwards in both *Poggendorf's Annalen* and the *Philosophical Magazine* for September 1861. Although Matthiessen calmly acknowledged Siemens' criticism of his mistake concerning the effect of impurities on mercury, he sought to rebut Siemens on almost every other point.[72] Stung by Siemens' derision at his supposedly groundless support for the absolute resistance units, Matthiessen performed a *volte face* by retrospectively declaring that he had 'always understood' determinations of resistance by the absolute method to be 'most accurate'. Matthiessen invoked the authority of Professor William Thomson as one whose opinion would be of such 'great weight' as to settle the question, citing a letter from him contending that there could 'scarcely be a doubt' that Weber's original absolute determinations were 'considerably within one half per cent. of the truth'.[73] As if to bring home the accuracy attainable with the absolute method, the editors of the *Philosophical Magazine* published a translation of Weber's original 1851 paper in the same issue.[74]

Matthiessen's biggest grievance against Siemens was the latter's contention that the gold–silver alloy was insufficiently stable to reproduce a resistance standard. He compared Siemens data on reproducibility with his own in the table that is reproduced here as Table 3.1.

Matthiessen strongly disagreed with Siemens' claim that these results showed that the mercury standard could achieve 'any degree of accuracy', and was certainly not persuaded that discrepancies in Siemens' results were

[72] Augustus Matthiessen, 'Some remarks on Dr Siemens's [2nd] paper "On standards of electrical resistance, and on the influence of temperature on the resistance of metals"', *Philosophical Magazine*, 4th series, 22 (1861), pp. 195–202, original version in *Poggendorf's Annalen*, 114 (1861), pp. 310–21.
[73] Of Weber's two more recent trials by completely independent methods (values 190.3 and 189.8) the two numbers differed by less than 0.14% from the mean; hence Thomson had apparently opined that it was 'scarcely possible' that it was 'wrong' by as much as 0.5%. Matthiessen, 'Some remarks', p. 195. Contemporary correspondence between Matthiessen and Thomson (although not directly on this specific point) is held in the Kelvin correspondence at University Library, University of Cambridge.
[74] Wilhelm Weber (trans E. Atkinson), 'On the measurement of electric resistance according to an absolute standard', *Philosophical Magazine*, 4th series, 22 (1861), pp. 226–40, 261ff. The original appeared in *Poggendorf's Annalen*, 82 (1851), p. 337. An editorial footnote had this comment: 'From the great scientific and practical importance which the determination of electrical resistance has of late acquired, it has been through advisable to give a translation of Weber's original paper published in 1851, containing the method of referring these resistances to an absolute standard.'

Table 3.1. *Matthiessen's Comparison of Reproducibility of Mercury and Alloy Resistance Standards*

Number of Tube	Siemens' Data on Mercury Resistance Ratio[75] (Calculated/Actual)	Matthiessen's Data on Hard-Drawn Alloys Resistance Ratio[76] (Calculated/Actual)
1	1.008	1 = 1.003
2	1.000	2 = 1.002
3	1.0008	3 = 0.988
4	0.992	5 = 1.004
5	0.994	6 = 0.997
6	1.005	7 = 1.003
		8 = 1.001

explicable by mundane temperature variations as Siemens had claimed. Focussing on the numbers themselves, Matthiessen noted that the range of variation was nearly identical in both sets of results; accordingly he had harsh words for his Prussian adversary:

Now on comparing the[se] values it will be seen that the maximum differences he finds in each case are the same. If therefore, in the opinion of M. Siemens, the gold silver alloy is useless as a standard, how much more must his mercury standard be so, when according to his own determinations with the same mercury tubes carefully picked from a large quantity, he does not arrive at a greater accuracy than I did with alloys made in *different* places by *different* persons, of *different* gold and silver, and drawn by *different* wire drawers.[77]

Siemens had not even shown that mercury had stable and consistent properties within the confines of the Siemens company's own testing rooms, let alone comparing samples produced independently at sites thousands of miles apart as Matthiessen had done. And whatever consistency of results had been achieved owed much, he averred, to Siemens' great selectivity in reporting data. Matthiessen asked rhetorically, if Siemens had been fully honest and open as he had been, was it not 'probable' that 'much greater differences' would have emerged in the mercury data? After all, Lenz, Becquerel, and he (Matthiessen) had shown that mercury was one of the least reliable metals as it was so generally prone to inconsistent performance. Given the recurrent difficulties in preparing mercury such as the need to clean glassware

---

[75] These figures are taken from Siemens, 'Proposal for a new reproducible Standard', p. 33.
[76] These figures are drawn directly from Matthiessen's 1861 paper 'On an alloy', pp. 108–9. Wire number 4 in that paper was only prepared in annealed form so does not appear in this list of the hard-drawn form.
[77] Matthiessen, 'Some remarks', pp. 196–7.

thoroughly, to chemically purify the metal, and to prevent incursion of further impurities, Matthiessen concluded that the mercury standard required far too much care and attention to be a 'useful and good one'.[78] At this stage then, Matthiessen implied that mercury was an untrustworthy substance, that the practices of using mercury in standards were not trustworthy, and that Siemens himself had probably been less than trustworthy in reporting his results.

After Matthiessen published this reply, he continued his critique of the Siemens position, working with his assistant, Dr C. Vogt, to show that even the minutest quantity of 'foreign' metal would 'materially' affect the conducting power of mercury. The publication of this in the pages of the *Philosophical Magazine* in early 1862[79] provoked a brief but pungent exchange in which the chemists were soon answered by Robert Sabine, who had been appointed by C. W. Siemens to the London branch of the company in 1858 and had transferred to the submarine telegraph section of the Siemens Berlin company in 1860–1.[80] Whilst undertaking research on the electrical conducting powers of metal amalgams, Sabine identified some serious calculation errors made by Matthiessen and Vogt in their criticisms of his employer's work.[81] Matthiessen and Vogt replied a few months later but directed the force of their critique directly at Siemens rather than at his deputy Sabine, deploying their own experimental data on amalgams to criticize Siemens' theoretical understanding of mercury's anomalous behaviour, and thus his competence to pre-empt significant sources of error in measurements involving mercury.[82] A major upshot of this exchange in the first half of 1862 was that Matthiessen and Vogt raised further doubts about Siemens' fitness to act as a reliable spokesman for the behaviour of mercury, thereby further eroding the relations between the Berlin and the London antagonists. This was to have major consequences for two new projects that ran concurrently with the squabble: the Siemens company's sale of commercial resistance coils calibrated in terms of the mercury unit, and the formation of a new BAAS Committee on Electrical Standards, of which Matthiessen was an original member.

[78] Ibid., pp. 197–8.
[79] Augustus Matthiessen and Carl Vogt, 'On the influence of traces of foreign metals on the electric conducting power of mercury', *Philosophical Magazine*, 4th series, 23 (1862), pp. 171–9, quote on p. 172.
[80] Robert Sabine (1837–85) moved to the Berlin branch of Siemens & Halske, ca. 1861, to work for Werner von Siemens on submarine cable technology until 1867, whence he departed to take charge of Sir Charles Wheatstone's work for the British Telegraph Company and married Wheatstone's daughter. See obituary in the *Proceedings of the Physical Society*, 6 (1886), appended to Proceedings of the Annual Meeting of 1885 at the end of volume, pp. 10–11.
[81] Robert Sabine, 'Some remarks on a paper by Dr Matthiessen', *Philosophical Magazine*, 4th series, 23 (1862), pp. 457–60.
[82] Alexander Matthiessen and Charles Vogt, 'Reply to Mr Sabine's "Remarks"' *Philosophical Magazine*, 4th series, 24 (1862), pp. 30–7.

3.5. THE BAAS COMMITTEE'S CONTEMPLATION OF MERCURY AND ALLOY STANDARDS

> He had also the faculty of availing himself of the powers of others by associating them with his work.
> Obituary notice of Augustus Matthiessen in *American Journal of Arts and Sciences*[83]

As has been well documented, the BAAS appointed a committee in September 1861 to investigate the establishment of units and construction of material standards for electrical resistance.[84] Consisting of Matthiessen along with Fleeming Jenkin, William Thomson, James Clerk Maxwell, Charles Wheatstone (Kings College, London), as well as the chemists W. H. Miller (Cambridge) and Alexander Williamson (University College, London), the committee moved quickly to endorse the Weber–Thomson scheme of an absolute unit of resistance.[85] What has not been quite so thoroughly documented is the debate about the constitution of the material resistance standard, and some commentators have overstated the extent to which there was a necessary opposition between the BAAS Committee's absolute *unit* and the use of mercury for calibrating the material *standard*. When Jenkin solicited the views of various 'foreign men of science' in 1862, the response he received to his circular letter was minimal. No replies were received from Joseph Henry in Washington DC, Polite in Paris, Jacobi in St Petersburg, Matteucci in Turin, Weber in Göttingen, Poggendorf in Berlin, Neumann in Königsburg, Fechner in Leipzig, or Edlund in Uppsala. The only three replies that Jenkin did receive were all in favour of adopting some compromise between the Weber absolute unit and the Siemens material standard, leaving Matthiessen as the only figure in 1862 opposed to any reconciliation.[86]

Gustav Kirchhoff, Matthiessen's erstwhile mentor at Heidelberg, observed that although the absolute unit more satisfactorily met the 'demands of science', the Siemens unit could be determined with greater 'accuracy'. And yet, he conjectured, it was not necessary to decide either way, for if the ratio between them were established with the highest 'attainable' accuracy, the parallel use of both units would be no more confusing than the dualism of expressing lengths in metres or in millimetres.[87] This was supported by Siemens' former assistant Esselbach, who presented evidence that the conversion

---

[83] 'Obituary: Dr Matthiessen', *American Journal of Sciences & Arts*, 3rd series, 2 (1871), p. 71.
[84] Lynch, 'History of the electrical units'; Smith and Wise, *Energy and empire*; Schaffer, 'Late Victorian metrology;' Hunt, 'The ohm is where the art is', offers a subtle and compelling revisionist account of the complex genesis of the committee and its aims.
[85] Hunt, 'The ohm is where the art is', pp. 57–60.
[86] See 'Provisional Report of Committee On Standards of Electrical Resistance', *BAAS Report*, 1862, Part 1, 'Appendix G – Circular Addressed to Foreign Men of Science', pp. 156–9.
[87] 'Provisional Report', *BAAS Report*, 1862, 'Appendix D – Professor Kirchhoff's letter', pp. 150–2, quote on p. 151.

process would be very simple indeed, 1.1 Siemens units being equal in magnitude to 10^{10} absolute units, such determination with the mercury unit being reproducible to within 0.25%.[88] Even Werner Siemens in Berlin was entirely amenable to the parallel harnessing of the two units, proposing that the mercury standard be ascertained in absolute units with the 'greatest possible accuracy'. Such a determination would suffice for all reasonable purposes because only 'very seldom' would there be a need for resistance to be expressed in absolute measure, namely, by a minority of physicists.[89]

Of the three letters received by the committee, only Siemens' expressed views about the material constitution that ought to be adopted by the BAAS standard, and unsurprisingly he proposed that mercury furnished the best metal for this purpose. Implicitly speaking against Matthiessen's earlier criticisms of his proposal, he recorded that the resistance of a spiral of mercury had remained 'constant' over six months without any need to change or repurify the mercury. He also noted that the threat of impurity encountered in bringing mercury into contact with other metals was of 'really less inconvenience' than was 'generally believed'; he had kept the same copper terminals in contact with mercury cups for over a week and detected 'not the slightest' change in resistance. Indeed, so reliable was this method that Siemens now claimed that independent reproductions of the mercury unit could be 'trusted' to within 0.05%. His hostility to the Matthiessen proposal was, however, undiminished, and he reiterated the judgements that the performance of a solid metal was too contingent on the vicissitudes of its 'molecular structure' and the problem of soldered end connections.[90] Moreover, the risk to the reliability through haphazard injury was much greater for wire standards: The breakage of a glass spiral containing mercury was infinitely to be preferred to the 'treacherous' results of a bruised wire.[91]

In three appendices to the Electrical Standards Committee's report for 1862, Matthiessen and Williamson presented very different conclusions. Their aggressive critique of mercury as a material from which to build standards borrowed heavily from the data and strategies adopted by Matthiessen in his two 1861 papers, notably repeating[92] his analysis of 1.6% variability in early Siemens mercury standards. They were also highly sceptical about

---

[88] 'Provisional Report [of the Standards Committee]', *BAAS Report*, 1862, 'Appendix F – Extracts from a letter addressed to Professor Williamson by Dr Esselbach', pp. 155–6, quote on p. 156. The value cited by Esselbach is apparently at odds with Siemens' judgement that the mercury unit was 2.5% greater than 10^{10} absolute units, but this would be explained by their respective use of mercury columns of different sizes.

[89] 'Provisional Report', 'Appendix E – Dr Siemens' letter', pp. 152–5, quote on p. 154.

[90] Ibid., pp. 154–5.   [91] Ibid., p. 153.

[92] Augustus Matthiessen and Alexander Williamson, 'On the reproduction of electrical standards by chemical means', *BAAS Report*, 1862, Appendix C of 'Provisional Report', pp. 141–50.

Siemens' claim to the committee that he could now achieve agreements to within 0.05%, attacking him in a footnote for neglecting to give details of the experimental 'precautions' he had taken to achieve this result. The clear implication was that Siemens might be self-deluded or deceitful on this point.[93] More brutally still, they cited evidence gathered by Jenkin from two sets of mercury-calibrated resistance coils, apparently from Siemens factories in London and Berlin, that had been displayed at the 1862 International Exhibition in London. Although each set was admittedly 'perfectly accurate' within itself, the sets were 1.2% adrift from each other.[94] Importantly, Matthiessen and Williamson did make some concessions to Siemens' case that the electrical conductivity of pure mercury did not depend on the history of its handling, and its conductivity changed little with temperature. In their conclusion, however, they simply issued blunt denials of Siemens' claims of the convenience and reliability of his method. And they now also doubted that the dimensions of Siemens glass tubes would remain unchanged as this was a calibration problem 'well-known' in regard to thermometer bulbs.[95]

Matthiessen and Williamson presented a strikingly more charitable account of the capacity of a hard-drawn silver–gold alloy to sustain stable and readily measured resistances, albeit one that reveals interesting tensions in their position. They cited four reasons to favour such alloys as a standard-bearing material and only one against. Whereas the conductivity of both pure metals could vary by up to several per cent, that of gold–silver alloys from different sources did not vary more than 1.6%; they did not acknowledge that this compared unfavourably with the 1.2% variation in the 1862 Siemens resistance coils. They presented explicit evidence for two of their other reasons: Their alloy's conductivity was barely affected by annealing (only 0.3%) and changed less with temperature than any other metal (6.5% between 0 and 100 °C compared with 8.3% for mercury). The remaining reason in favour of the alloy was the somewhat speculative claim that the homogeneity and 'molecular condition' of gold–silver alloys were 'always the same'. Ironically, this point was juxtaposed with their admission of one reason that told against the choice of gold–silver alloy: The problem was that the conducting power of this alloy 'might alter' with age as the properties of alloys were more prone to long-term secular change than were pure metals.[96]

---

[93] Ibid., p. 145. [94] Ibid., p. 147. [95] Ibid., p. 150.

[96] The extraordinariness of this combination of claims is clear from an earlier assertion of a crucial *similarity* between this gold alloy and pure gold from which they inferred that it was 'more than possible' that their gold–silver alloy would be as stable as the gold used in commercial jewellery. More circumspectly, they contended that this alloy would probably be immune to such environmentally induced changes as made brass and silver become brittle and crystalline over time. Ibid., pp. 149–50.

106   *The Morals of Measurement*

Notwithstanding that admitted difficulty, Williamson now concurred with Matthiessen that the best material embodiment of the BA unit was a section of gold–silver alloy. Tellingly, though, they recognized that their conclusion might well not compel universal assent, admitting that further research needed to be undertaken by three or four 'scientific men and electricians' to compare the resistance of pure mercury with that of the gold–silver alloy in order to resolve the 'many disputes' on this subject.[97] Indeed, the overall Committee's report drawn up by Fleeming Jenkin was equally open to either metal, noting that both methods 'seem susceptible of considerable accuracy'. Although the Standards Committee had not yet officially decided to follow Matthiessen's line, it did accept his judgement – against Siemens' testimony[98] – that the mercury in a Siemens standard would have to be 'continually' changed to prevent corrupting amalgamations with contiguous metals. It nevertheless welcomed the further advice from Siemens on the possible construction of an *absolute* mercury standard and professed enthusiasm to his offer to assist in establishing the accuracy with which such a standard could be reproduced.[99]

In March 1863 Werner Siemens was busy with both political matters and managing company operations in Berlin and Russia, planning with his brother William the foundation of the first Siemens cable-making factory in London.[100] Thus it was not Siemens but his assistant Sabine who publicly replied to the BAAS Committee's 1862 report, claiming in the *Philosophical Magazine* that the 1.6% error lambasted by Matthiessen was due solely to earlier errors in using measuring apparatus.[101] [Only much later did Siemens himself explain that the coils referred to as 'Siemens, London (1862)' by Matthiessen and Jenkin were in fact on display at the 1862 Exhibition only for historical reasons as prototypes constructed in 1859 for use in laying the Red Sea and Indian Ocean submarine cables.[102]] According to Sabine's newly refined method of calculating resistance from the geometrical properties of glass spiral tubes, the agreements between mercury measurements turned out to be consistent to within 0.05% after all – just as Siemens had claimed in his letter to Jenkin. In claiming exactness for this method, Sabine cited data to six significant figures, citing no quantitative estimates of error

---

[97] Ibid., p. 150.
[98] Siemens wrote to Jenkin, 'Nor does mercury change its resistance in the least by standing in the air. This I have proved by keeping a [glass] spiral six months filled without changing the mercury, and I found its resistance to be constant', 'Appendix E', BAAS Report, 1862, p. 154.
[99] 'Provisional Report', BAAS Report, 1862, p. 133.
[100] Werner was by now member of the Prussian parliament – having co-founded the German Progressive Party in 1861; Feldenkirchen, *Werner von Siemens*, p. xix.
[101] Robert Sabine, 'On a new determination of the mercury unit of electrical resistance in Dr Siemens' laboratory', *Philosophical Magazine*, 25 (1863), pp. 161–74, footnote on p. 161.
[102] Siemens, 'On the question of the unit of electrical resistance', p. 333.

but offering detailed analysis of the care employed to remove all possible sources of error.[103] The incursion of British protocols of measurement reporting into the Berlin Siemens company was mirrored by the inflitrations of coils calibrated to the new 'third' determination of the Siemens mercury unit into the international telegraphic practice of resistance measurement. According to Sabine, the Siemens unit had already found employment by 'many physicists' and 'very extensively in telegraph measurements', both in England – presumably at least in the London Siemens works – and on the continent.[104]

In its report for the 1863 meeting of the BAAS, the Electrical Standards Committee accepted Sabine's results as having 'conclusively proved' that mercury resistances could be reproduced to within 0.05%.[105] Indeed, the BAAS Committee's relations with the Siemens company were not manifestly antagonistic at this stage. Later it revealed that it had allowed the Siemens and Halske company in Berlin to reproduce copies of the BAAS's 1863 determination of an absolute resistance in a set of 1–10,000 units to meet the request of the superintendent of the government telegraph lines in India.[106] After the 1863 report appeared, these relations cooled considerably, however, seemingly as a result of Matthiessen's continued hostility to the use of mercury. The 1863 report had noted that Matthiessen was convinced that several solid metals were 'equally fitted' for the purpose of reproducing resistances and was 'disposed' to put his conviction to experimental proof by securing the concordance of results from a number of independent observers.[107] And yet in 1864 it transpired that the committee had not done this. Rather, it had transferred all responsibility on this question to Matthiessen – working with Charles Hockin, a young Fellow of St John's College Cambridge. The results for the Siemens mercury proposal were thus somewhat partisan.

After six months of laborious experiments on various materials including mercury, lead, and gold–silver alloy, this pair produced measurement data that they claimed had been accomplished by exercising 'great' care. The unusual amount of care they expended fulfilled the degree of precaution required to be sure of 'trustworthy' results, and involved recourse to 'extraordinary precautions' not taken when only 'ordinary' care was shown. Such distinctions enabled them to explain away the rather awkwardly disparate results in Siemens' first publication on mercury columns (his 1860 paper) and Matthiessen's first trials in which a gold–silver alloy was used. The use of only ordinary care had led to Siemens' results that had a degree of accuracy of 1.6%, and Matthiessen's results that had a comparable figure of 1.5%.

---

[103] See especially Sabine, 'On a new determination of the mercury unit', p. 169.
[104] Ibid., p. 162.
[105] 'Report of the Committee on Electrical Standards' BAAS Report, 1863, Part 1, pp. 122–3.
[106] 'Report of the Committee on Electrical Standards', BAAS Report, 1864, Part 1, p. 345.
[107] 'Report of the Committee on Electrical Standards', BAAS Report, 1863, Part 1, p. 122.

Table 3.2. *Results of Matthiessen and Hockim*

Metal	Number of Wires Tested	Maximum Discrepancy in Conducting Power (as a Fraction of Total)	Probable Error*	Probable Error[†] (%)
Silver	3	0.0014	0.00052	0.052
Copper	3	0.0011	0.00021	0.021
Gold	4	0.0005	0.00011	0.011
Lead	4	0.00054	0.00006	0.006
Gold–silver alloy	5	0.00073	0.00001	0.001
Mercury	3	0.00151	0.00009	0.009

* This appears to be the probable error expressed as a fraction.
[†] This is an extrapolated column to show the probable error as a percentage.
Source: BAAS Electrical Standards Committee, *BAAS Report* 1864, p. 348.

By contrast, the much more concordant results attained by Sabine in 1863 and by Matthiessen and Hockin in this paper involved 'great' care.[108] From the way they presented the results of their most recent research, however, Matthiessen and Hockin tried to convey the impression they had exercised still *greater* care than Sabine, securing somewhat closer agreement for almost all the metals used than Sabine's 0.05% (see Table 3.2).

Given that, with 'great care' Matthiessen's gold–silver alloy accomplished the greatest degree of reproducibility, Matthiessen and Hockin moved rather swiftly to the conclusion that this material would produce the best results even if only ordinary care were used. They concluded that anyone who could not afford the time or expense necessary for great care, could produce their own reproduction of the BA absolute resistance by taking a metre of gold–silver alloy weighing 1 g and using 0.5995 of that metre as equal to the required unit.[109] Jenkin's report for the BAAS Standards Committee in 1864 not only confirmed these conclusions, but also made unprecedentedly harsh remarks about the 'great difficulty' in dealing with mercury. It compared the two Berlin units of 1862 attacked by Jenkin and Matthiessen in earlier reports; two values for a commercial coil produced by the Siemens company in 1864, and two specially prepared by Matthiessen. Because the discrepancy among results amounted to over 1%, Jenkin concluded that no trustworthy standard existed that strictly fulfilled Siemens' definition. It was somewhat disingenuous of Jenkin therefore to tell readers that the Committee did not 'desire to express any opinion' but just hoped to draw attention to the 'great

---

[108] Augustus Matthiessen and Charles Hockin 'Appendix C – On the reproduction of electrical standards by chemical means', *BAAS Report*, 1864, Part 1, pp. 352–67, quotes from Part 1, pp. 352–3.
[109] Matthiessen and Hockin, 'Appendix C', pp. 366–7.

discrepancies' in realizations of the mercury unit. In fact, the Committee's case against the mercury unit could hardly have shown less charity: Much was made, for example, of Siemens' use of a questionable figure for the specific gravity of mercury of 13.557, the standard figure for this having been established by Committee member Balfour Stewart as 13.595. When the table previously discussed (i.e., Table 3.2) was constructed, these two different values for specific gravity were used to maximize the apparent discrepancy between values calculated for the Siemens commercial unit and for Matthiessen's own construction.[110]

Why had the Committee's report become so hostile to the use of mercury to calibrate resistance units? One obvious reason concerns the long-term stakes in getting the BA unit into the working lives of resistance users. Whilst there was still no officially sanctioned copies of the BA unit yet available to purchase, the Siemens 1864 unit was apparently becoming embedded in the working routines of many telegraph engineers.[111] As the BAAS Committee's report primly noted, 'Defective systems are daily taking firmer root.' Practical standards of resistance were therefore 'urgently' required, and the Committee felt pressed to come to a decision on this matter. The BAAS Committee thus adopted the interim strategy of enthusiastically promoting the Matthiessen–Hockin specification for an alloy reproduction of the BA unit previously discussed – and speculated openly that 'authentic copies' might soon be issued at a cost of about £ 2 10s each.[112] Clearly the committee had been sounding out its future market, for it claimed to have received assurances that its units would be adopted readily in the United Kingdom, India, Australia – and even in the Germanic states where the Siemens unit already had such an easy market. Jenkin even went so far as to say that it believed the unit would be 'accepted in America' and in 'many other parts of the world', although no reply had been received to their enquiries about its prospective adoption in France. The very fact that such assurances had been sought is indicative of the pressure the Committee felt from competing units,[113] and by thus publicizing such assurances they may well have swayed some readers into following their (prospectively self-fulfilling) predictions of the inevitable universality of the BA unit.

Ironically, it transpired that the BAAS rejected both Matthiessen's original proposal of a gold–silver alloy for the standard as well as the Siemens' mercury construction, opting instead for a cheaper alloy of silver–platinum.

---

[110] 'Report of the Committee on Electrical Standards, BAAS Report, 1864, Part 1, pp. 348–9.
[111] For evidence that, apart from the Siemens and Halske set of BA units sold to the Indian Telegraph service, 'Messrs Elliott had sold some copies of the coil to those unwilling to wait for the Committee's final experiments, but the Committee refused to certify these as being correct', see ibid. pp. 345 and 348.
[112] Ibid, p. 348.
[113] Evidence subsequently presented of the geographically limited distribution of the BA unit will show that their concern was by no means ungrounded.

But the die was effectively cast. Although Werner's commercial partner, his brother Charles, was by now a member of the BAAS committee, the primary means of embodying and reproducing the BA resistance unit was by use of an alloy standard and or use of a Wheatstone bridge to effect comparative copying.

3.6. 'DR MATTHIESSEN HAS BEEN OPPOSED TO MERCURY': THE ACRIMONY OF COMMERCE 1865–1866

The Committee's barbed attacks on Siemens' advocacy of mercury in the BAAS Committee 1864 Report received no immediate direct reply. The two Siemens brothers were mostly preoccupied with attempts to lay the submarine Cartagena–Oran cable on behalf of the French government – and the subsequent collapse of both attempts. This outcome was financially catastrophic for the Siemens brothers and damaged both the company's prestige and international relations. Nevertheless, it was well recognized at the time that the cable ruptures owed rather more to mechanical problems with the cable technology and oceanographic misfortune rather than to the techniques of electrical tests in which the Siemens Company had rather more specialist expertise.[114] In any case, it is not obvious that the Siemens company would have needed to respond directly to BA Reports: The readership of such potentially damaging criticisms on the Siemens' commercially issued standard might not have included a large enough number of telegraph engineers to warrant a formal reply. A reply from Siemens came only once members of the BAAS Committee began taking aggressive steps to publicize the rival BA units, copies of which became commercially available in early 1865 after Maxwell and Hockin had made apparently secure measurements of the absolute unit.[115] One tactic employed in promoting the BA unit was to expound the value of the absolute system, but two potentially much more effective techniques were also in play: Denigrating the integrity of the popular rival Siemens unit and casting aspersions on the trustworthiness of Siemens and his fellow workers.

In February 1865, Committee Secretary Fleeming Jenkin actually took the extraordinary step of *advertising* the BA unit in a letter to the editors of the *Philosophical Magazine* in February that year. Jenkin expressly hoped its editors and readers would share their reasons for adopting the Weber–Thomson absolute system of measurement and would 'assist' in ensuring that their resistance unit was generally adopted.[116] By sending Jenkin the sum of

---

[114] Siemens, *Inventor and entrepreneur*, pp. 147–57.
[115] Schaffer, 'Late Victorian metrology', pp. 27–8.
[116] Fleeming Jenkin, 'Electrical standard', letter to editors of the *Philosophical Magazine*, 29 (1865), p. 248.

£2 10s – the cost indicated in the previous year's report – readers could in fact directly secure from him their own boxed platinum–silver coil calibrated to the BA unit. This advertising was repeated in his follow-up paper at the Royal Society in April 1865[117] in which he revealed that having sent out twenty free copies of the new standard and now expected that it would be 'gladly accepted' throughout Britain and the colonies: a geographical specificity that implicitly conceded the continent of Europe to the Siemens unit and other rivals. Jenkin claimed that the 'only obstacle' to the wider adoption of the BA unit was the difficulty of explaining to potential users how the resistance of a wire could have the dimensions of velocity, and he accordingly offered an accessible account of this point.[118] He then explained how many more difficulties attended the use of the mercury unit. Having worked as Siemens' assistant seven years previously, he politely acknowledged that Dr Siemens merited special recognition for the 'great care' with which the mercury-calibrated coils issued by his company had been made.[119] And he conceded that four observers – two of them 'practical engineers' – had now independently shown that the BA and the mercury units could be mutually calibrated within a few hundreds of a per cent.[120] Yet twice in his paper Jenkin hinted darkly that the mercury-calibrated coils were not what the Siemens' company claimed. He suggested that the unit had changed in value by about 2% since the first issue, and inaccurate specific gravity data made it 'doubtful' whether even the most recent standard 'truly' represented the Siemens definition.[121]

Augustus Matthiessen was much less circumspect than Jenkin in his contemporaneous attack on the 'so-called' mercury units that he published in the same British, German, and French journals used on previous occasions.[122] The ostensible purpose of his paper was to publicize the BA unit both by showing how it could be used accurately to quantify the specific resistance of various metals. Tellingly, it repeated Jenkin's message about how to secure a copy of these units, claiming for them (without explicit calculations)

---

[117] Fleeming Jenkin, 'Report on the new unit of electrical resistance proposed and issued by the committee on electrical standards appointed in 1861 by the British Association', *Proceedings of the Royal Society*, 14 (1865), pp. 154–64, and reproduced in *Philosophical Magazine*, 29 (1865), pp. 477–86. The subsequent citations are from the version published in the *Philosophical Magazine*.

[118] Jenkin, 'Report', pp. 484–6.   [119] Ibid., p. 480.   [120] Ibid., p. 484.

[121] Ibid., p. 479, footnote 'II II' and p. 484.

[122] Augustus Matthiessen 'On the specific resistance of the metals in terms of the B.A. unit (1864) of electric resistance, together with some remarks on the *so-called* mercury unit', *Philosophical Magazine*, 29 (1865), pp. 361–370, and also in German as 'Ueber den specifischen Leitungswiderstand der Metalle, bezogen auf die von der BRITISH ASSOCIATION angenomme Widerstandseinheit, nebst einigen Bemerkungen über die sogenannte Quecksilbereinheit', *Annales de Physique et Chimie*, 125 (1865), pp. 497–509.

112   *The Morals of Measurement*

Table 3.3. *Matthiessen's Table*

Resistance Measured	'True Value in B.A. Units'
Siemens (London) 1862 unit	0.9723
Siemens (Berlin) 1862 unit	0.9605
Siemens 1864 unit	0.9534
Mean	0.9620
Hockin & Matthiessen 1864	0.9619

Source: Matthiessen, 'On the Specific Resistance of the Metals', *Philosophical Magazine*, 1865, pp. 364–5.

an accuracy of 0.01%.[123] This overrode the gold–silver protocol for the home-made unit issued in the previous year's BAAS Report: Now that the absolute unit could be purchased *ready-calibrated* from Jenkin, Matthiessen hoped that no one would actually *try* to reproduce for themselves a solid-metal unit – not even by using the trustworthy gold–silver alloy. From that he immediately launched into an attack on the Siemens unit by challenging the veracity of the Siemens camp. Given the persistently incorrect use of specific-gravity data and the discrepancy in the early London and Berlin mercury units, Matthiessen presented a table (reproduced as Table 3.3) to support his claim that *no true mercury unit* had yet been issued.[124]

Why, asked Matthiessen, were there such significant 'differences' between successive determinations of the mercury unit? Had not the readers of Siemens' and Sabine's papers been led to think that the reproduction of the mercury unit was the 'most simple thing possible', requiring little trouble or expense?[125] With increasing hysteria, Matthiessen asked these questions:

[D]oes anyone for a moment suppose that a standard [mercury] measure can be reproduced with a *little trouble*? Would it not take any observer weeks to reproduce with accuracy a resistance which can be relied on? Must not such an observer check and recheck every determination he makes? Must not he be sure all his instruments are graded accurately? and he can only be sure of this if they are carefully tested by himself, or by perfectly reliable persons; and, what is of very great importance, must he not be sure that he deals with an absolutely pure metal?[126]

To clinch his claim, Matthiessen even cited Werner's own brother and fellow BAAS Committee member, C. W. Siemens, in admitting that his experience

---

[123] Matthiessen, 'On the specific resistance of the metals', p. 369.
[124] Matthiessen read Siemens as implying that the higher results attained for the specific gravity of mercury by Regnault, Kopp, and Stewart were 'all incorrect'. Matthiessen explained: '[O]ne is irresistably led to the conviction that it is utterly impossible that those distinguished observers can all have made such mistakes.' Matthiessen, 'On the specific resistance of the metals', pp. 364–5.
[125] Ibid., p. 367.   [126] Ibid., p. 367.

in reproducing the mercury unit had taught him the 'great difficulty' of such operations.[127] And if in Dr Siemens' skilled hands errors could occur, how much more were errors to be 'feared' in efforts by those who had not such refined apparatus as that of the Siemens at their command?

At this juncture of the debate the central issue was whether Siemens and his co-workers had forfeited the status of trustworthiness by giving unreliable and misleading claims about the performance of mercury.[128] Werner von Siemens replied in the *Philosophical Magazine* and *Poggendorf's Annalen* of 1866 with a robust defence.[129] Conceding that he had adopted the wrong value for the specific gravity of mercury, he confidently claimed that the consequent 0.3% error could nevertheless be eliminated simply by raising the temperature specification at which the unit represented the Siemens standard resistance from 0 to 10.5 °C. If that change of specification were implemented, the hundred or so standards that Siemens had shrewdly presented *gratis* to physicists, technologists, and institutions across the continent and the UK would thus agree with the third Siemens determination of the mercury unit in 1863–4 to within 0.05%.[130] Thus Siemens could claim that there was little in the critique of Matthiessen and Jenkin to bother users of the Siemens resistance standards, especially given that the next corrected Siemens resistance unit would be available from 1866. In any case, he claimed, the 1864 Siemens commercial unit was 'very generally employed' and met all 'practical wants' in telegraphy.[131]

Given that Siemens was so confident about the present and future success of the mercury-calibrated unit, it is important to ask why he bothered to reply in considerable detail to Jenkin and Matthiessen. As Olesko has emphasized, Siemens was caustic about the Committee's bizarrely optimistic calculation of a 0.1% error in the BA unit, arguing that these actually amounted to 1.4% if subjected to a full standard error analysis.[132] Siemens was more perturbed, however, that Matthiessen and Jenkin had criticized his results by 'attacking the trustworthiness' of his labours,[133] throwing a 'false colouring' on his (company's) resistance determinations. They had simply failed to properly understand the historical succession of ever-improved units from the Siemens atelier.[134] Now that all the existing Siemens commercially issued resistance coils agreed with each other to within 0.05%, the discrepancies between the range of Siemens units found by Matthiessen arose simply from

---

[127] Ibid., p. 368.   [128] Ibid., p. 369.
[129] Siemens, 'On the question of the unit of electrical resistance', pp. 325–36.
[130] Ibid., p. 332.
[131] Ibid., pp. 330–1. See Latimer Clark and Robert Sabine, *Electrical tables and formulae for the use of telegraph inspectors and operators*, London, 1871, p. 6.
[132] Olesko, 'Precision, tolerance, and consensus', pp. 129–30; Siemens, 'On the question of the unit of electrical resistance', p. 328.
[133] Siemens, 'On The Question Of The Unit Of Electrical Resistance', p. 330.
[134] Ibid., p. 334.

the latter's incompetence in handling mercury. He had failed to follow the protocols recommended by Siemens, namely, not adding drops of mercury to the tube,[135] but dipping the tubes into a trough. In pressing the ends together with his bare fingertips, Matthiessen had thereby diminished the quantity of mercury in the tube, thus upsetting the reliability of calculations made on the amount of mercury used. Siemens contended that differences between their results stemmed from this instance of the 'little care' exercised by Matthiessen.[136]

Matthiessen's reply published in the same (May 1866) issue of the *Philosophical Magazine*[137] suggested that Siemens had not 'carefully read' his reports and papers[138] and straightforwardly denied that there was any significant material difference in the way that he and Siemens had conducted their experiments. And in reply to Siemens' accusation of his 'lack of care' in his measurements, he would simply 'leave it to others' to judge whether his experiments with Hockin had been carried out with as 'great a care' and with 'as many precautions' as those of Siemens and Sabine. Matthiessen thus left it diplomatically unclear as to whether differences in their respective results were due to Siemens' incompetence and misrepresentation or to the inherently unreliable nature of mercury.[139] A few months later, Fleeming Jenkin leaped into the fray in the *Philosophical Magazine* for September 1866 with a lengthy reply covering many issues.[140] He refused to mediate between Siemens and Matthiessen, claiming the latter's reputation was 'too high' for anyone uncritically to accept Siemens's contrary claims on the reproducibility of mercury measurements. Dr Matthiessen was 'opposed to mercury', and that was sufficient reason for Jenkin to adopt the same position.[141]

Jenkin did also recognize that Siemens had an international reputation in electrical measurement, and he was especially concerned that his former employer's claims had been given 'fair consideration' by the BAAS Committee.[142] Yet he asserted that Matthiessen and his collaborators had not been careless in their experiments with mercury as Siemens had contended. The very names of the members of the BAAS Committee were a 'guarantee that care would be taken' in whatever experiments they undertook, and moreover, that the names of the Committee at large guaranteed that no results would be adopted unless they were considered trustworthy. Such a representation of the Committee's integrity was evidently important

---

[135] Siemens, 'Proposal', pp. 28–9.
[136] Siemens, 'On the question of the unit of electrical resistance', pp. 334–5.
[137] Augustus Matthiessen, 'Note on Dr Siemens' paper "On the question of the unit of electrical resistance"', *Philosophical Magazine*, 4th series, 31 (1866), pp. 376–80.
[138] Matthiessen, 'Note on Dr Siemens' paper', pp. 376–7.   [139] Ibid. p. 380.
[140] Fleeming Jenkin, 'Reply to Dr Siemens' paper "On the question of the unit of electrical resistance"', *Philosophical Magazine*, 4th series, 32 (1866), pp. 161–77.
[141] Ibid., pp. 168–9.   [142] Ibid., p. 161.

because the absolute pedigree of the BAAS unit had thus far not been sufficient to displace the Siemens mercury unit in the market place. Alleging that no 'large' English telegraph company used the BA unit, nor any in France, Jenkin grudgingly admitted that, owing to the 'undoubtedly excellent' manufacture and convenient arrangement of Siemens resistance coils, this unit had been used at large. But this extensive use owed nothing to the 'excellence' of his unit's definition: It was merely that people had ordered coils from the 'most celebrated' firm in Europe and took what was given them – the miles of copper wire before 1860 and the mercury units thereafter.[143]

Jenkin then went on to stress how much the Siemens unit was prone to gross errors and discrepancies when brought into practical realization, the production of it requiring all of Dr Siemens' skill, 'if even he can do it'.[144] The difference between the Siemens value of the BA unit as 1.0486 mm of mercury of 1-mm cross section at 0 °C and Matthiessen's of 1.0396 mm – implicitly of around 1% – was, inferred Jenkin, a serious problem. What was needed was a concordance of independent results to within 0.01%, although he offered no reasons why that figure should be the threshold of reasonable agreement. Given that these values had been identified as 'discrepant' after both parties had taken 'all possible care' in their measurements, the only recourse was to suspend judgement as to which value was nearest the truth. How then could this discrepancy be explained? Jenkin was reluctant to identify the problem in mercury itself for this substance was one of the various metals still being tested by the Committee in long-term trials to see which could furnish the most invariable embodiment of the BA unit. Instead, Jenkin argued that the difficulty was that no reliable *method* had yet been identified for reproducing resistance values with mercury. Symmetrically, the same applied to Matthiessen's claims for the gold–silver alloy: Until Siemens or some other 'competent' observer had attained thoroughly concordant results with that material, it too could not be relied on as a means of reproducing a resistance measurement standard.[145]

And with that diplomatic defusing of the conflict by Jenkin, the overt manifestation of the metals controversy came to an end.

There are a number of reasons why Siemens had neither the time nor immediate interest to reply to Jenkin. In 1866 Werner was preoccupied with the death of his wife, Mathilde, and in the following year as a member of the new parliament, he threw himself vigorously into the politics of a new Prussian War. Later he busied himself with developing the powerful possibilities of the self-exciting dynamo for electrical generation that he had announced contemporaneously with Cromwell Varley and Charles Wheatstone in Britain in 1866–7.[146] As we shall see in later chapters, the

---

[143] Ibid., p. 163.   [144] Ibid., p. 164.   [145] Ibid., pp. 168–9.
[146] Feldenkirchen, *Werner von Siemens*, p. 87; Percy Dunsheath, *A history of electrical engineering*, London: Faber & Faber, 1962, pp. 109–10.

commercial possibilities of electric lighting were indeed to prove very profitable to the Siemens company in the 1880s and enhanced the reputation of the Siemens name in ways that eclipsed all earlier embarrassments over the calibration of the mercury unit. Coils of that continued to sell well on the continent despite all the controversy previously discussed.[147]

For their part, by the summer of 1866 the various members of the BA Committee could relax a little after the second transatlantic cable had been successfully tested and laid (probably) by use of the BA's own resistance units. As Hunt has observed, this lent credibility and saleability to the unit in cables laid by British companies in the British Empire.[148] And by 1867 the BAAS had acquired further evidence that vindicated Matthiessen's claims about the difficulties of reproducing mercury standards. Hockin had checked the consistency of pairs of standards made from different metals by Matthiessen in 1865 and found discrepancies in the mercury construction that he attributed to the troublesome use of nitric acid to pre-clean the tubes holding the mercury. The faulty mercury tube was abandoned and a new one set up, only to be abandoned soon after.[149] The 1867 Report of the BA Committee, however, reveals that it had accepted the duarchy[150] at least of the mercury and absolute units.[151] Significantly, it had approved the use of a cheap portable resistance measurer calibrated in both BA and mercury units, which C. W. Siemens had recommended from his experiences in employing it on the Siemens European landlines.[152] We might plausibly infer that this device was endorsed by the Committee as a means of familiarizing many users of the Siemens mercury unit with the equivalent values of the BA absolute unit.[153] This was a shrewd move because the Vienna International Conference on Telegraphy adopted the Siemens mercury unit as its conventional standard for operations in 1868.[154]

---

[147] For overall company sales figures, see appendices in Feldenkirchen, *Werner von Siemens*, pp. 161–173.

[148] Hunt, 'The ohm is where the art is', p. 61.

[149] Charles Hockin, 'Appendix III, 'Comparison of the B.A. Units to be deposited at Kew Observatory, Report of the Electrical Standards Committee', *BAAS Report*, 1867, Part 1, p. 483. See also committee report in *BAAS Report*, 1870, Part 1, pp. 14–15. No mention of them is made in R. T. Glazebrook, 'Report of the electrical standards committee', Appendix I, 'On the values of certain standard resistance coils', *BAAS Report*, 1890, Part 1, pp. 98–102.

[150] I thank Jack Morrell for suggesting the use of this term in preference to 'dualism.'

[151] 'Report of the Committee on Electrical Standards', *BAAS Report*, 1867, Part 1, pp. 474–522.

[152] Ibid., Werner von Siemens, Appendix III 'On a resistance measurer', pp. 479–81; complemented by Fleeming Jenkin's 'On a modification of Siemens' resistance measurer', ibid., pp. 481–3.

[153] 'Report of the Committee on Electrical Standards', 1867, p. 475.

[154] Olesko, 'Precision, tolerance, and consensus', p. 139.

3.7. RESISTING MERCURY: THE UNRESOLVED AFTERMATH OF THE METALS CONTROVERSY

> The electrical properties of metals and their alloys have been studied with great care by MM. Matthiessen, Vogt, and Hockin, and by M.M. Siemens, who have done so much to introduce exact electrical measurements into practical work.
>
> James Clerk Maxwell, *Treatise on Electricity and Magnetism*, Vol. 1, 1873[155]

No complete consensus was accomplished during the ensuing years of the nineteenth century as to whether mercury or an alloy was the most trustworthy metal to embody a resistance standard.[156] In 1866, the principal antagonists returned to their own particular localized concerns, still brooding over the debate in ways that were directly manifested again only decades later. It was doubtless as Matthiessen's private pupil in 1868 that the young Alexander Muirhead imbibed the distrust of mercury that he avowed to a Board of Trade Committee on Electrical Standards in 1891 (see Chapter Introduction). Siemens' vociferous complaints about Matthiessen's conduct in his 1892 autobiography were matched only by his gloating about Lord Rayleigh's (apparent) vindication of his claim about the reproducibility of mercury standards to within 0.01%.[157] One contemporary hint of the acrimony can still be found, however, in Robert Sabine's 1867 volume *The Electric Telegraph*, dedicated to C. W. Siemens. After tracing the histories of both the BA and the Siemens unit, Sabine criticized the BAAS Committee for having lost sight of its 'dignified' position as an impartial jury, certain unnamed members showing so much 'apparent animism' that they could be suspected of having some 'personal feeling' in the matter.[158] It does not require a great interpretive leap here to identify this as an allusion to the fulminations of Augustus Matthiessen.

Ironically, rather less controversy surrounded the occasion of Matthiessen's suicide in October 1870. He took prussic acid in his laboratory following accusations that he had indecently assaulted a young male assistant, an allegation he denied in his suicide note. An obituarist in *The Times* suggested that Matthiessen had a 'high character' and 'world wide reputation', and accordingly there could be 'no foundation whatsoever' for the charges against him. It emphasized not only his blameless character, but

---

[155] J. C. Maxwell, *Treatise on electricity and magnetism*, London, 1873, Vol. 1, p. 416; 1891, 3rd edition, p. 496.
[156] By contrast, simple sociological models of scientific controversy might lead us to expect that such disputes would be marked by a definitive outcome, whether produced as a broad consensus or by a definitive 'core-set' of practitioners determining authoritatively which party had won. See H. Collins, *Changing order: Studies on scientific induction and replication*, London: Sage, 1985, pp. 142–5, 154–5.
[157] Siemens, *Lebenserringerungen*, pp. 166–7, 264.
[158] Robert Sabine, *The electric telegraph*, London, 1867, p. 339–40. By then Sabine had left Berlin to become superintendent to Charles Wheatstone's telegraph company in London.

also the 'trustworthy character' of his results, emphasizing that the laws derived from his research on copper and its purification were indeed now in 'constant' use by telegraphic engineers.[159] A writer for the Chemical Society more elliptically reported the incident in terms of the 'strain' put on his brain by overwork, hinting at the pressures of managing his disability.[160] Given this and the events of the previous decade, it is striking that one of Matthiessen's colleagues on the BAAS Committee took some trouble to uphold the reputation of both sides in the erstwhile metals dispute. In his *Treatise on Electricity and Magnetism*, Maxwell lavished praise on Matthiessen's work on the resistance of copper and alloys, but also maintained that mercury showed considerable stability and uniformity as a conductor. In contrast to Matthiessen, he acknowledged that 'great care' had been exercised by the Siemens brothers in measuring the specific resistance of this metal.[161] Maxwell then pointedly declined to specify which was the more trustworthy metal for the purpose of embodying standards. And he was not alone in refusing to adopt a partisan stance on this matter.

By the time Maxwell's *Treatise* was published, C. W. Siemens had arranged for the London branch of his Anglo–German company to sell coils calibrated to both BA and Siemens units.[162] This sales policy clearly supports Hunt's scepticism about the claim of one BAAS Committee member in 1868 – Latimer Clark – that the BA unit was by then 'universally adopted'.[163] The Committee's report two years later observed merely that its unit had been ratified over a 'great portion of the globe' and copies placed in 'the hands of electricians in various parts of the world'.[164] More telling still is the evidence of the *Electrical Tables & Formulae For The Use Of Telegraph Inspectors And Operators* that Clark co-wrote with Sabine in 1870. Despite recommending the BAAS's productions as theoretically the most 'rational

---

[159] See inquest report in *The Times*, 8 October 1870, p. 5, column 5. This was not a reference to his work on alloys, but rather to his extensive 1860 studies on effects of temperature and impurity on the conductivity of copper that were drawn on by the Atlantic cable committee of enquiry. In his article on Matthiessen for the *Dictionary of National Biography*, Phillip Hartog wrote 'Matthiessen bore a high personal character among his contemporaries'.

[160] As Olive Anderson has noted in her study of Victorian suicide, it was rare for men to take their own lives when accused of sexual misconduct. Matthiessen's obituaries in October 1870 maintained a strongly exculpatory tone or silently passed over the circumstances of his fate. Olive Anderson, *Suicide in Victorian and Edwardian London*, Oxford: Clarendon, 1987, p. 196. 'Obituary of Augustus Matthiessen', *Proceedings of the Chemical Society*, 25 (1871), pp. 615–17.

[161] Maxwell, *Treatise*, 3rd edition, 1891, Vol. 1, pp. 497–8.

[162] See citation of this in Fleeming Jenkin's Cantor Lecture No. IV, 'Electrical tests', in Fleeming Jenkin (ed.), *Reports of the Committee on Electrical Standards*, London, 1873, pp. 225–6.

[163] Hunt, 'The ohm is where the art is', p. 61, referring to Latimer Clark's *Elementary treatise on electrical measurement*, London, 1868, p. 43.

[164] 'Report of the Committee on Electrical Standards', *BAAS report* 1870, Part 1, pp. 14–15.

and concordant system', Sabine and Clark defined the practical manifestation of the now-christened BA 'ohm' as a prism of mercury of 1-mm^2 cross section and 1.0486 m long – the definition offered by Siemens but opposed by Matthiessen in 1866. For obvious territorial reasons, whereas the BA unit was 'generally adopted' in Britain and 'elsewhere' (presumably in British colonies), the Siemens mercury unit was used where that company had installed major landlines or submarine cables: Germany, Russia, and the Mediterranean. More revealing, however, is the report of Clark and Sabine that in all French submarine cable work, resistance coils were 'adjusted' to the mercury unit. The Swiss company, Bréguet and Digney, had readjusted its coils to 1/10 of the mercury unit, and even the Varleys working for the British-based E & I Telegraph company had now abandoned their British statute mile of copper wire in favour of coils adjusted to 25 mercury units.[165]

This ubiquity of the Siemens mercury-based products was not the only problem facing protagonists of the BA resistance unit. The trustworthiness of the 1865 absolute determination, on which that unit was based, was undercut by James Joule's use of it in 1866–7 to redetermine the mechanical equivalent of heat by electrical methods. The value Joule thereby secured, 783 ft lb/°F, was 1.4% greater than his previous determination by frictional methods (772 ft lb/°F). Although this discrepancy was as great as any that the BAAS Committee had attacked in Siemens' publications on mercury, the Committee nevertheless glossed over this inconvenient result, concluding that Joule's experiments removed 'all fear of any serious error' in determinations of the mechanical equivalent or in the BA standard.[166] Ironically, later observers decided that Joule's frictional method was the more trustworthy – and indeed that this very difference between his results represented the *error* in the BA's first determination of its absolute unit.[167] The reasons for this judgement emerged after the reconstituted BAAS Committee admitted in 1881 that 'discrepant results' had shown the 1865 unit to be adrift by around 1.2%.[168] In addition to this, the discrepancies between various solid-metal and alloy reproductions of the BAAS unit set up by Matthiessen in 1865 undermined the credibility of British attempts to argue against the mercury-embodied standard at the 1881 Paris congress.[169] Although that meeting adopted the

---

[165] Robert Sabine and Latimer Clark, *Electrical tables & formulae for the use of telegraph inspectors and operators*, London, 1871, pp. 6–7. Certainly too, mercury units were used on board the Siemens ship *Faraday* that, throughout the 1870s, competed with the vessels owned by John Pender's companies in laying cables across the Atlantic; Feldenkirchen, *Werner von Siemens*, pp. 101–5.
[166] 'Report of the Committee on Electrical Standards', *BAAS Report*, 1867, Part 1, p. 475.
[167] Sir Frederick Abel, 'Presidential address', *BAAS Report* 1890, Part 1, pp. 1–9. Abel thus attributed to Joule the first, albeit indirect, determination of the 'true value' of the BA Unit.
[168] 'Report of the Committee on Electrical Standards', *BAAS Report*, 1881, Part 1.
[169] Lagerstrom, 'Universalizing units', pp. 5–8.

BA absolute unit, it followed the entrenched practice of the Vienna Telegraph Congress of 1868 to embody this in the standard form of a mercury column, now 106 cm long and 1 mm² in cross-section area at 0 °C.

After three years of further debate, the BAAS Committee formally abandoned its commitment to the virtues of Matthiessen alloy. At the 1884 BAAS meeting in Montreal, the BAAS Committee nominally agreed to adopt the Paris protocol as a form of 'legal ohm', albeit (yet) without any Parliamentary sanction. As previous historians have noted, a further redetermination of the absolute ohm as 106.3 cm of mercury in 1888 brought considerable derision, not least for the further proliferation of units it generated; this, indeed, to some, vindicated the critics of those who were critical in principle of mercury-based standards.[170] Throughout much of this period, Matthiessen was not forgotten for his name was preserved eponymically as the standard for absolutely pure copper.[171] He also remained part of the BAAS heroic pantheon, as for instance when Sir Frederick Abel related the history of the electrical standards in his Presidential Address at Leeds in 1890.[172] Ironically, it was a discussion on this topic at that BAAS meeting which prompted one engineering journal to comment that new electrolytic techniques made it possible to secure copper that was, according to Matthiessen's criterion, 102% [sic] pure. The label of Matthiessen 'pure' copper, it concluded, now conveyed 'less meaning than ever'.[173] And this problem had a wider significance in relation to the problem of pure metals.

The preceding comment in the *Electrician* for 3 October 1890 was made in response to puzzled questions from its readers as to why mercury had been chosen as the pure metal out of which the anticipated new British legal standard for resistance would be made. Surely everyone knew that a column of mercury was known to be 'difficult' to construct, 'troublesome' to set up, and so much more 'inconvenient' to use than a solid wire. Twenty-five years after Matthiessen had set up the first solid-wire embodiments of the BA unit, an *Electrician* editorial noted that both pure metals and alloys had turned out to be insufficiently reliable for this purpose. Researches by Glazebrook and others presented at the Leeds meeting had shown that the very act of freezing metals to 0°C for the purpose of intercomparison substantially altered their resistance. And even with alloys little affected by temperature change, it

---

[170] As James Swinburne noted rather scathingly in 1888, the so-called 'legal ohm' had no special status in the British legal system; James Swinburne, *Practical Electrical Measurement*, London, 1988, pp. 22–24.

[171] Maxwell, *Treatise*, 1873, Vol. 1, pp. 416–18, 1891 (3rd edition, 1891, pp. 496–9); Kempe, *Handbook of electrical testing*, 2nd edition, 1881, p. 266, cites Matthiessen's criterion as being that a 'foot-grain' of pure copper at 24 °C had a resistance of 0.2262 ohms – this value being 'the most trustworthy' yet obtained.

[172] Abel, 'Presidential address', 1890, p. 9.

[173] Anonymous, 'Notes', *Electrician*, 25 (1890), p. 600; 'Reviews', *Electrician*, 30 (1895), p. 676, referring to W. A. Price, *The Measurement of Electrical Resistance*, Oxford, 1895.

transpired that no two wires cut even from the same alloy specimen had sufficiently similar electrical properties. Moreover, alluding to Matthiessen's recent dethroning as arbiter of copper purity, the journal's writer noted that not even solid metals could be made consistently pure to a well-defined and meaningful protocol.[174] Implicitly, this left mercury as the only possible candidate metal for a standard that was unharmed by freezing and susceptible to unproblematic criteria of 'purity'. Yet in its reporting of the BAAS Section A meeting the previous month, the journal had noted that one eminent physicist on the BAAS Electrical Standards Committee of eight years of experience in measuring the resistance of mercury had raised old doubts about the reproducibility of mercury resistances. This was none other than Lord Rayleigh.

At the Section A meeting in question, Richard Glazebrook had presented the entire range of nineteen extant determinations of what length of a mercury column should be taken to express the absolute ohm, starting from Lord Rayleigh's 1882 determination of 106.24 cm. Although that determination by Rayleigh had arguably resolved the prevailing 3% disagreement on this matter in the early 1880s, there was in 1890 still a considerable range of values, the extremes being H. F. Weber's notoriously low 105.84 cm of 1884 and the 1890 figure of Salvioni and Wuilleumeier of 106.33 cm. Glazebrook started by discounting seven results that he considered unreliable from the point of view of experimental construction or performance (such as H. F. Weber's) and a further five which were also probably doubtful, he found a mean of 106.28 cm from the remaining seven trustworthy outcomes. However, by correcting Lord Rayleigh's values with an arithmetical refinement and then tactically reintroducing three of the five 'doubtful' determinations, he arrived at a value of 106.30 cm to within 1 or 2 parts in 10,000. This highly selective claim agreed quite astonishingly closely with most the recent determination by the Oxford-trained John Viriamu Jones – 106.307 ± 0.012 cm – and this value was accorded the invaluable concurrence of US visitor Henry Rowland.[175]

Such claims provoked two concerned remarks from Lord Rayleigh. He suggested first that too much importance had been attached to the fifth significant figure, and asked what difference it would really make if in fact the value of the mercurial ohm could be established to within 1 part in 100,000, as was implied by the results of Glazebrook and Jones. 'What use', Rayleigh pragmatically enquired, 'could we make of our knowledge?'[176] Accordingly

---

[174] Anonymous, 'Notes', pp. 599–600. The name Matthiessen is erroneously spelled quasi-phonetically as 'Matheson'.

[175] Richard T. Glazebrook, 'Recent determinations of the absolute resistance of mercury', *The Electrician*, 25 (1890), pp. 543–6.

[176] Anonymous, 'British Association – Discussion on electrical units', *Electrician*, 25 (1890), p. 558.

the BAAS Committee resolved that the length of the mercury column should be cited to only four significant figures as 106.3 cm. A more searching question raised by Lord Rayleigh, however, was why mercury should be the metal chosen for this purpose at all? As an experienced measurer of mercury's resistance he was in a strong position to argue that electrical researchers and practitioners would not find it 'convenient' to set up a mercury tube to contain the standard, and 'still less' would mercury be employed for the purpose in a factory. To make any use of a practical determination of the ohm, said Rayleigh, the electricians 'must have a wire'.[177] This was precisely the point with which the *Electrician* of October 1890 disagreed,[178] as it would seem at first sight did the other electrical luminaries at the BAAS 1890 meeting: Sir William Thomson, J. J. Thomson, W. E. Ayrton., Oliver Lodge, and John Hopkinson all publicly agreed that the absolute ohm for practical purposes was to be specified as the resistance of a column of mercury 106.3 cm in length and 1 mm^2 in section at a temperature of 0 °C.[179]

Rayleigh, Ayrton, Hopkinson, and William Thomson were also among those concurrently appointed to the Board of Trade Committee on Electrical Standards[180] first convened to discuss the creation of a new legal standard for resistance, current, and potential difference in January 1891. Yet, whilst acting on this committee, they entirely reversed the line they had adopted at the Leeds meeting. All were members of the BAAS Electrical Standards Committee, and they now followed the scepticism of their 1860s predecessors in regard to mercury columns. Whilst they agreed that 106.3 cm of mercury was a legally feasible definition of the ohm, they declined to make this the definitive *British* definition. The committee concurred with Hopkinson, a veteran professional witness in technical litigation, that for a permanent standard to be of value in legal proceedings it would have to be constructed from a solid metal of an uncontested (and incontestable) form. If the ohm were defined as the resistance of a 'piece of wire' under certain specified conditions, this would furnish something very 'direct' for lawyers to appeal to, expediting court business more effectively than definitions open to indefinite disputation – as was implicitly the case for mercury.[181] Accordingly, when the Committee agreed draft proposals to be put to its witnesses, these specified that the resistance standard should be embodied in a solid

---

[177] Anonymous, 'British Association – Discussion on electrical units', p. 558.
[178] Anonymous 'Notes', pp. 599–600.
[179] 'Report of the Electrical Standards Committee', *BAAS Report*, 1890, Part 1, pp. 95–138, eps. p. 97.
[180] Committee Membership: Lord Rayleigh and Sir William Thomson for Royal Society; W. E. Ayrton and John Hopkinson for the IEE; George Carey Foster and Richard Glazebrook for the BAAS, and William Preece and Edward Graves for the Post Office.
[181] *Report of Board of Trade Committee on Standards for the Measurement of Electricity*, Parliamentary Papers, 1891, p. 4.

metal – although tellingly no agreement was reached about *which* metal this should be.[182]

When witnesses appeared before the Committee from February to May 1891; none dissented from the virtues of the 'solid' ohm. Silvanus P. Thompson (representing London County Council) said that he made no 'exact' experiments' on the subject, only some 'rough' ones a few years ago, but he had found 'great difficulty' in getting the value of two columns of mercury to agree with each other.[183] The electrical manufacturer R. E. B. Crompton, speaking on behalf of the London Chamber of Commerce did not even raise the possibility of mercury, simply agreeing 'very heartily' with William Thomson's somewhat leading question on whether the material standard should be constructed out of 'solid metal'.[184] Professor of Electrical Engineering at University College London, John Ambrose Fleming, announced that he regarded that a 'carefully made' wire of platinum–silver alloy to be the most 'permanent thing' possible for the purpose.[185] And the most explicitly hostile testimony against the use of mercury for the standard came from a witness who had been Augustus Matthiessen's pupil just over twenty years previously, Dr Alexander Muirhead. Whilst accepting that it might be 'expedient' to define the quantitative value of the ohm in terms of a mercury column, Muirhead felt it was nevertheless an 'unnecessary elaboration' to do: Few 'practical men' would 'take the trouble' to set up such resistances. In words that could almost have been taken directly from Matthiessen's quarrels with Werner von Siemens three decades previously, he made this statement:

I doubt whether a very close agreement would result if different persons were to set up standards of resistance in mercury for themselves. I would prefer a definition in terms of some pure solid metal or some definite alloy . . . I do not think it will be possible to get mercury standards set up by different people, to agree within [o].3 per cent, the difference between 106 and 106.3.[186]

Muirhead's claim was not interrogated by any of the Committee members present, not even by Glazebrook, who had argued at the Leeds BAAS meeting in the previous year that the mercury ohm *was* in fact reproducible to within a few ten thousandths of a per cent! Indeed the residual scepticism about the stability of mercury led to the final report of the Committee of 23 July 1891 specifying that considerations of 'practical importance' made it 'undesirable' to adopt a mercurial construction for a material standard.[187]

The 1894 Act signed by Queen Victoria epitomized the international agreement of the Chicago 1893 Congress in stipulating that the *primary* denomination of the resistance standard would be a mercury column. Yet the evasion

---

[182] Ibid., pp. 4–9. [183] Ibid., S. P. Thompson evidence, q. 310.
[184] Ibid., R. E. B. Crompton evidence, q. 12. [185] Ibid., J. A. Fleming evidence, q. 76–8.
[186] Ibid., Alexander Muirhead, evidence, q. 123–4. [187] Ibid., p. vi.

of the mercury definition was still subtly maintained. This was accomplished by the wording of the practical schedule specified that the *secondary* reference standard to be kept for daily comparisons at the Board of Trade Standardising Laboratory in Whitehall was in fact to be a *coil of insulated wire* at a temperature of 15.4 °C. There was – remarkably – still no indication, however, of what metal would be used to make this wire.[188] Over a decade later scepticism about mercury lingered at the St Louis International Electrical Congress in 1905, by which time Glazebrook was in charge of determination of the mercury ohm at the new National Physical Laboratory (NPL). The NPL staff claimed to have accomplished agreement of 1 part in 100,000 between their mercury ohm and that of the Physikalische Technische Reichsanstalt in Berlin.[189] Yet at a meeting of the IEE, William Ayrton doubted that even the fourth significant figure was certain, and his friend and colleague S. P. Thompson complained that 'nobody' seemed to be able to build a mercury ohm twice over 'exactly alike'.[190] Thus the historian seeking communally shared agreement of what constituted numerical criteria of 'reasonable agreement' between mercury-based measurements of electrical resistance might find candidate thresholds as evasive and problematic to quantify as the behaviour of mercury itself.

3.8. CONCLUSION

Mapping the changing terms and language with which the metals controversy was conducted 1861–6 and in its aftermath show us how complex, contingent, and perspectival are the shifting patterns of trust in measurements that involve a material substrate of contested qualities. Although in such instruments as the thermometer or barometer, mercury was for most purposes a mundane and ubiquitous fluid in the 1860s, judgements of its appropriateness for furnishing a resistance standard were affected by factors that changed contingencies. Werner von Siemens' trust in mercury owed much to his *relative* distrust of solid metals from his experiences with the Jacobi copper unit in the 1850s. Augustus Matthiessen opposed mercury initially on the grounds of its chemical susceptibility to impurity and his contrasting expert advocacy of alloys as the most stable substrates for electrical measurement.

---

[188] Ayrton and Mather, *Practical electricity*, pp. 489–90.

[189] For contemporary developments in State standards management, see M. Eileen Magnello, *A century of measurement: An illustrated history of the National Physical Laboratory*, Bath: Canopus, 2000.

[190] See IEE discussion on International Electrical Congress at St Louis in 1904, *Journal of the Institution of Electrical Engineers*, 35 (1905), pp. 3–42. Although an IEC meeting in 1933 proposed that the unit of electrical conductance be christened the 'Siemens', it was not internationally accepted as such until SI units were formally ratified in 1971.

Later, however, Matthiessen's opposition to mercury became further entrenched in response to Siemens' sardonic attacks on his competence in handling mercury. Yet he persuaded fellow members of the BAAS Committee to share his outright rejection of the trustworthiness of mercury only when the commercial success of the Siemens mercury unit threatened to preempt the universal acceptance of the BA unit. That explains why Fleeming Jenkin, hitherto agnostic about the mercury issue, worked with Matthiessen in the period 1864–5 to cast maximum doubt on the reproducibility of the Siemens unit, challenging the trustworthiness both of mercury and of the Siemens staff who claimed to speak on its behalf. Only when the acrimony became publicly very sour indeed in 1866 did Jenkin backtrack somewhat and agree once again to suspend judgement about the trustworthiness of mercury and Siemens himself. Nevertheless, a tradition of scepticism about the suitability of mercury for embodying the ohm was born and resurfaced in variant forms for the next forty years among later members of the BAAS Electrical Standards Committees.

None of these factionalized judgements about the trustworthiness of mercury can be accounted for simply by a referral to the numerical evidence evinced in measurement activities. At certain key points, some measurements on mercury columns evinced the same or greater numerical consonance than those involving other metals. And yet this was not the deciding matter. Discrepancies between measurements on alloys were not sufficient for Matthiessen and later Jenkin to distrust their own results, whilst closer agreements among mercury data were not sufficient for them to trust it as the basis of a material standard. Their interpretations made of what constituted sufficient reasonable agreement were formulated from somewhat pre-judged positions on the (un)trustworthiness of the metals and/or practitioners concerned. The issue of care here does some explanatory work in accounting for the apparent inconsistencies in the position of the BAAS Committee. It conceded in 1864 that, with 'great' care, measurements on mercury could be nearly as easily reproduced as those on silver–gold alloy or lead. Yet it then characterized the degree of care required in working with mercury to achieve sufficiently accurate results as too problematically great for it to be worth the attentions of instrument-makers. The degree of care deployed in a measurement does, indeed, as Schaffer suggests (Chapter 2), account for judgements about the (possible) accuracy of that measurement. Yet it does not explain all that there is to be explained in the case of Matthiessen's and Jenkin's comments on the attainability of accuracy with mercury.

Insofar as issues of trust and distrust were linked to notions of care in judgements about the putative accuracy of measurements, there was a fourfold heterogeneity and considerable asymmetry of judgement. Trusting the work of others' measurements meant trusting them in all of the following respects: their honesty in reporting their own quantitative results and procedures; the quality of skill, labour, and care exercised in undertaking measurements; the trustworthiness of the substances on which they relied in making measurements; and their reliance on the trustworthiness of measurement work undertaken by third parties. When these areas were judged to be securely managed – or at least if no challenges had been made on this point – the measurement activities of individuals or groups could be judged as trustworthy or otherwise, and attributions of accuracy made accordingly. Yet it took only *one* of these areas of trust to be thrown into doubt for attributions of accuracy to dissolve. Then again, quite which of these areas might be held to blame was ambiguous and subject to what might be called interpretive 'flexibility'. When Siemens and Matthiessen disputed each other's measurement results, there was a recurrent discursive slippage between expressing distrust in the measurer, in the measurement practices, in the metal undergoing measurement, or in the inappropriate use of research. These kinds of (dis)trust were of course interrelated because an attack on the trustworthiness of a metal as a reliable material for a resistance standard was usually also an attack on the judgement of anyone who represented that metal as a trustworthy substrate for a measurement (see Chapter 1).

To the extent that the metals controversy was about a direct conflict of trust among individuals, I can now suggest some further dimensions to Schaffer's claims about the relations between trust and precision. Schaffer has argued that the attribution of precision to a measurement hinged on cultures of 'communal trust' and was thus a consequences of the strength of the social relations between separate and complex institutions.[191] Such a claim might seem tautological insofar as those measurers who worked together in a community were *de facto* those who trusted each other to make appropriate judgements about the precision of numerical results anyway. Insofar as it is not a tautological claim, then we find in the case of the metals controversy that the 'communities' involved in trust relations were somewhat fluid and could indeed be nucleated into opposed camps around the *distrust* generated only between specific individuals. The Germanic community, Siemens employees (including Britons), and many European telegraphists came to trust Werner von Siemens on the mercury question, whereas the BAAS Committee and many telegraphists in the British Empire came to support Matthiessen's judgement on the value of alloys. Then, after Matthiessen died and Siemens became preoccupied by other projects in electrotechnology, the matter of whether one trusted mercury or not became a more overtly nationalistic matter, especially at the International congresses on electricity. Then again it was also a somewhat context-dependent matter: As we shall see in Chapter 6, many of the British specialists who opposed the use of mercury in resistance

---

[191] Simon Schaffer, 'Accurate measurement is an English science', in M. N. Wise (ed.), *The values of precision*, Princeton, NJ: Princeton University Press, 1995, pp. 135–172; see quote on p. 164.

standards were quite happy to employ Ferranti domestic electrical meters whose whole operation hinged on the trustworthiness of mercury. In that case, we shall see that trust in the Ferranti mercury motor meter was very context dependent: It owed much to being a more attractive alternative to Edison's distinctly consumer-unfriendly and little-trusted rival electrolytic meter. Such was the mercurial and somewhat context-dependent nature of trust in mercury.

4

Reading Technologies: Trust, the Embodied Instrument-User and the Visualization of Current Measurement

> Professor Ayrton's galvanometer will be a great help to us electric light engineers: we are greatly in want of a trustworthy galvanometer to be carried down to the various installations as we fix them.
>
> David Chadwick, discussion at the STEE, London, 1881[1]

> I myself began with great faith in the ordinary electrical instruments; but, after taking readings sometimes with one and sometimes with another instrument, I began to lose it.
>
> Peter Willans, discussion at the IEE, London, November 1891[2]

Why did late nineteenth-century electricians decide to trust particular measurement instruments in preference to others? Was it a particular faith in the theory that informed the design of their electromagnetic mechanism? Was it their trust in the instrument-maker's effectiveness in enacting reliable designs for the devices? Or was it the fidelity with which users could take quantitative readings from performances yielded by the instrument? We need to consider these sorts of questions to recover how evaluations were made of the trustworthiness of the new species of 'direct-reading' current-measuring instruments that came onto the market in the 1880s. In contrast to the idealist tradition common in the history of science that treats instruments as essentially theory-laden devices, I consider measurement instruments from the perspective of users as technologies that had to be *read*. As in the previous chapter, my aim here is to show that pressing problems of trust were generated by the recalcitrance and fallibility of instrumental readings. The variety of practitioners' responses to these concerns show that no simple consensus existed in the 1880s and 1890s about the relationship between trustworthiness and the design of what I call 'reading technologies'. To throw this discussion into relief, I analyse the *bodily* nature of instrumental practice to show that trustworthiness was not the only important issue in taking

---

[1] From discussion of William E. Ayrton and John Perry, 'A portable absolute galvanometer for strong currents, and a transmission dynamometer', *JSTEE*, 10 (1881), pp. 156–64, discussion on pp. 165–70, quote on p. 169.

[2] From discussion of H. R. Sankey and F. V. Andersen, 'Description of the standard volt- and ampere-meter', *JIEE*, 20 (1891), pp. 516–34, discussion on pp. 534–90, quote on p. 573.

readings: Devices otherwise judged to be trustworthy could be rejected if they proved bodily awkward, inconvenient, or demeaning to read in daily practice.

The starting point of my discussion is the way that historians of science have long considered scientific instruments to be, in Gaston Bachelard's famous phrase, 'reified' expressions of scientific theory.[3] Designers of the nineteenth-century current-measurement instruments certainly used Ørsted's and Ampere's interpretation of magnet–current interactions to quantify the magnitude of a current by its deflection of a wire, magnet, or piece of iron. And this certainly made gauging currents less painful for experimenters than using their own bodies as test conductors and more convenient than using standard cells, resistances, and Ohm's law. It must be conceded Bachelard's thesis of 'instrument as reified theory' has long been influential as a counter to simplistic positivist assumptions of the supposedly 'neutral' role of instruments in experiment. Yet it needs to be supplemented by an account of how the *same* theoretical principles could be embodied in instruments that served very different roles for quite distinct audiences. Bachelard's thesis gives little scope for understanding, for example, how electrical instruments devised for such diverse purposes as measurement, exploration, manipulation, magnification, refinement, display, entertainment, pedagogy, and persuasion could be constructed in significantly different ways with exactly the same corpus of electrical theory.[4] Whilst the design of galvanometers, electrodynamometers, or ammeters is an undeniably theory-loaded enterprise, the making and *using* of them involved much more than abstract theoretical knowledge. There were, as we shall see, many practical problems in ascertaining the magnitude and direction of a current from the displacement of a needle, light-spot, or torsion-head to which historians have previously given little attention.

To put the point most directly: Apprehending the theories that had been reified in such instruments was neither necessary nor sufficient for a user to obtain a trustworthy current reading from it. Instrument-users also expected an effective means of *reading* the instrument's registrations so as to minimize error and uncertainty. Much of this chapter is about the context-specific 'reading technologies' developed for current-measuring instruments in late Victorian physics, telegraphy, and dc electrical engineering. Whereas previous histories have, in a Bachelardian vein, focussed on the internal

---

[3] Gaston Bachelard, *La nouvelle esprit scientifique*, Paris: Press Universitaires de France, 1934, p. 17. (Trans A. Goldhammer as *The new scientific spirit*, Boston: Beacon Press, 1984); idem, *Intuitions atomistiques*, Paris: Boivin et cie, 1933, pp. 140–1. See discussion in Graeme Gooday, 'Instrument as embodied theory' in Arne Hessenbruch (ed.), *Readers' guide to the history of science*, London: Fitzroy Dearborn, 2000, pp. 376–8.

[4] Albert van Helden and Thomas L. Hankins have plausibly argued that it is not possible to provide any unified characterization of all past instruments that is both satisfactory and inclusive. See A. van Helden and T. L. Hankins, 'Introduction: Instruments in the history of science', *Osiris*, 2nd series, 9 (1993), pp. 1–6.

mechanisms used to transform currents into material deflections, I look at debates on how users should obtain meaningful and trustworthy quantitative readings by ocular encounter with these deflections. An act of instrument reading could be a more or less heterogeneous act, undertaken by calculation with analytical formulae, the manual implementation of a balancing condition, or directly reading from a pre-calibrated dial – perhaps with the multiplication of a simple scaling factor. Although the first two embody conventional 'absolute' and 'null' notions of Victorian measurement, as discussed in Chapter 2, the watching of needles or light-spots in *direct-reading* devices deserves consideration as new kinds of measurement altogether. Such new reading technologies greatly expedited the process of measurement. They nevertheless also posed new problems of trust in the construction and calibration of instruments insofar as they mediated a changing – and not entirely unproblematic – division of labour among designers, manufacturers, and users. They also raised related complex questions about what sort of instrumental reading technology was most appropriate for a particular type of user with a given purpose in a specific context. Given the mobile and metallic nature of the instrument medium to be read, my treatment of such questions will draw rather more heavily from Lisa Gitelman's account of reading and writing technologies than from Latour's account of the reading of paper 'inscriptions' produced by self-registering devices.[5]

In Section 4.1 I show how previous histories of the galvanometer and the ammeter have focussed on their internal mechanisms, usually devoting insufficient attention to the reading technologies of these instruments. Then I examine how the reading of instruments to within certain levels of sensitivity or accuracy was the critical feature to be considered in designing them to be used in diverse contexts. Instead of there being a *universal* device for current measurement, any given instrument was employed only by some users or in only some contexts depending on the constraints of environmental conditions (Section 4.2) and/or temporal schedules (Section 4.3). Following that, I consider a central design feature of a reading technology: the form of scales calibrated against which an instrument's registrations were gauged. I examine why William Ayrton and James Swinburne disagreed over the trustworthiness of instrument scales calibrated with equally spaced markings and the merits of making instruments easy to read by that means. Finally, in Section 4.5 I draw these issues together by looking at attempts by Peter Willans, supplier of steam-engines to many early electrical-generating companies, to establish a consensual means of measuring a wide range of currents in tests of

---

[5] See the account of the technological reading of a meter in Lisa Gitelman, *Scripts, grooves, and writing machines: Representing technology in the Edison era*, Stanford, CA: Stanford University Press, 1999, p. 10. Compare with the account of self-registering laboratory devices in Bruno Latour, *Science in action*, Milton Keynes, England: Open University Press, 1987, pp. 64–70.

commercial efficiency. Although the Siemens electrodynamometer and later the D'Arsonval reflecting light-spot galvanometer were both proposed by Willans and his staff as effective universal devices, I examine how neither offered a reading technology that was accepted and trusted by all clients and commentators.

Before that, I critique the extant literature on the history of current-measuring devices, drawing on themes from recent historical scholarship on instruments.

4.1. 'INTERNALIST' HISTORIES AND THE NEW HISTORIOGRAPHY OF INSTRUMENTS

Traditional accounts of nineteenth-century current-measuring instruments focus on the emergence of the three principal internal mechanisms for generating deflections from currents: moving-magnet, moving-iron, and moving-coil. This taxonomy of devices is adopted by Dunsheath, Stock, and Vaughan and Lyall in offering what is – in more than one sense – a predominantly 'internalist' historiography.[6] They identify the origins of such mechanisms, circa 1820, in Ørsted's observations of how a current-carrying wire deflected a magnet, Ampère's work on the mutual attraction and repulsion of two such wires, and Arago's demonstration that a coil of current-carrying wire behaved like a magnet.[7] These effects became especially useful to those seeking to measure currents in submarine telegraphy from the 1850s and electric-light engineering three decades later. Practitioners across Europe and the USA tried various ways of devising instruments that used these effects in a combination of deflecting and controlling mechanisms to represent the direction and magnitude of a passing current. Instrument historians typically refer to only one point in the history of reading technologies, in contrast to their detailed discussion of internal mechanisms. This point is that mirror techniques of projecting a moving light-spot onto a distant scale external to an instrument, as developed in the mid-nineteenth century, enabled much more sensitive readings of currents than the use of a metal needle travelling over the scale or dial within an instrument itself.[8] For an historical account

---

[6] It is a moot point as to whether, as Dunsheath has claimed, new kinds of instruments were *necessitated* by the demands of commercial electric supply. Thomas Hughes implicitly disagrees, making no reference to ammeters or voltmeters in his account of Edison's system-building activities in the early 1880s. Percy Dunsheath, *A history of electrical engineering*, London: Faber & Faber, 1962, p. 305. See Thomas P. Hughes, *Networks of power: Electrification in Western society, 1880–1930*, Baltimore: Johns Hopkins University Press, 1983.

[7] Dunsheath, *A history of electrical engineering*, pp. 53–65; Stock and Vaughan, *The development of instruments to measure electric current*, London: Science Museum, 1983, pp. 7–9. For a traditional theory-centred account, see A.W. Humphreys, 'The development of the conception and measurement of electrical current', *Annals of Science*, 37 (1937), p. 164–78.

[8] Kenneth Lyall, *Catalogue No. 8: Electrical and magnetic instruments*, Cambridge: Whipple Museum, 1991.

of instrument *usage* we shall see that this point demands rather more than just a passing comment.

The tangent galvanometer was the standard 'moving-magnet' device in a Victorian laboratory. It consisted of a magnetic needle, with attached dial pointer, suspended at the centre of the coil of wire carrying the current to be measured; its reading technology consisted of a 360° compass scale against which the deflections of the pointer could be gauged. Following various early German and British forms developed in 1820–1, the canonical version was developed in 1837 by Claude Pouillet, Professor of Physics at the Sorbonne, who deployed a geometry of large circular coils from which he derived an analytical 'tangent formula' (subsequently discussed) to yield absolute values of current. In ensuing decades the redesign and refinement of tangent galvanometers was a distinctly international endeavour.[9] Stock and Vaughan note, for example, that variants of the tangent galvanometer were produced by Hermann von Helmholtz and Jean-Mothée Gaugain in the early 1850s in attempts to reduce the experimental errors to which application of Pouillet's 'tangent law' was prone. James Clerk Maxwell noted in 1873 that Helmholtz had made the Frenchman's device a more 'trustworthy' instrument by adding a second deflecting coil to create a deflecting field (assumed in Pouillet's formula) that was more uniform than Pouillet had accomplished in his single-coil instrument.[10]

An alternative tradition of constructing instrument mechanisms used a moving-coil as a more stable electromagnetic substitute for an iron magnet. In this kind of instrument, the reading technology was normally a needle attached to the coil arranged so that the coil deflections would cause the needle to travel over an instrument dial. Its origins are conventionally traced to instruments made by William Sturgeon in 1836 and Cromwell Varley's device patented for use in telegraphic work in 1856.[11] The definitive form of a moving-coil is normally attributed, however, to the British-born Edward Weston. He patented moving-coil ammeters in the USA in 1886 and 1888,

---

[9] The origins of this instrument are conventionally traced to Schweigger at Halle and Poggendorff at Berlin, who followed up reports of Ørsted's experiments in 1820 by developing a form of 'multiplier' that coiled the wire in many turns to amplify the deflection. The name galvanometer was probably applied to a similar device developed by James Cumming in Cambridge ca. 1821. See Stock and Vaughan, *The development of instruments*, 1983.

[10] This was important because the 'tangent' law was premised on the effective uniformity of the field acting on the magnet. With single-coil instruments this assumption rested on an approximation that was feasible only if the magnet was very much smaller than the deflecting coils; Maxwell, *Treatise on electricity and magnetism*, 2 Vols., Oxford: 1873, Vol. 2, p. 318; 3rd edition, 1891, p. 356. For a more detailed discussion of the tangent galvanometer and illustrations, see Graeme Gooday, 'The morals of metering, constructing and deconstructing the precision of the electrical engineer's ammeter and voltmeter', in M. Norton Wise (ed.), *The values of precision*, Princeton, NJ: Princeton University Press, 1995, pp. 239–82, esp pp. 243–8.

[11] Stock and Vaughan, *The development of instruments*, p. 27.

although their use before the early twentieth century is probably somewhat exaggerated by both Dunsheath and Stock and Vaughan. For example, the Deprez D'Arsonval moving-coil device produced in France in 1881 was much copied in the UK in both direct-reading and mirror-reflecting forms for at least a decade after 1887 (see the following discussion). This was not merely an extension of the Varley device, but also an adaptation of the third tradition of moving-iron devices produced by fellow Frenchman Marcel Deprez in 1880.

The third tradition of deflection mechanisms used the tendency of iron to move from weaker to stronger magnetic fields in ways that could be harnessed, again, to a needle-and-dial-based reading technology. The genealogy of how moving-iron devices were developed for the electrical-lighting industry in the hands of Kohlrausch, Deprez, Ayrton, and Perry is subsequently discussed.[12] Before doing so, however, I should reiterate the point made earlier about the importance of industrial audiences for new kinds of current-measuring devices. Dunsheath and Stock and Vaughan recognized that the greatest development of electrical-measuring devices in the nineteenth century was closely tied to the needs of engineers in large-scale enterprises of telegraphy and electric lighting. Although that helps in explaining the general growth of current-measurement devices, these authors rarely explore the reasons why individuals or companies developed the *particular* devices when they did. To understand that particular characteristic of an instrument's history I would argue that we need to relate its development to particular socio–economic or disciplinary rationales.[13] The origins of the Siemens electrodynamometer is an exemplary case in point (Figure 2).

This instrument was based upon a cross-coiled device developed by Wilhelm Weber in 1841 as a test of Ampere's theory of the mutual attraction of current-carrying wires, although its role was soon inverted to become a standard device for measuring currents absolutely. It was fitted for the non-instantaneous null method of reading currents: The passage of current would generate a mutual attraction between the coils, and the user then had to counterbalance this by rotating a counter-rotating torsion-head through a certain angle. Users then referred to a calibration chart to read off the

---

[12] Ibid., pp. 34–6.
[13] For an antidote to this see Crosbie Smith and M. Norton Wise, *Energy and empire*: *A biographical study of Lord Kelvin*, Cambridge: Cambridge University Press, 1989, pp. 649–798, for William Thomson's development of instruments in collaboration with Whites of Glasgow. For a business history of instruments, see Mari Williams, *The precision makers: A history of the instruments industry in Britain and France, 1870–1939*, London: Routledge, 1994. For the wider European perspective see R. Fox and A. Guagnini, *Laboratories, workshops, and sites: Concepts and practices of research in industrial Europe, 1800–1914*, Berkeley: Office of the History of Science and Technology, University of California at Berkeley, 1999; also published as special issues of *Historical Studies in the Physical and Biological Sciences*, 29 i (1998) and 29 ii (1999).

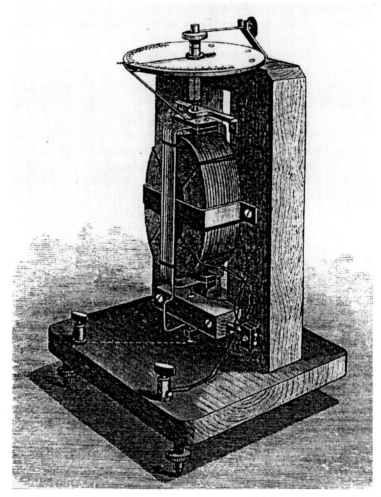

Figure 2. A Siemens electrodynamometer, ca. 1884, with housing removed to show the two vertical coils: The larger was fixed in position, the smaller suspended perpendicular to it dipping into mercury cups so it could rotate in response to attractive forces when a current passed through it. The user then rotated the spring-loaded torsion-head at the top of the device until the instrument's zero-position was restored, and then used a calibration chart to convert the angle of head rotation into a reading of current. (*Source:* James E. H. Gordon, *A Practical Treatise on Electrical Lighting*, London, 1884, p. 34.)

current from the size of the angle required for achieving the null condition. Why did the Siemens company choose to produce this particular design and a long series of commercial successors from 1877? As it was not clear then whether alternating- or continuous- (direct-) currents would predominate in the electrical industry, Siemens produced the dynamometer as a device that could (uniquely) measure both. And just as the mercury resistance unit had been integral to the Siemens submarine cable business in 1860 (see Chapter 3), this dynamometer was an integral part of Werner's plans to launch the company into large-scale commercial production of dynamos and lighting equipment in 1878.[14] Such was its popularity and reliability that its use persisted even when direct- and instantaneous-reading instruments became available in the following decade.

The Siemens dynamometer's unique flexibility in measuring not only ac and dc currents but also voltage and hence electrical-power consumption might be described as an instance of 'functional flexibility'.[15] The dynamometer was one of quite a number of instruments which were developed so that their deflection mechanism could be harnessed to more than one kind of purpose and thus to more than one kind of 'reading' practice. Not addressed explicitly in internalist historiographies, this was evidently a quality that mattered to both makers and users of instruments. Although as Peter Willans found in 1891 it was not a universal instrument and could not serve all current-measuring needs in all situations (Section 4.5), the Siemens dynamometer was certainly valuable for the busy, mobile, and financially constrained practitioner who wanted a single instrument to perform several different tasks. Similarly, the equally ubiquitous mirror galvanometer developed by William Thomson in 1858 could be used quantitatively to measure small currents and qualitatively to read submarine telegraph signals.[16] Landbound telegraph workers by contrast used differential galvanometers, such as Cromwell Varley's 1859 form, both to measure currents and to compare resistance by the null method of balancing.[17] A more sophisticated version

---

[14] Wilfried Feldenkirchen, (trans. B. Steinbrunner), *Werner von Siemens: Inventor and international entrepreneur*, Columbus, Ohio: Ohio State University Press, 1994, pp. 105–10. Given Siemens' reluctant acquiescence in the absolute system of resistance measurement in 1881, it is rather ironic that the dynamometer was used by a number of practitioners for the *absolute* measurements of dynamo efficiency – see Section 4.5. Stock and Vaughan, *The development of instruments*, p. 40. In 1890, a new direct-reading form of this device was patented in the UK by the English-born Edward Weston, a device which Stock and Vaughan claimed superseded the Siemens form: British/Patent number 12221, 1890.

[15] Compare with the term 'interpretive flexibility', coined by Pinch and Bijker to highlight the diverse reasons of different users for accepting a technology. See Trevor Pinch and Wiebe Bijker, 'The social construction of facts and artefacts: Or how the sociology of science and the sociology of technology might benefit each other', in Wiebe Bijker, Trevor Pinch, Thomas P. Hughes (eds.), *The social construction of technological systems*, Cambridge, MA: MIT Press, 1987, pp. 17–50.

[16] Silvanus P. Thompson, *Life of Lord Kelvin*, 2 Vols., London: 1910, Vol. 1, pp. 347–9.

[17] Varley revived Antoine Becquerel's differential instrument for comparing the conductivity of metals and used it to locate the position of faults in telegraph cables. Stock and Vaughan, *The development of instruments*, pp. 10–17. See Chapters 2 and 5 for the theory of the electrical 'balance'.

of this device was produced by Latimer Clark and promoted rather self-servingly in his *Elementary Treatise on Electrical Measurement* of 1868 as a measuring instrument that he claimed could meet all the needs of telegraph inspectors and operators.[18] A similar wide range of options was offered by the Siemens 'Universal Galvanometer' available from 1873, although this device was evidently not 'universally' used.[19]

The persistent proliferation of both multi-purpose current-measurement devices was also characteristic of all single-purpose devices in their embodiment of any of the three forms of deflecting mechanism discussed earlier. Dunsheath shows that whilst some early forms of moving magnet devices were abandoned, the development of more stable magnets and electromagnets meant that all three main mechanisms remained in use, no single form being generally favoured over others during the period covered by this book.[20] This closure-free development is an important feature of that any historiography of such instruments must explain. Taking a cue from George Basalla's evolutionist historiography, we see that the main explanandum for the historian of current-measuring instruments is the *diversity* of instruments – not the closure towards a uniquely popular form much discussed by the first generation of social-constructionist scholarship.[21] One important factor in this persistence of diversity was the prevalence of commercial attempts to patent distinctive designs. This potentially highly lucrative practice in intellectual property often drove innovators to seek distinctive technological differentiation by means of unprecedented novelty, albeit often in elaborate and cunning variations on a theme to avoid convergence on a single form. James Swinburne thus commented in his *Practical Electrical Measurement* of 1888 that permanent magnets were then little used in ammeters as they were unpatentable through widespread prior usage. Swinburne noted that each maker preferred 'to push something else', that is, something that *could* be

---

[18] These functions included measuring currents, sending and receiving telegraph signals, testing the potential difference of batteries, measuring resistances, ascertaining the specific conductivity of copper, and locating faults on telegraph lines and cables. See Latimer Clark, *Elementary treatise on electrical measurement*, London: 1868.

[19] See Anonymous, 'Siemens universal galvanometer', *The Telegraphic Journal*, 2 (1873–4), pp. 46–8.

[20] Dunsheath, *A history of electrical engineering*, p. 314.

[21] George Basalla, *The evolution of technology*, Cambridge: Cambridge University Press, 1989. Basalla borrows from the neo-Darwinian language of environmental 'selection' processes to explain diversity of technological form, whilst taking care to signal a clear disanalogy between organic and technological evolution; ibid., pp. 2–3. For early social constructivism that took final closure of technological form as its principal explanandum, see Pinch and Bijker 'The social construction of facts and artefacts'. For a more dynamic view of incessant mutual shaping between technology and society see Wiebe Bijker, and John Law (eds.), *Shaping technology/Building society*, Cambridge, MA/London: MIT Press, 1992; Wiebe Bijker, *Of bicycles, bakelite and bulbs* Cambridge, MA/London: MIT Press, 1995.

patented.[22] Stock and Vaughan thus hint that various individuals sought to win patents for means of overcoming problems in the reliability and convenience of existing devices, notably Edward Weston in his classic refinement of the moving-coil mechanism.[23] Insofar as the makers of such instruments used them in their own localized company operations (see Section 4.5) and insofar as customers often developed a particular loyalty to their favoured makers, the process of patent-driven and financially underwritten diversification was resistant to a 'closure' process.

In relation to another of Basalla's major arguments, designers and makers were often forced[24] to *adapt* their devices to extend their usage to the uncongenial material circumstances of new contexts. Although keeping the same sorts of deflecting mechanisms, designers introduced some radical innovations into the *reading* technology of some instruments. Both Thompson and Stock and Vaughan, for example, relate the genesis of William Thomson's first mirror–galvanometer in 1857–8 to solve the problem of detecting the tiny electric signals expected in the 1858 transatlantic submarine telegraph. Noting that a new highly sensitive kind of reading technology was required for any visually detectable representation of the signals, Thomson borrowed Poggendorf's magnification technique of attaching a small mirror to the back of a tiny moving-magnet mechanism reflecting a light beam off this mirror onto a projected screen[25] (Figure 3). But there were practical limits to the functional flexibility of Thomson's instrument insofar is it became chaotically unreadable on board a rolling cable-laying ship. Thomson's marine galvanometer, patented at the same time as the original, was more robust in several respects. In reference to the massive iron casing used to shield the mechanism from the disturbing effects of the ship's magnetism, as it had a robust fibre suspension to avoid disturbance from the ship's movements, it was nicknamed the 'ironclad' (after the contemporary armoured Naval vessel).[26] A land-bound variant of the mirror galvanometer developed with multiple current-carrying coils by Thomson in 1863 was for several decades thereafter used in the laboratory as a standard device for 'sensitive'

---

[22] James Swinburne, *Practical electrical measurement*, London, 1888, p. 25.
[23] British patent number 3059, 1856. Stock and Vaughan, *The development of instruments*, p. 20.
[24] For an account of forced vs. free moves see Andrew Pickering and A. Stephanides, 'Constructing quaternions: On the analysis of conceptual practice', in Andrew Pickering (ed.), *Science as practice as culture*, Chicago: University of Chicago Press, pp. 139–67.
[25] Stock and Vaughan, *The development of instruments*, pp. 23–4. Thomson's patent specification number 329, 1858.
[26] Smith and Wise discuss only the later marine version of this instrument; *Energy and empire*, pp. 668–70. For correlated problems of using microscopes on board ship, see Graeme Gooday, 'Instrumentation and interpretation: Managing and representing the working environments of Victorian experimental science', in B. Lightman (ed.), *Victorian science in context*, Chicago: University of Chicago Press, 1997, pp. 409–37, esp. pp. 423–6.

Figure 3. Integrated 'convenience' version of the Thomson mirror galvanometer specially built by the Elliot Brothers for the BAAS Electrical Standards Committee. By physically connecting scale and galvanometer together, this version apparently saved 'much trouble in adjustment' – a characteristic of previous forms in which scale and galvanometer were supplied separately for the user to orientate before taking any readings. A light behind the screen was directed at a tiny mirror in the galvanometer, whose movements were traced by the movement of a reflected light-spot along the scale. (*Source*: George Chrystal, 'Galvanometers', in *Encylopedia Brittanica*, 9th edition, 1879, p. 51.)

measurements of small currents, its 'astatic'[27] mechanism shielding its operations for nearby disturbance (see further discussion in the next section).[28]

The historical importance of looking at reading technologies is made most evident by showing how the study of these helps us to recover the *bodily* nature of instrumental practice that is not accessible by focussing solely on the history of deflection mechanisms. The ineluctably bodily nature of

[27] The 'astatic' system, is an arrangement of opposed pairs of needles developed to counteract the effects of fluctuations in the geomagnetic field by Nobili at the Florence Museum in 1823, drawing on Ampere's prior work at the Ecole Polytechnique. See Stock and Vaughan, *The development of instruments*, pp. 12–13.

[28] Stock and Vaughan, *The development of instruments*, pp. 23–4; Lyall, *Electrical and magnetic instruments*, items 13–18; James E. H. Gordon, *A physical treatise on electricity and magnetism*, 2 Vols., London: 1880, Vol. 1, pp. 238–9. For dealing with much larger currents, however, Thomson and his colleagues on the BAAS Committee used a Weber electrodynamometer in determining the absolute ohm in 1863–4 (see Chapter 2). The authority Thomson derived from the practical success of his various galvanometers earned him a special seven-year extension on his 1858 patent rights by Parliament in 1871. See Thompson, *Life of Lord Kelvin*, Vol. 1, p. 349, footnote. On the authority accruing to developers of new instruments, see Albert van Helden, 'Telescopes and authority from Galileo to Cassini', *Osiris*, 9 (1994), pp. 9–29, and Jan Golinski, 'Precision instruments and the demonstrative order of proof in Lavoisier's chemistry', *Osiris*, 2nd series, 9 (1993), pp. 30–47.

instrument usage is a topic that has been addressed by both Hankins and van Helden and more recently by Shapin and Lawrence. They show that the performance of instruments is intimately linked to the bodily conduct and disposition of those who use them. This is especially relevant in cases in which the bodily configuration of workers might be constrained by the material and temporal features of their working environment.[29] Otto Sibum has shown, for example, that there were two important reasons why James Joule's early work on the 'mechanical equivalent of heat' could not initially be replicated: Other experimenters lacked both his unique brewery-trained ability to read sensitive thermometers and also his bodily techniques of shielding thermometers from corporeal heat.[30] Electrical-lighting engineers, by contrast, were often impatient with instruments that needed such contrived conditions and bodily precautions. Given the celerity with which they had to work in the power station and their constant need for mobility and panoptical awareness of more than one instrument, some engineers objected to the use of D'Arsonval reflecting galvanometers. As we shall see in Section 4.5, such devices required them to become preoccupied by and positioned as 'spot-watchers' – a self-indulgent activity held by James Swinburne to be fit only for the laboratory.

From the preceding discussion we can infer that instrument reading for the late Victorian was a complex material and bodily practice far removed from the world of paper texts in which Bruno Latour locates in his tendentious account of 'metrology'. In *Science in action* Latour arbitrarily excludes from the category of 'instrument' all but self-registering technologies that automatically generate 'inscriptions' on paper. He then traces the movement of such paper inscriptions around the networks of technoscience to explain how late twentieth-century technoscientists universalized their results. Such a reductively papyrocentric account of instrumental work cannot explain much about nineteenth-century electrical-measurement practice, however.[31] For that kind of practice, paper documents were no more than contingent

---

[29] Van Helden and Hankins, 'Introduction: Instruments in the history of science'. For further discussion of this sort of issue, see Christopher Lawrence and Steven Shapin (eds.), *Knowledge incarnate: Human embodiments of natural knowledge*, Chicago/London: University of Chicago Press, 1998.

[30] Otto Sibum, 'Reworking the mechanical value of heat: Instruments of precision and gestures of accuracy in early Victorian England', *Studies in History and Philosophy of Science*, 26 (1995), pp. 73–106. See Gooday, 'Instrumentation and interpretation', for further analysis.

[31] Latour narrowly specifies instruments as being 'inscription devices', specifically only those which provide a tangible textual output to be deployed directly in a research publication. He explicitly disavows the conventional historical notion of an instrument as that which has a 'little window' that allows someone to take a reading from a dial, although he does concede that thermometers, watches, and Geiger counters were once 'instruments' in his sense of furnishing readings directly cited in technical papers, Latour, *Science in action*, pp. 67–9. Elsewhere he contends that making technologies out of science is not going from a paper world to a messy greasy concrete world, but rather from paperwork to 'still more paperwork'. Ibid., pp. 227ff, esp. p. 253.

auxiliaries to the skill-laden interaction of body with metal. Undeniably, calibration charts were required by users of the Siemens electrodynamometer, D'Arsonval devices, and other instruments that were not pre-calibrated; equally much paper was required for calculating the final results of measurements obtained from a tangent galvanometer. Otherwise, though, the definitive characteristics of instrument reading were bodily proximity and scrutiny of non-textual resources, and often active bodily intervention to adjust instrument settings. Alas for the historian, such non-textual procedures left behind fewer recoverable traces than a Latourian paper-based practice might have done. Nevertheless there are just enough traces left for the historian to reconstruct something of that practice.

As unhelpful as Latour's account of how instruments are read is his claim that instruments are of themselves immutable, sufficiently black-boxed for their *context-independent* operation in the universalizing practices of technoscience. In the preceding discussion of reading technologies for Thomson galvanometers we saw that at least some current-measuring instruments were characteristically designed for *context-specific* usage. Far from being black-boxed, these devices required users to adjust their operations and to take visual readings from a reading mechanism in ways that very much depended on the vicissitudes of context. In any case Latour himself ironically undermines his own claim about the importance of 'black-boxed' instruments when introducing his thesis on the 'extramural' laboratory.[32] This thesis maintains that to get instruments to work outside the laboratory of origin, the rest of the world needs to be rearranged, 'literally papered over' with continuous calibration chains and pre-prepared 'landing strips', to reproduce the conditions of the laboratory.[33] This is clearly an admission that some products of the laboratory never manage to be immutable and need the advantageous setting of the well-managed laboratory to work properly. Yet such an explanation of how instruments work in electrical technology is unsustainable: Weather-beaten telegraph huts and dynamo-filled power stations palpably could not be turned into laboratory-like settings.[34] Even with sufficient accurate calibrations for laboratory purposes, an ammeter placed in the chaotic environment next to an operating dynamo could yield results several per cent adrift.

In Sections 4.2 and 4.3 we shall see that some current-measuring instruments were arguably made with the quasi-Latourian assumption that working environments would be adapted – if not exactly 'papered over' – to furnish ideal conditions for reading their performance. Importantly, this was not

---

[32] Bruno Latour, 'Give me a laboratory and I will raise the world', in Karin Knorr-Cetina and Michael Mulkay (eds.), *Science observed: Perspectives on the social study of science*, London: Sage, 1983, pp. 141–70, and Latour, *Science in action*, p. 249.

[33] Latour, *Science in action*, p. 253.

[34] Latour himself admits the existence of this resistance in his discussion of the failed attempt to build a solar village at Frangocastello in Crete in 1980, Latour, *Science in action*, pp. 248–9.

Reading Technologies 141

just a matter of material architecture, but concerned also the bodily deportment and disposition of human operatives and traffic. By contrast, though, we will see that other instruments for reading currents were designed so that they could (ideally) withstand the disturbing forces in their likely working environment and indeed be read without a need for the reconfiguration of that environment to be more congenial. We shall also see that such alternatives had important implications for whether an instrument-user sought to read their devices to the highest degree of accuracy.

4.2. SENSITIVITY VERSUS ROBUSTNESS: GALVANOMETER ACCURACY IN THE WORKING ENVIRONMENT

Some months ago a test had to be made of a cable miles from any large town, on a beach of loose sand exposed night and day to the full force of the trade winds. Inside the hut, which was in a perpetual state of vibration, tests with mirror galvanometer[s] or the suspended needle were impossible, and an expedition therefore had to be made to the other end of the cable to find a still hut where reliable observations could be made. Had the testing apparatus been so constructed as to be entirely independent of the cable hut, this second journey would have been saved.

Engineer, 'The Manufacture of Electrical Instruments', Telegraphic Journal, 1873[35]

Neither William Thomson's original reflecting galvanometer nor the tangent galvanometer could have been considered as 'immutable mobiles'. Translated into the terms of their Victorian users, these devices were not expected to be capable of supporting accurate measurement in any and every situation. Anguished contemporary testimony on the difficulty of using the sensitive mirror galvanometer in a wind-buffeted beach hut can be gleaned from the anonymous engineer's letter to the *Telegraphic Journal* in 1873. In the quote that opens this section, he contended that this instrument was inadequate for the purpose as it seemed to him to have been constructed only for the 'favourable' still conditions of the physical laboratory. He suggested somewhat testily that instrument-designers ought to adapt these devices so that readings could be taken from them in the conditions to which distant telegraph stations were subject; such designers should accordingly visit, or interview 'working electricians' like himself to find out what these conditions were. But no response was forthcoming from instrument makers or anyone else in subsequent issues of the *Telegraphic Journal*. Given the engineer's suggestion that the galvanometer apparatus should be put 'neatly' into a box and made so as to operate independently of cable hut and trade winds, a judicious colleague might have recommended to him that he use a Thomson marine galvanometer which was constructed in just such a fashion. Thomson had, after all, designed this later variant in direct response to

[35] 'Engineer', 'The manufacture of electrical instruments', *Telegraphic Journal*, 2 (1873), p. 14.

his personal experience of the difficulty of receiving signals on board rolling cable ships.[36]

The anonymous engineer's point was, nevertheless, an important one. As Post Office electrician Harry Kempe noted in articles for the *Telegraphic Journal* and successive editions of his *Handbook on Electrical Testing*, there were considerable challenges in trying to take readings from a standard Thomson mirror galvanometer even indoors, far away from the outdoor extremes of inclement tropical weather. Whereas users of the 'ironclad' marine galvanometer could rely on the instrument's robust reading technology to avoid external disturbance, Kempe told users of the original Thomson galvanometer that the responsibility for excluding sources of disturbance fell on them. He thus offered advice on both the bodily discipline required for spot-watching and the care required in managing the material environment.[37] In addition to excluding all dampness from nearby surfaces,[38] Kempe noted in the *Telegraphic Journal* for July 1874 that, to get any readable results, the instrument should be set up on a 'very firm' table in a basement storey. It was 'almost useless' to test with it in an upper room as the least vibration sent the spot of light 'dancing and vibrating to and fro', making it impossible to read: The mirror-galvanometer was not even sufficiently immutable to be used with equanimity anywhere within the same building.[39]

Kempe also warned users to discipline their indulgence in personal accessories. Whilst no threat was posed by stationary iron near the instrument, he advised experimenters to remove any iron keys or knives from their pockets because, if the user moved about too much, these would 'very much affect' the performance of the galvanometer.[40] The kind of table on which work

---

[36] Ibid. At least one form of the ordinary mirror galvanometer available by 1880 was situated on a tripod stand to work in rough conditions. See Gordon, *A physical treatise*, Vol. 1, p. 238. I thank Colin Hempstead for pointing out that the title of Gordon's 'practical' treatise shows deference to Maxwell's *Treatise on electricity and magnetism* of 1873. Gordon was one of Maxwell's students at the Cavendish Laboratory in Cambridge, 1876–8. For Gordon's career, see Anonymous, 'Obituary: James Edward Henry Gordon', *The Electrician*, 30 (1893), pp. 417–8.

[37] Harry R. Kempe, *Handbook of electrical testing*, London, 1876; 2nd edition, 1881, p. 33; all subsequent citations are from the 2nd edition as this is the most widely available edition of the book. Maxwell also wrote somewhat allusively about 'disturbing agents' in his lecture to the 1876 South Kensington conference on apparatus. James Clerk Maxwell, 'General considerations concerning scientific apparatus', *Catalogue of the special loan collection of scientific apparatus at the South Kensington Museum*, London, 1876, pp. 1–21, esp. pp. 1–2. See also Gooday, 'Morals of metering', pp. 278–9.

[38] Some contemporary instruments employed sulphuric acid inside the body of the instrument instead for removing moisture. See Gordon, *Physical treatise*, Vol. 1, pp. 37 and 39.

[39] Harry R. Kempe, 'Resistances and their measurement. XI. The Thomson galvanometer', *The Telegraphic Journal and Electrical Review*, 2 (1873–74), pp. 241–4, quote on p. 243 (numbering of XI was accidentally omitted from the article's title but fits the overall sequence).

[40] Harry R. Kempe, 'Resistances and their measurement. XII', *The Telegraphic Journal and Electrical Review*, 2 (1873–74), pp. 266–8, quote on p. 266.

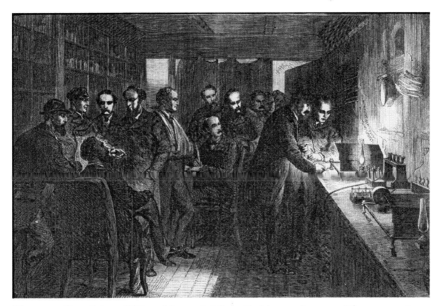

Figure 4. Electricians Willoughby and Oliver Smith stand over a Thomson mirror galvanometer in the test-room on board the Great Eastern, accompanied by a dozen eminent shipboard witnesses. They await a signal to confirm that the 1865 cable recently retrieved from the ocean bed had been reconnected in readiness for full commercial operation. As the senior of the two, Willoughby stands more upright to oversee the operation. (*Source*: Robert Dudley, 'Testing the Recovered Atlantic Cable', *Illustrated London News*, 13 October 1866, p. 357.)

took place was also of some significance as it was most convenient for users of the galvanometer to stand at a table 3–4 ft high to observe the spot of light moving over the dial at near eye-level. Although Kempe described this mode of using the galvanometer as a matter of the experimenter's 'fancy' in measurement work, this is the bodily position in which users of the mirror-galvanometer were portrayed in the most famous depiction of its employment in 'reading' telegraph signals on the transatlantic cable of 1866[41] (see Figure 4). We shall see in Section 4.5 that, whilst this sort of bodily disposition was popular among telegraphists, it was not congenial to those accustomed to the rather different bodily regimes of mechanical engineering.

Whilst the Thomson marine galvanometer, like the Siemens electrodynamometer, had a deflection mechanism and reading technology sufficiently

---

[41] In 1866 the *Illustrated London News* published a picture of two clerks using mirror galvanometers to transmit and receive telegraph messages from the Great Eastern in Valencia, South West Ireland. The instruments are represented in idealized conditions – a stable environment and no human movement around the mirror galvanometers. The image is reproduced in Smith and Wise, *Energy and empire*, p. 671.

robust to give accurate readings in virtually any indoor or outdoor environment, 'robustness' was not a quality sought in all galvanometers. Although the marine galvanometer was robust enough to give reliable results in inclement circumstances, not even a highly skilled user would normally be able to use it to get readings of the high 'degree of accuracy' (later known as 'precision') required for most other purposes of telegraphy or physics. Victorian practitioners engaged in delicate testing work or determination of material properties cherished galvanometers that had a deflection mechanism and reading technology that were sensitive enough to yield readings to a higher degree of accuracy. As we shall see, there was an important ambiguity about the extent to which the accuracy and the degree of accuracy attainable with an instrument were more closely dependent on the construction of the instrument or on the skill of the user.

That being said, accuracy was not necessarily synonymous with sensitivity – certainly not with *any* arbitrarily great degree of sensitivity. Anyone seeking to understand the relation between accuracy and sensitivity in a galvanometer's construction would have found different sorts of advice available from two major sources. One was the analytical survey of galvanometers in Volume 2 of James Clerk Maxwell's *Treatise on Electricity and Magnetism* of 1873 that won warm praise from a reviewer in the *Telegraphic Journal*.[42] Focussing on tangent and sine galvanometers that exemplified his treatment of the magnetic effects of circular currents in the preceding chapter, Maxwell argued that such devices could be constructed primarily either for accuracy *or* for sensitivity. To be a 'sensitive' galvanometer, the device would have to be constructed with a magnetic field as *intense* as possible to give maximum deflection for any given current. This generally meant having coils very close to the moving-magnet so that they could effect maximum influence on its orientation. By contrast for accuracy, the so-called 'standard' galvanometers would be constructed with a magnetic field as *uniform* as possible to minimize uncertainties in subsequent calculations of the value of the current. As Maxwell put it, standard galvanometers were constructed so that the dimensions and relative positions of their fixed components could be accurately known and so that the variation in the position of the moveable parts introduced the 'smallest possible error' into the calculations of current.[43] Maxwell implied that there was a *forced* decision between these two forms because the most intense field available for a given configuration was very non-uniform.[44]

Maxwell's opposition of sensitivity to accuracy in galvanometer construction was focussed specifically on the case of absolute instruments from which calculations of current would be made. His advice accordingly differed from accounts on devices that were calibrated empirically, such as those often

---

[42] Anonymous, 'Notice of books, a treatise on electricity and magnetism. By James Clerk Maxwell MA', *Telegraphic Journal*, 1 (1872–3), pp. 145–6.
[43] Maxwell, *Treatise*, Vol. 2, 1st edition, p. 313.   [44] Ibid., pp. 286–98.

employed in telegraphic work. This is apparent in the articles published soon after Maxwell's *Treatise* by senior Post Office engineer Harry Kempe in the *Telegraphic Journal*, of which Kempe was a co-founder.[45] These articles were edited and republished in his *Handbook of Electrical Testing*, 1876, and read by many engineers in subsequent editions. In expounding his view that instrumental accuracy was premised on sensitivity rather than opposed to it, Kempe devoted much attention to the Thomson reflecting galvanometer not discussed in Maxwell's analysis. In his introduction to a lengthy discussion of this instrument's virtues Kempe wrote:

The accuracy with which measurements can be made depends chiefly upon the sensitiveness of the galvanometer employed in making those measurements. The Thomson reflecting galvanometer supplies this requisite sensitiveness, and is the instrument which is almost invariably employed when great accuracy is required.[46]

According to Kempe, Thomson had designed the most commonly employed version of this device to meet the requirements for both sensitivity and accuracy. For the purposes of sensitivity its coils had been arranged to elicit a 'maximum effect' on the magnet out of a 'minimum quantity of wire'. Yet, for the purposes of accuracy, he also managed to contrive the angular response of the needle to the current to follow a tangent law. For the very small deflections typically employed with the Thomson device, this was usually treated as approximating a linear law. Kempe's dissolution of the opposition between accuracy and sensitivity was accomplished in part by the magnifying reflection technology that enabled a tiny sensitive movement to be amplified sufficiently to be measured with accuracy by the movement of a light-spot against a calibrated screen several feet away.[47] Significantly, users had to intervene in the instrument to adjust its sensitivity for the particular purpose in hand by raising the magnet up or down to be nearer or further from the deflecting coils as required. To attain the condition of 'maximum sensitiveness', Kempe advised users to reverse the direction in which the magnet was suspended so that it opposed the Earth's field, the magnet's orientation being rendered less stable and thus more amenable to deflection.[48] Yet there was, he warned, a danger of adjusting the instrument *too* sensitively: In that condition the spot would respond to any 'slight external action' and never settle down to a final exact position that could be read with confidence. In this extreme condition, sensitivity once again became antithetical to accuracy – although for reasons quite distinct from Maxwell's account.

[45] Harry Kempe (1852–1935) studied at Kings College, London, 1867–70, joined the engineering branch of the Postal Telegraph Service in 1871 to work as assistant to William Preece, succeeding him as Principal Technical Officer in 1900 and as Electrician in 1907, and retired in 1913. [Obituaries], 'Harry Kempe', *JIEE*, 114 (1935, p. 522, and Donard de Cogan, 'Kempe, Harry Robert' (1852–1935), *Dictionary of National Biography Missing Persons*.
[46] Kempe, *Handbook of electrical testing*, p. 24.
[47] Ibid., pp. 24–5.   [48] Ibid., pp. 32–3.

The multi-form Thomson mirror galvanometer was probably the most widely used device for long-distance telegraph testing and was accordingly the instrument most thoroughly discussed in Kempe's *Handbook of Electrical Testing*. Yet it was a relatively expensive patent-laden instrument that was affordable only to enterprises as wealthy as Post Offices and telegraph companies. For others the cheaper tangent galvanometer would have been the obvious instrument to use, especially as one could be constructed out of basic laboratory resources. The major problem, however, was that to give trustworthy readings it needed an environment that was even more disturbance-free than the Thomson device. For example, C. V. Boys recalled that, as Frederick Guthrie's assistant at the South Kensington in the 1870s, he hardly ever found the tangent galvanometer to be a 'convenient' instrument. It was peculiarly 'unsatisfactory' in the Science Schools building on Exhibition Road because the steel structure of the ceilings and floors caused the Earth's magnetic field to have values which might have been only half that outside the building and seemed to be different in every place inside the building.[49] This was a real problem because the trustworthiness of the tangent galvanometer hinged on having a reliable figure for the strength of the Earth's magnetic field which provided the instrument's controlling force.[50] The accuracy of current values calculated from galvanometer deflection readings was thus highly vulnerable to localized variation in terrestrial magnetism as was endemic in Boys' working environment.

Even if users of the tangent galvanometer could find a room that was free of iron, they still had to establish their local magnetic field at the particular time of measurement. Unless they had access to a magnetometer and were expert in its use, they would have had to trust data taken by other observers for a given metropolitan location. Thus, for example, the English 1874 translation of Friedrich Kohlrausch's *An Introduction to Physical Measurements*[51] cited a table (reproduced here as Table 4.1) of *forecasts* of the values of H[52] extrapolated from existing data.

Kohlrausch emphasized that such a table could be relied on only if all iron objects and 'especially' long iron conductors were removed from the neighbourhood.[53] Furthermore he observed that further extrapolations would

---

[49] Charles V. Boys, 'The Physical Society of London, 1874–1924', *Proceedings of the Jubilee Meetings*, March 20–22. (special number, 1924), pp. 21–2.

[50] Gooday, 'The morals of metering', p. 243.

[51] Friedrich Kohlrausch, *An introduction to physical measurements*, 1874 (1st English edition translated from 2nd German edition by T. H. Waller and H. R. Procter), p. 155.

[52] Kohlrausch and his translators use the symbol T.

[53] Kohlrausch, *An introduction to physical measurements*, p. 234, Table 22. In the 3rd English edition of 1894, this was supplemented by the advice that the place in question 'may be tested as regards constancy of H' and compared with the 'out of doors' value by a standard magnetometer technique. Ibid., 1894 (3rd English edition translated from 7th German edition,) London, p. 277.

Table 4.1. *Kohlrausch's Forecasts of the Values of H*

Location of Value	Value of H in 1870	Value of H in 1875	Value of H in 1880
London	1.78	1.80	1.82
Göttingen	1.850	1.860	1.888
Darmstadt	1.91	1.93	1.95
Zürich	2.00	2.00	2.04

Source: F. Kohlrausch (trans. T. H. Waller and H. R. Procter), *An Introduction to Physical Measurements*, 1874, p. 155 (from 2nd German edition of 1873).

have to be made according to latitude and longitude from these civic locations and allowance made for an increase of about 0.004 units per year.

Thus although the Earth's magnetic field provided a usefully 'natural' element to the measurement of current by use of the tangent galvanometer, its very dependence on such a contingently variable quantity militated against its use for measurements to the highest degrees of accuracy. When Professor William Anthony lectured to the Franklin Institute in January 1888 on the 'great' tangent galvanometer he employed at Cornell University to measure currents up to 300 A, one 'great difficulty' was that the Earth's magnetic field varied there even from hour to hour. He had now arranged for this device to be mounted in a specially contrived building, constructed entirely without iron and placed 'at a distance' from all other buildings likely to cause disturbance. With the relevant appliances for determining 'natural' changes in the Earth's field, he claimed that the 'highest class' of work could be done there. Indeed, he argued further that the Cornell tangent galvanometer could serve as an invaluable standard device against which other instruments could be compared. From this expert position Anthony contended that a tangent galvanometer could be used *only* in places with such a special large-scale construction. They could not possibly be used in a dynamo station or a workshop: He had observed that a large magnet or dynamo even 'many feet' away could change the Earth's field by as much as 100%.[54]

Although in the mid-1880s William Thomson did arrange for the manufacture of a tangent galvanometer with an extra-strong controlling magnetic field generated by an electromagnet, this device was one of his commercially unsuccessful creations. As Anthony's wisdom had it, no amount of practical or discursive ingenuity could make this device reliable to any more than to within 5%–10% next to whirling dynamos in the power-station.[55] On the

---

[54] William Anthony, 'Electrical measurement, especially as applied to commercial work', *Journal of the Franklin Institute*, 3rd series, 125 (1888), pp. 56–72, esp. pp. 60–1.
[55] Gooday, 'The morals of metering', esp. regarding judgements of Robert Paul, p. 248.

basis of lengthy practical experience, the aristocratic James Swinburne reversed his early advocacy of such devices in *Practical Electrical Measurement* of 1888. There he claimed that the 'troubles' of tangent devices that allegedly arose from dynamo fields had been greatly exaggerated.[56] But by the time he lectured the ICE on this subject four years later, Swinburne regarded the only 'survivor' of the tangent galvanometer to be Lord Kelvin's lamp counter, a device used at some distance from dynamos for a self-explanatory purpose. Not all agreed, however: Chief Post Office electrician William Preece suggested that, if Swinburne were to 'cast his eyes' around, he would find at that moment more tangent galvanometers in practical use in electrical engineering than any other kind of current-measuring devices. Swinburne replied sharply that if the Post Office used such a large number of tangent galvanometers for its telegraphic work, he would recommend that it should 'alter that practice' – that is, give it up altogether.[57]

Not all obeyed Swinburne's injunction, especially those who were long-habituated in using tangent instruments, and they never came to trust later innovations in direct-reading instruments for heavy-duty current measurement. Even if devices such as the ammeter were immune to local variations in magnetic fields, their complex deflection and reading technologies required much more trust in the ingenuity of the instrument-designer and instrument-maker than did the simply made compass-based tangent galvanometer. Among these was the instrument-manufacturer Sydney Evershed, who continued to use his own form of tangent galvanometer into at least the early 1890s.[58] Such die-hards as Evershed were well accustomed to the non-instantaneous nature of the calculation-based practice of reading a current by using a tangent galvanometer. As we will see in the next section, however, many other practitioners were much more concerned about the temporal constraints of reading instruments.

4.3. TEMPORAL CHARACTERISTICS OF CURRENT-MEASUREMENT PRACTICES

The lamp-light falls on blackened walls.
And streams through narrow perforations;
The long beam trails o'er pasteboard scales,
With slow decaying oscillations.
Flow, current! flow! set the quick light spot flying!
Flow, current! answer, light spot! flashing, quivering, dying.

---

[56] James Swinburne, *Practical electrical measurement*, London: 1888, p. 28.

[57] James Swinburne, 'Electrical measuring instruments', *Minutes of Proceedings of the Institution of Civil Engineers*, 110 (1891–2), pp. 1–32, discussion on pp. 33–64, quote on p. 9, and further discussion on pp. 57–61.

[58] Evershed in discussion with Sankey and Andersen, 'Description of the standard volt- and ampere-meter', 1891, discussion p. 569.

J. C. Maxwell, first stanza of 'A Lecture on Thomson's Galvanometer' [Delivered to a Single Pupil in an Alcove with Drawn Curtains], *Nature*, 16 May, 1872[59]

The modern culture of instruments that give *instantaneous* direct readings from pre-calibrated scales makes it easy to acquiesce in the (naive realist) notion that such readings are transparent glimpses of the world, needing no interpretive skill or intervention. Accordingly, there is a certain amnesia about why many Victorian physicists and telegraph practitioners considered it to be a positive advantage that their galvanometers were *not* pre-calibrated in units of current and accordingly were far from instantaneous in their operations. To understand this, it should be borne in mind that it was financially and pragmatically advantageous for users to be able to undertake as many measurements as possible with just a few instruments, altering the sensitivity and hence calibration to meet the diverse conditions of each task. This involved the distinctly time-consuming manipulation of the instrument's deflections, often a prolonged wait for the needle or light-post to settle down, and then calculation of results. Hard-working measurers could at least be sure that they had effectively rested at least most of the trustworthiness and accuracy of their readings on the integrity of their *own* labours. And, indeed, if they were leisured natural philosophers like James Clerk Maxwell, they could linger over the slowly decaying oscillations of the Thomson galvanometer and ponder the analogy with the echoing trumpet calls of Tennyson's poem 'The Splendour Falls on Castle Walls', of which Maxwell's poem was a skilled parody (see the quote that opens this section).[60]

There were, however, limits to the tolerance of prolonged measurement readings, especially among the commercial fraternity. Whereas nineteenth-century determinations of resistance standards in which Maxwell participated took months and years of greatly exacting care to undertake (Chapter 2), the day-to-day work of testing telegraph lines and cables was a commercial enterprise that did not accommodate quite such leisurely attention to minutiae. This might not be obvious from a first glance at Kempe's detailed analysis in the *Handbook of Electrical Testing* of how maximum accuracy was to be accomplished in instrument reading. However, his detailed analysis of how to determine the degree of accuracy for a measurement reading was aimed at expediting the important business of detecting and locating faults on telegraph lines and cables. Kempe invoked the notion of 'figure of merit' to quantify the 'degree of sensitiveness' for a galvanometer's particular internal resistance and magnet configuration. The figure of merit represented the

---

[59] Cited in Thompson, *Life of Lord Kelvin*, Vol. 1, p. 349. For a full analysis of this poem in the context of galvanometer reading see Graeme Gooday 'Spot-watching, bodily postures and the '"practised eye"': The material practice of instrument reading in late Victorian electrical life', in Iwan Morus (ed.), *Bodies/Machines*, Oxford: Berg, 2002, pp. 165–195.

[60] See Gooday, 'Spot-watching, bodily postures and the 'practised eye'.

current required for producing a unit deflection on the instrument's scale; according to Kempe, this also defined the 'degree of accuracy' attainable in readings taken from it. For a typical galvanometer of 5000 ohms resistance with a figure of merit of 10^{-9} units of current per scale deflection, this meant that the degree of accuracy to which it could be read was 10^{-9} units.[61]

Kempe made this inference by assuming that the smallest change an observer could discern in the movement of a galvanometer light-spot was just one scale division. This point was heavily criticized by later commentators who suggested that the maximum degree of accuracy attainable would not solely be determined by the instrument's figure of merit: Skilled observers might reliably be able to read to a *fraction* of a scale division.[62] Nevertheless, by equating the figure of merit with degree of accuracy, Kempe was not thereby *assuming* that all electricians possessed identical abilities in reading galvanometer scales; nor was he suggesting that a galvanometrist trained to a great acuity of vision could not attain a higher degree of accuracy. Rather, Kempe's aim seems to have been to suggest the degree of accuracy in readings that electricians could cite with *certainty* for their work. By adopting this convention, they could make a quick and standardized judgement about what sensitivity was needed in associated apparatus for resistances to be calculated to within a required degree of accuracy – commonly 1 part in 100. This degree of accuracy was essential for speedily locating and mending faults in telegraph lines and cables that were hundreds of miles long – the topic that dominates much of the latter part of Kempe's *Handbook*.[63]

Kempe's instructions on reading a tangent galvanometer and the practical limits to which users could read its angular degree scale were tellingly

---

[61] For the mirror galvanometer a single figure of merit could typically be specified along its entire scale: Given the very small magnitude of angular deflections involved, it was conventional to consider it as effectively being a proportional instrument – at least to within an acceptable limit of error. For non-proportional instruments, a figure of merit could be meaningfully specified for only very short regions of the scale. Kempe, *Handbook of electrical testing*, 1881, pp. 37–8, 53. The weber was renamed the ampere at the Paris Congress of 1881.

[62] Kempe's analysis was criticized in 1890 by William Ayrton and his associates as being insufficient to compare the figures of merit for different kinds of current-measuring instruments. The figure of merit depended in fact on a complex of factors, including the skill of the experimenter, and Ayrton et al. devised a criterion for measuring the factor of merit for instruments arranged to have the same period of oscillation for its suspended system as this was the most useful for other purposes. William E. Ayrton, Thomas Mather, and William.E. Sumpner, 'Galvanometers', *Proceedings of the Physical Society of London*, 10 (1888–90), pp. 393–434, discussion on pp. 414–9.

[63] For Kempe's comments on the limits of 'certainty' in the possible degree of accuracy attainable, see his *Handbook of electrical testing*, p. 58; see also pp. 112–13, 123–4. Kempe offers a worked example to show how galvanometer indications read with 'great accuracy' could enable a fault in a 1000-knot telegraph cable to be located to within 1 knot; ibid., pp. 289–90.

different. He recommended that they adjust this instrument so that its deflection for a given current fell in the region of 40°–50°. This was the range, he explained, over which its deflections responded most sensitively to changes in current and thus afforded the highest degree of accuracy; accordingly, users were enjoined to make their figure of merit calculations for this range. Importantly, Kempe suggested that they should be able to report scale deflections to the nearest $\frac{1}{4}°$ with an associated maximum error also of $\frac{1}{4}°$. Why so? Kempe relied on the notional normal user to be able to judge when a needle was exactly midway between two scale markings ($\frac{1}{2}°$) and thus could tell to within about $\frac{1}{4}°$ whether the needle was nearer to a scale marking or to the adjacent halfway point.[64] Thus we see that different measurement technologies were bound up with different reading techniques and different assumptions about the bodily competence of users to differentiate between scale markings.

Entirely unspecified in Kempe's account, however, is the amount of time required for taking a reading. Neither the mirror galvanometer nor the tangent galvanometer could be used to take an immediate or instantaneous reading of currents. This was not just because of the need to refer to paper calibration charts to perform measurements. Rather it was because the light-spot on the former and the needle on the latter would take time to settle to the 'final' position that yielded the reading. The mirror galvanometer in particular was pinpointed by Maxwell as being prone to sensitive dwindling oscillations, configuring its user into being a patient 'spot-watcher' waiting for the exasperating light beam to settle to its ulitmate destination. This was an experience that inspired Maxwell's satirical poetic borrowing from the fading bugle call echoing around the Irish valley in Tennyson's famous poem 1848 'The Splendour Falls on Castle Walls'.[65] Whilst perhaps aesthetically pleasing for Maxwell, such prolonged dainty attenuation did not suit those in the world of commerce. Indeed, it was with this problematic phenomenon in view that Kempe contrasted galvanometer sensitivity not with accuracy but with the pragmatic virtue of 'deadbeatness'. For those users who would not wait for light-spots to settle, and were prepared to sacrifice their instruments' sensitivity to short-term fluctuations in current, Thomson developed

---

[64] As Kempe explains, if a current of 0.01 webers gave a deflection of 50°, the reading for a 1° deflection would be a current of 0.01 × tan 1°/tan 50° = 0.000146 webers, Kempe, *Handbook of electrical testing*, 1881, pp. 37–8 and 54–5.

[65] In Tennyson's original, the last two lines of the first verse read:

Blow, bugle, blow, set the wild echoes flying,
Blow, bugle; answer, echoes dying, dying, dying.

This was apparently Tennyson's rendition of the experience in Killarney in 1848 of hearing a bugle call echo eight times. See J. D. Jump (ed.), *Tennyson: In Memoriam and other poems*, London: Dent, 1974, pp. 71–2 and 222.

an alternative 'deadbeat' form of galvanometer. In this adapted device, optimal air-damping was employed to ensure that the needle moved without oscillation – and as quickly as possible – to its final position.[66] Clearly there were temporal constraints as well as spatial limits that required the Thomson mirror galvanometer to be adapted to the diverse needs of users in their diverse contexts.

This same range of techniques of reading and the tactic of deadbeatness was adopted initially by the designers of the new instruments developed for the stronger currents used in electrical lighting. The non-linear Deprez galvanometer developed in 1880 was issued with a calibration chart, and the portable and allegedly proportional galvanometer of Ayrton and Perry, produced in 1881, was issued with a calibration constant akin to the mirror galvanometer's figure of merit. A new practice of reading was soon developed, however, to deal with the somewhat different temporal constraints of lighting engineering that emerged in the mid-1880s. In the instruments they developed between 1882 and 1884, the need to measure currents *quickly* led them to develop the practice of producing instruments which would allow readings to be taken effectively instantaneously from the dial.[67] Just a few seconds of inattendance to an ampere-meter (later the ammeter) might allow current surges to damage circuit wiring or dynamo machinery, and neglect of a voltmeter might lead to drops in circuit potential that angered customers who resented temporary dimming or permanent damage to their lights.[68] And in his *Practical Treatise on Electrical Lighting* of 1884, J. E. H. Gordon was highly critical of the Siemens dynamometer he had been using in preparing the first public ac installation at the Paddington railway station. The 'great objection' was that its time-consuming null operation meant it could not be used to read momentary variations of the current, such as during critical moments when lights were about to break down or were in the process of doing so. The only advantage he saw in the dynamometer was that it could measure alternating currents, but even then not 'extremely' accurately.[69]

The extent to which high-speed direct-reading had become orthodoxy a decade later might be gauged from the response to Swinburne's lecture on electrical measuring instruments at the ICE in 1892. He declared of ideal instruments in the electric-power industry, notably the burgeoning world of power-stations, that all should be direct-reading. By this he meant 'the index should point to the pressure or current' on a dial, and there should

---

[66] Kempe, *Handbook of electrical testing*, 1881, pp. 34–7.
[67] William E. Ayrton and John Perry, 'Measuring instruments used in electric lighting and transmission of power', *JSTEE*, 11 (1882), pp. 254.
[68] Rookes E.B. Crompton, *Reminisences*, London: 1928, p. 147; J. E. H. Gordon, 'The development of electric lighting', *Journal of the Society of Arts*, 31 (1883), pp. 778–87, discussion pp. 787–91.
[69] James E. H. Gordon, *A practical treatise on electrical lighting*, London: 1884, pp. 35–7, 44–7.

be no 'turning of buttons on the tops' as for the Siemens device, or reference to calibration tables. Unlike the electrician at the telegraph station, the power-station engineer in particular would have to monitor many different instruments simultaneously. For that practitioner, Swinburne pronounced that all instruments should have 'bold and visible scales', with vertical dials, so that it was unnecessary to go up 'close' to the switchboard to take a reading.[70] Nobody in the enusing discussion contested this view. As we shall see, however, these were about the only remarks in Swinburne's paper that seemed to have been entirely uncontroversial.

One particularly colourful feature of this ICE lecture was Swinburne's attacks on the futility and untrustworthiness of attempts by William Ayrton and others to give equal spacing to the scales of direct-reading instruments. To understand the basis of Swinburne's attack, I next consider the different engineering traditions from which he and Ayrton drew in their preferences in instrument reading.

4.4. PROPORTIONALITY VERSUS TRUSTWORTHINESS: CONSTRUCTING THE DIRECT-READING AMMETER

We still want for commercial use a really good ammeter for measuring large currents, one that we can trust as we trust our scales for weighing merchandise, to give correct indications for weeks and months and years.
    Professor William Anthony, Electrical Measurement at Franklin Institute, January 1888[71]

In 1881, two professors at Finsbury Technology College, William Ayrton and John Perry, proposed that quick and accurate current readings were best accomplished by using galvanometers with a 'proportional' mechanism. Whilst Ayrton at least adhered strongly to this position into the 1890s, Swinburne disagreed and was their most voluble opponent on this matter. Their highly divergent views were manifested several times in the period 1888–92 when Ayrton and Swinburne publicly vied for authority and credibility in the field of electrical measurement. A self-styled acolyte of Sir William Thomson, W. E. Ayrton had been trained in experimental physics at University College, London, and Glasgow University, and had been engaged in telegraphic work and training in England, India, and Japan.[72] It was presumably from this close familiarity with the use of the mirror galvanometer in many contexts since the late 1860s that he acquired his long-term conviction that instruments were most conveniently read with proportional calibrated and equally divided scales. This was the form of scale reading that he and Perry

[70] Swinburne, 'Electrical measuring instruments', p. 2.
[71] Anthony, 'Electrical measurement', pp. 56–72.
[72] For Ayrton's familiarity with Thomson and the mirror galvanometer, see William E. Ayrton, 'Kelvin in the Sixties', *Popular Science Monthly*, 27 (1908), pp. 259–68, especially pp. 265–7.

proposed for their first ampere-meter in 1881 and all its successors. They emphasized that calibration with scales of equally spaced markings facilitated faster and easier reading than did non-linear alternatives. This was especially important when deflections fell between scale markings: Interpolated readings could easily be gauged by eye from the relative distance of the needle from adjacent scale markings. By contrast, (non-linear) markings that increased or decreased in separation along their scale did not lend themselves to simple proportionate approaches to interpolated readings.

By contrast, Swinburne's position was that veracity of the readings mattered more fundamentally than convenience. Trying to design or make a device that had a strictly proportional reading was theoretically very difficult, and Swinburne considered that it was never actually realized in practice with sufficient accuracy for the engineer's purposes. According to him, the trustworthiness of the deflection mechanism was likely to be seriously compromised by contrived and complex compensation techniques that attempted – but usually failed – to render scales reliably equally spaced. For Swinburne the crucial matter was whether the instrument's mechanism furnished a trustworthily unvarying relation between current and deflection. The most likely archetype for Swinburne's approach was the Bourdon pressure gauge that was becoming increasingly ubiquitous in contemporary steam technology. This device used the unfurling of a pre-stressed metal tube under pressure to give reliable but non-proportional steam-pressure readings on boilers.[73]

Ironically, Swinburne was probably introduced to this pressure gauge by his mechanics teacher at Clifton College in the early 1870s – none other than John Perry[74]; Swinburne would then also have encountered these or similar non-linear steam gauges regularly during his mechanical engineering apprenticeship in Manchester locomotive shops and then in the Tyneside shipyards.[75] Whilst there is no extant direct evidence of Swinburne's early experiences of pressure gauges in engineering, his later writings clearly indicate the nature of his commitment to mechanical engineering instrumentation. In the introduction to his 1888 *Practical Electrical Measurement* he defined his subject as instruments designed for use in the workshop as opposed to 'mere laboratory practice'.[76] And in his lecture to the ICE four years later he declared approvingly that electrical engineers' methods of measurement were becoming ever more like those of the mechanical engineer: Although

---

[73] For discussion of the Bourdon pressure gauge, see J. Stock, 'Gauge, pressure', in Robert Bud and Deborah Warner (eds.), *Instruments of science*, London/New York: Garland, 1998, pp. 273–5.

[74] Perry described the Bourdon pressure gauge as a 'very exact device' that was 'almost universally employed for all purposes' in his textbook on the subject; John Perry, *Elementary treatise on steam*, London: 1874, pp. 186–7.

[75] F. A. Freeth, 'James Swinburne, 1858–1958', *Biographical memoirs of the Royal Society*, 5 (1959), pp. 253–64.

[76] Swinburne, *Practical electrical measurement*, p. 111.

they were departing from those of the physics laboratory, they were not necessarily therefore less accurate.[77]

There was one specific practical reason why Ayrton and Perry, among others, were most interested in developing new kinds of current-measuring devices in the 1880s. Whereas potential differences could be gauged by skilled used of other technical means, this was very difficult to accomplish for currents. When Swinburne first installed a lighting system in Antwerp for the Swan company in December 1881, he could easily judge the voltage load on a dynamo by using the 'touch' of a filament lamp to judge its heat output.[78] J. E. H. Gordon recommended that engineers trust instead to their visual faculties to ascertain the potential drop across a lamp, comparing its brightness with one connected to a circuit of known voltage. In 1886 S. Z. de Ferranti used twenty-four 100-V lamps to gauge the potential across the circuits at the Grovesnor Gallery ac station in central London: no existing voltmeters could safely register 24,000 V.[79] By contrast, there were no comparatively safe bodily means of gauging currents. Electrical engineers doubtless soon learned the injurious – and occasionally lethal – effects of becoming accidentally connected to lighting circuits and that equipment could be damaged by large currents that did not render wiring visibly hot. Mrs J. E. H. Gordon reported of her husband's inventive trials for the Telegraph Construction and Maintenance Company in early 1883 that, by the time a current generator could be heard to 'scream', the omens were not good. By the following day, the remaining fragments of the machine could be picked up with a shovel.[80]

The first instruments for current measurement used by lighting engineers employed a moving-iron mechanism, but deployed 'permanent' magnets strong enough to shield the mechanisms from the deleterious effects of nearby dynamos.[81] The first such device was Marcel Deprez's 'fishbone' galvanometer, produced in 1880.[82] Ayrton and Perry soon criticized this at theSTEE for requiring users to get current 'readings' from a calibration chart. They proposed that the magnet pole pieces in such instruments be moulded so as to give deflections proportional to current. Users of their 'portable absolute galvanometer for strong currents' could, they claimed, just multiply scale

---

[77] Swinburne, 'Electrical measuring instruments', 1892, p. 1.
[78] J. Swinburne [50-year jubilee of the IEE], *JIEE*, 60 (1922), p. 422.
[79] Gertrude de Ferranti and Richard Ince, *The life and letters of Sebastian Ziani de Ferranti*, London: 1934, pp. 54–5. I thank Neil Brown for explaining Ferranti's purpose here; see Chapter 6 for further discussion of Ferranti's domestic meters.
[80] Mrs J. E. H. Gordon, *Decorative electricity*, London: 1891, 2nd edition, 1892 pp. 161–2. Compare this with J. E. H. Gordon's account in his piece, 'The development of electric lighting', *Journal of the Society of Arts*, 31 (1883), pp. 778–87, discussion on pp. 787–91, quote on pp. 781–2.
[81] Swinburne, *Practical electrical measurement*, pp. 25–6.
[82] See Anonymous, 'Marcel Deprez's galvanometer for strong currents', *Nature* (London), 22 (1880), pp. 246–7.

deflections by a calibration 'constant' to get a reading.[83] Their novel device, soon renamed the 'ammeter', was evidently quite popular: At a meeting of the STEE in May 1882, J. E. H. Gordon commended it for being deadbeat in operation, registering fluctuations in current more rapidly and clearly than rival devices.[84] In October, two electrical engineers wrote to the *Electrician* praising this 'excellent little current measurer', albeit suggesting improvements in it for peripatetic work.[85] But at the time, Ayrton and Perry showed the robustness and readability of this device and its new sibling 'voltmeter' by attaching one each to the handlebars of their new battery-powered electric tricycle (see Figure 5). Whilst riding it through the City of London at the illegal speed of nearly nine miles an hour, Ayrton calculated its power consumption by multiplying the instrument's readings and constants together, presumably in his head.[86] In February 1883, rather more sedate publicity was given to the ammeter in James Shoolbred's sympathetic review of the ammeter alongside five other devices for current measurement.[87] Delivering a paper, 'On the Measurement of Electricity for Commercial Purposes', to the STEE, Shoolbred argued that no other device had the ammeter's full range of user-centred advantages of portability, proportionality, deadbeatness, and invulnerability to disturbance from nearby electrical machinery.

By then, Ayrton and Perry had systematized the mass production of their ammeters with sufficient consistency that, depending on the position of the commutator, the calibration constant was generally either 1/5 or 2 amperes

---

[83] Gooday, 'The morals of metering', pp. 250–2. Two such calibration constants were required for instruments that had a commutator to allow for use over two ranges (e.g., 1–10 and 1–100 wb), and such constants were checked and adjusted probably at least daily. Ibid. pp. 251–8; Ayrton and Perry, 'A portable absolute galvanometer', pp. 156–6. These devices were manufactured by Patterson & Cooper in 1881, whereas later instruments were produced by Ayrton's erstwhile employers of 1878–9, Latimer Clark Muirhead and Co, and then later still by Ayrton's and Perry's own Acme company: See 'The Acme Electrical Works', *The Electrical Engineer*, 3 (1889), pp. 110–12.

[84] Discussion of Ayrton and Perry, 'Measuring instruments used in electric lighting and transmission of power', *JSTEE*, 11 (1882), p. 289. Ayrton had helped Gordon in preparing his *Physical treatise* for publication, and Gordon returned the gesture by giving a sympathetic representation of Ayrton's instruments. Gordon says in the preface, 'Mr Ayrton has read the whole of Part 1 in proof, and most of Part III.' Gordon, *Physical treatise*, Vol. 1, p. vi.

[85] The letter was written 2 October 1882 by L.Probert and Alfred Soward, *Electrician*, 7 (1882), p. 500.

[86] See note in *Nature* (London), 27 (1882), p. 19; Anonymous, 'The electric tricycle', *Telegraphic Journal and Electrical Review*, 11 (1882), pp. 394–5 with illustration; P. J. Hartog. 'Prof. W. E. Ayrton, F. R. S. A biographical sketch', *Cassier's Magazine*, 22 (1902), pp. 541–4, quote from p. 543.

[87] James Shoolbred, 'On the measurement of electricity for commercial purposes', *JSTEE*, 12 (1883) pp. 84–107, discussion on pp. 107–122. The other devices cited were two from the Siemens company (dynamometer and Obach tangent galvanometer), Deprez's fishbone galvanometer, Thomson's graded galvanometer (adapted from the tangent galvanometer), and Major Cardew's apparatus for comparing a known current with the unknown current.

Figure 5. Ayrton's and Perry's electric tricycle in November 1882. Developed to show the motive power of the secondary battery cells underneath, it has an ammeter attached to the right steering arm and a voltmeter attached to the left. Reading from these enabled the rider (typically Ayrton) to 'calculate at any moment' the horsepower expended in propelling the tricycle. A light bulb was attached to each arm to illuminate both instrument dials and the roadway. [*Source*: 'The Electric Tricycle', *Telegraph and Electrical Review*, 11 (1882), p. 394.]

per degree of deflection to simplify the mental multiplication needed to get a reading.[88] Yet, even with this, mishaps occurred when workers using several ammeters simultaneously had confused the calibration constants of the instrument with those of another. Moreover, as they noted to the Physical Society in January 1884, the device presumed too much skill on the part of the largest constituency of ammeter readers: skill in mental calculation 'not always predominating' among the men in charge of lighting circuits. Ayrton

[88] Gordon, *Practical treatise*, pp. 46–7.

and Perry thus made their hitherto merely proportional instruments completely direct-reading. Ammeter dials were now to be manufactured with scale markings in amperes, so that current values could now be 'immediately read off' to an accuracy of around 1%. And by means of adjustable calibration knobs, users could take the final responsibility for ensuring that the ammeter's readings were trustworthily calibrated, even if the strength of the deflecting permanent magnet changed from day to day.[89] Within a few years this format was adopted by manufacturers of almost all commercial current-measuring instruments, as can be seen in the models advertised in Swinburne's *Practical Electrical Measurement* of 1888.[90]

These Ayrton and Perry permanent-magnet ammeters and voltmeters were widely used, but it soon became common knowledge that these devices did not always furnish dial deflections that faithfully represented the size of currents generating them. Perry conceded in 1885 that very short-term 'memory' effects – later dubbed 'hysteresis' – in rather less than permanent magnets could cause errors of up to 37% in certain parts of their range.[91] Writing in the *Telegraphic Journal* in 1887, Swinburne emphasized rather differently that the 'very serious faults' of these instruments stemmed from attempts by Ayrton and Perry to make them proportional, which, 'needless to say', was impossible to accomplish with 'any accuracy'. In reshaping the magnet pole pieces to achieve this end, they had made the controlling field so strong and left so little space for the moving mechanism that the instruments soon got so hot that its deflections then became unreliable. Moreover, even with the very limited success at proportionality that they had achieved in their instrument, the instruments of Ayrton and Perry had rather short scales extending over only about 45° deflection, so Swinburne contended that their scale markings were too cramped together for readings to be taken with 'very great accuracy'.[92]

By the time Swinburne was writing, Ayrton and Perry had responded to these irremediable problems, not by abandoning the troublesome attempt at proportionality, but rather by developing an alternative means of accomplishing it by means of their patented 'proportional spring'.[93] In their 'spring'

[89] William E. Ayrton and John Perry, 'Direct-reading instruments, and a non-sparking key', *Proceedings of the Physical Society of London*, 6 (1884), p. 59.
[90] See advertising in the end leaves of Swinburne, *Practical electrical measurement*, 1888.
[91] See Perry's comments in a discussion on Peter W. Willans, 'The electric regulation of the speed of steam-Engines, and other Motors for Driving Dynamos', *Minutes of Proceedings of the Institution of Civil Engineers*, 81 (1885), pp. 166–89; ibid., following discussion on p. 197.
[92] James Swinburne, 'Practical electrical measurement', *Telegraphic Journal and Electrical Review*, 20 (1887), pp. 492–3; Anthony 'Electrical measurement', p. 61; Swinburne, *Practical electrical measurement*, pp. 29–30. Similar criticisms were repeated by Swinburne to the ICE in 1892, although he observed then that large numbers of very similar instruments were still being produced in France by Carpentier, Swinburne; 'Electrical measuring instruments', p. 9.
[93] British patent number 2156, 1883.

ammeter of 1884–5, they borrowed Kohlrausch's 1876 solenoidal technique of attracting a thin iron core downwards into a current coil, deploying their patent 'magnifying spring' to unfurl (allegedly) in proportion to the core's vertical movement and rotate an attached pointer over a horizontal dial. Publicizing this new instrument in both the Royal Society *Proceedings* and *Nature*, they showed how a new phosphor bronze metal could make springs that rotated in proportion to extension up to 270°. Ayrton's survey of ammeters, published in his 1887 textbook *Practical Electricity*, certainly presented the spring ammeter as though it had solved almost all problems to which earlier and rival devices were prone. He conceded only the inconvenience that it did not respond at all to currents less than 20% of the maximum reading (the threshold at which the iron core became saturated) and thus was not calibrated for the bottom end of the scale. Proportionality and long range had thus been accomplished but only at the expense of sensitivity for smaller currents – a typical trade-off in instrument construction borne of the dilemmas facing instrument-designers.[94]

The spring ammeters of Ayrton and Perry seemed to have attained at least as great a popularity as their earlier magnetic siblings, but some practitioners were still not satisfied. In his series of articles for the *Telegraphic Journal* of 1887, Swinburne was generally rather more sympathetic to these devices than he had been to their permanent-magnet predecessors. The spring ammeter, he opined, was 'most ingenious', 'exceedingly pretty', and 'free from many errors', being easy to read, usable close to a dynamo, and prone to no more than 1% secular drift. Yet, given how 'very largely used' these instruments were in industry, Swinburne was again emphatically critical of Ayrton's and Perry's attempts to make their instruments proportional. Whilst the spring ammeter he held in front of him on his desk was 'right' at two points at either end of its scale, its readings were as much as 8% too high in the middle of its range. He regretted that he had not been able to secure from instrument-makers the latest range of ammeters so that he could systematically check the trustworthiness of all other instrumental calibrations in this regard.[95] By way of transatlantic comparison, it is notable that, in his 1888 lecture at Cornell previously discussed, Professor William Anthony reported from his experience of testing a wide range of commercial instruments that he had been unable to get such spring ammeters to be mutually consistent to within even 5%. Accordingly he declared that a commercial current-measuring instrument was still needed that could be 'trusted' as much tomorrow as it was today.[96]

Swinburne swiftly re-edited his articles into the book *Practical Electrical Measurement*, published by Alabaster and Gatehouse for a readership of lighting engineers, complementing thereby Kempe's *Handbook of Electrical*

---

[94] William E. Ayrton, *Practical electricity*, London: 1887, pp. 386–91.
[95] Swinburne, 'Practical electrical measurement', pp. 536–7.
[96] Anthony, 'Electrical measurement', pp. 56–72.

Testing produced by the same house for telegraphic audiences. Ayrton and Perry had by then undertaken major revisions to improve the proportionality of their spring ammeters and voltmeters,[97] and Swinburne agreed that the most recent instrument was much better. He reported that its designers had even apparently received an 'exceedingly complimentary' letter from 'no less an authority' than H. F. Weber confirming its accuracy.[98] Despite acknowledging that their spring ammeter was still a 'remarkable' instrument in his ICE lecture on 'Electrical Measuring Instruments' in April 1892, Swinburne was otherwise unremittingly harsh about the calibration techniques of Ayrton and Perry. He criticized certain unnamed designers for having spent years devoting 'unlimited ingenuity' to making instruments with equally divided scales – a feature of 'no value', especially if other crucial design advantages had been 'sacrificed' in this process. Ayrton took umbrage at this and in the discussion that followed Swinburne's lecture complained of being 'derided'. Swinburne retorted that the instrument should not be 'tampered' with to make it fit the scale, but the scale engraved to 'suit the instrument'. Instrument maker Sydney Evershed agreed, maintaining that Swinburne need not be so concerned to counter Ayrton's strategy, because for the past five to six years all the major manufacturers he knew had calibrated every device 'without assuming it obeyed any simple law at all'. This practice of making instruments with non-linear calibrations became the convention thereafter. Ayrton's and Perry's efforts at promoting a simplified practice of reading had been little heeded by instrument-makers, on whom such a practice would have placed an enormous burden of technological ingenuity.[99] (Figure 6)

In Section 4.5 we will see how issues of accuracy, trustworthiness, design, and bodily deportment all merged together in a debate at the IEE in London in 1891. This concerned the D'Arsonval reflecting galvanometer as a possible universal current-measuring device. Further issues emerged, such as the nature of the trust relation between manufacturers and clients, especially when the latter insisted on witnessing tests of electrical performance. Once again we find that the commercial practitioners gathered there did not unreservedly welcome Ayrton's fulsome expertise in galvanometer technology.

4.5. IRONIES OF READING INSTRUMENTS: PROPORTIONALITY AND SPOT-WATCHING

We do not think that engineers who have once worked with a good d'Arsonval galvanometer will ever part with it in favour of instruments with pointers. Readings can, of course, be taken to greater accuracy with the spot than a pointer. The

---

[97] British patent number 7802, 1887. [98] Swinburne, *Practical electrical measurement*, p. 47.
[99] Swinburne, 'Electrical measuring instruments', pp. 1–32; see also pp. 1–3, 6, 33, 42–3, 57. See range of available instruments listed as exhibited at the ICE monthly reception, cited in ibid., pp. 66–8.

sensitiveness of the reflecting galvanometer is absolute . . . [and it] is easily made to cover the whole of the requirements of a shop by means of a single instrument.
Captain H.R. Sankey and R.V. Andersen, discussion at IEE, November 1891[100]

From the late 1880s onwards, much attention was paid in the British electrical engineering community to the remarkable properties attributed to the Deprez – D'Arsonval galvanometer (Figure 7). The 'galvanomètre aperiodique de MM. Deprez et D'Arsonval' first appeared in France in 1882 as a result of a close collaborative effort. Jacques Arsène D'Arsonval had taken Deprez's 1880 fishbone galvanometer and inverted its operation so that the deflection was generated by the more sensitive movement of the current-carrying coil rather than the 'fishbones.' This was amplified by a mirror arrangement of the form used in the Thomson reflecting instrument, and the whole set-up was shielded from external magnetic disturbances by a vertical iron tube which took the place of the iron fishbones. Its French designers presented it as an *'aperiodique'* [deadbeat] device that could measure currents *'d'un intensité extrêmement faible'*[101] [of extremely weak intensity]. Contemporary texts refer to two versions of this device, and these prima facie fit James Swinburne's taxonomy of laboratory versus workshop instruments. The more recently developed direct-reading 'industrial' form was praised by Swinburne in his 1888 *Practical Electrical Measurement*, as being perfectly deadbeat. He advised prospective users not to depend, however, on the direct-reading scale supplied by the manufacturer: They should instead calibrate it for themselves. Swinburne was much less charitable about the original D'Arsonval instrument, contending that such reflecting devices were fit only for the laboratory and were 'not suitable' for work in power engineering, especially not in power-stations.[102] Accordingly he could not recommend to his readers the policy adopted by his erstwhile employer, R. E. B. Crompton, the eminent Chelmsford dynamo manufacturer, for using this device in testing the performance of incandescent lamps.[103] Other people in Britain, however, were showing a strong interest in adapting the French galvanometer to use in electrical engineering.

On 8 February 1889, *The Electrical Engineer* published a piece on the new instruments under development at the testing department of Ayrton's and Perry's Acme electrical factory, recently relocated to Kentish Town. This department was superintended by H. B. Bourne, Ayrton's South Kensington

---

[100] Sankey and Andersen, 'Description of the standard volt- and ampere-meter', p. 589.
[101] Marcel Deprez and Jacques D'Arsonval, 'Galvanomètre aperiodique', *Comptes Rendus del 'Academie des Sciences*, 94 (1882), pp. 1347–50, quote on p. 1347. Deprez seems to have worked on its highly non-linear performance to produce a device for heavy-duty engineering which offered direct readings proportional to intensity. Marcel Deprez, 'Sur un galvanomètre à indications proportionelles aux intensités', *La Lumiere Electrique*, 14 (1884), p. 401.
[102] Swinburne, *Practical electrical measurement*, pp. 49–50.   [103] Ibid., pp. 29.

Figure 6. A Johnson & Phillip's hot-wire ammeter ca. 1910 illustrates the non-uniform divisions of scales typical of instruments used in early twentieth-century electrical engineering. The extra needle to the right records the highest current registered. (*Source*: Rankin Kennedy, *The Book of Electrical Installations: Electric Light, Power, Traction and Industrial Machinery*, 3 Vols., London, n.d., ca. 1910–5, Vol. 2, p. 26.)

student, and it followed up instrumental researches by Ayrton and his students at the City & Guilds Central Institution. The article reported that a new and original 'universal instrument' was being fitted up there that could measure currents from 0.1 to 600 A and potential differences for 1–500 V. Although no details were published, it is clear that this device was an adapted form of the Deprez–D'Arsonval reflecting galvanometer. At a meeting of the

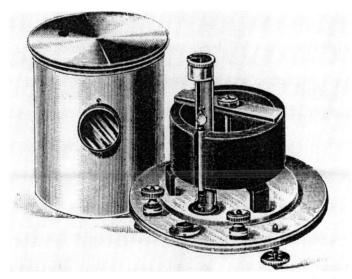

Figure 7. An Ayrton and Mather version of the Deprez–D'Arsonval galvanometer manufactured by R. W. Paul ca. 1902 for delicate use in Wheatstone bridge measurements of resistance. The housing is removed to show the internal construction of a ring-shaped permanent magnet in the air-gap of which is suspended a cylindrical coil carrying a circular mirror for reflecting a light beam onto a distant scale. [*Source*: Rankin Kennedy, *Electrical Installations*, 5 Vols. (originally 4), London, 1902, Vol. 2, p. 1.]

Physical Society of London two weeks later, Ayrton explained the attempts he and his students had undertaken to bring proportionality to the notably non-linear D'Arsonval instrument. Their efforts to overcome a peculiar discontinuity in their moving-coil's response to currents were persistently thwarted for months until Bourne exercised his 'dexterity' in suspending the coil. From that point the various settings of the galvanometer showed a proportionality that agreed with a Thomson standard current balance to within 1/5% for every current range between 0.01 to 600 A.

From this vantage point Ayrton and his students were able to attack the trustworthiness of a rival current claimant to a universal measuring device, the widely used Siemens dynamometer. They showed that calibration charts for five such instruments were unreliable owing to the distortion of their springs.[104] After another year's researches with his assistants Thomas Mather and William Sumpner comparing the figures of merit of various kinds of galvanometer, Ayrton came back to the Physical Society in January

---

[104] William E. Ayrton and John Perry, 'Electrical measurement', *Electrical Engineer*, 3 (1889), pp. 175–7, 189–92. Notably, they had investigated the D'Arsonval reflecting galvanometer with the aim of solving the problem of how a laboratory could use a single inexpensive instrument to register currents very accurately over the range of the circuit instead of having to purchase a dozen Thomson standard current balances at a cost of £200–300.

1890. There he presented data to show that the proportional D'Arsonval galvanometer had the highest 'sensibility' of all. When modified to their specifications, they claimed it could register the constancy of a current to within a 'very small fraction' of a per cent.[105] Yet, as we shall see, this device did not gain acceptance as a universal instrument as Ayrton had hoped. In the following year, the contentiousness of the Deprez–D'Arsonval reflecting galvanometer was clear in a debate on the electrical testing procedures of steam-engine manufacturers Willans and Robinson at the IEE in November 1891.

At that time the Willans and Robinson company had a strong reputation, its high-speed steam-engines were being adopted by many of London's recently launched electric supply companies.[106] Eleven years earlier, Peter Willans had set up business with his commercial partner (Mark) Robinson at Thames Ditton near Hampton Court in 1880 to make steam-engines for ships and locomotives. He soon followed the advice of R. E. B. Crompton in 1883 that there was an unmet demand for high-speed engines in the electric-lighting industry. And not long after Willans and Robinson began producing specialist low-noise machines for electrical lighting on railways, the company won contracts in 1884 to supply four such engines for the electrical illumination of Buckingham Palace. The company's reputation for economy and reliable testing methods were enhanced by two award-winning papers that Willans presented to the ICE in 1885 and 1888.[107] The latter concerned trials on the efficiency of a wide range of steam-engines; as a major user of Willans engines, Crompton considered that this paper established Willans' methods of measuring horsepower and steam and water consumption to be the 'most perfect and accurate' ever adopted.[108]

[105] Ayrton, Mather, and Sumpner, 'Galvanometers', pp. 414–29. Swinburne was in the audience for that paper and reiterated that he 'saw no great advantage in making practical instruments proportional'; ibid., p. 434. The eponymous Ayrton–Mather galvanometer became a standard laboratory instrument manufactured by the Cambridge Scientific Instrument Company for several decades thereafter. See William E. Ayrton and Thomas Mather, 'Galvanometers', *Philosophical Magazine*, 42 (1896), pp. 442–6; Ibid., 'Galvanometers [Third Paper]', *Philosophical Magazine*, 46 (1898), pp. 349–79 [also published in the *Proceedings of the Physical Society of London*, 16 (1899), pp. 169–204].

[106] These were only displaced by the higher-efficiency Parsons steam-turbine after Willans' premature death in 1892.

[107] Peter Willans, 'The electrical regulation of the speed of steam engines, and other motors for drivng dynamos', *Proceedings of the ICE* (hereafter *PICE*), C.E., 81 (1885), pp. 166–89, discussion on pp. 190–232, won the ICE Telford prize, and Peter Willans, 'Economy trials of a non-condensing steam-engine: Simple, compound and triple', *Minutes of the PICE*, 93 (1888), pp. 128–88, won the ICE Watt Medal and Telford Premium. For a biography of Willans see the obituary in *PICE*, 111 (1892), pp. 395–8 and *DNB Missing Persons*.

[108] Crompton in discussion of Sankey and Andersen, 'Description of the standard volt- and ampere-meter', p. 564. For the trials, see John Perry, *Steam engine*, London: 1899; James Alfred Ewing, *The steam engine and other heat engines*, 2nd edition, Cambridge: 1897, pp. 162–4. For Gordon's early use of Willans engines, see Robert H. Parsons, *The early days*

In 1887 Willans had commissioned his electrical staff, Captain Sankey and F. V. Andersen, to extend such tests to the efficiency with which the company's engines converted steam power into electrical power when driving his customer's dynamos. To do this they had to find instruments with an unprecedented combination of qualities. They needed a voltmeter that could operate over 1/1000 of a volt to 700 V and an ammeter that could measure 1/40 of an ampere to 1,100 A with errors no greater than 1/5%. Moreover, they sought devices that could attain such 'high accuracy' whilst being robust enough to work alongside working dynamos. Equally important, they needed instruments that *all* dynamo manufacturers could trust to produce reliable readings, and do so instantaneously in a way that enabled customers to stand next to company staff in the test-room to corroborate the results of the efficiency tests. Choosing the right instruments for the Thames Ditton test-room was thus a tricky decision as Sankey and Andersen explained at the start of their IEE paper in November 1891. Judgements of the accuracy and trustworthiness of various instruments were somewhat moot when they began their investigations in 1887. It was widely maintained that even the best available ammeters and voltmeters could not be relied on to give the degree of accuracy of a fraction of a per cent that Willans had obtained in steam-consumption measurements. Moreover, all such instruments 'laboured under the disadvantage' in that they were calibrated elsewhere, so that results attained with them were dependent on *other people's* measurements to a degree Sankey and Andersen held to be 'undesirable'.[109] The problems of trust in instruments and unseen persons thus came to the forefront of discussion.

Sankey and Andersen adopted a form of D'Arsonval galvanometer (such was the partisan abbreviation by then commonly used) only after two years of trials on a wide range of commercial voltmeters and ammeters. Whilst they found the best instruments gave measurements to within $\frac{3}{4}$% when placed 'sufficiently far away' from the engines and dynamos, these discrepancies rose to 3%–5% when the instruments were placed next to the dynamos under test. And notwithstanding manufacturers' claims to the contrary, such errors palpably changed when a new dynamo was started or stopped.[110] In his later discussion of their paper, Willans himself expressed the problem of existing

---

of the power station industry, London: 1940, pp. 45, 73, 78; Crompton, *Reminiscences*, pp. 89–94; Dunsheath, *A history of electrical engineering*, p. 148. Ayrton's colleague in civil engineering at the City & Guilds college in South Kensington later said, 'However purely practical the object Mr Willans had in view, his experiments were made in the spirit of true scientific research. No trouble was too much to secure accuracy to the last decimal, no possible cause of error was so trivial that its investigation was rendered unnecessary.' Alfred Unwin, 'The James Forrest lecture: [The experimental study of steam engines]', *Minutes of PICE*, 122 (1895), p. 179.

[109] Sankey and Andersen. 'Description of the standard volt- and ampere-meter', p. 516.
[110] Ibid., p. 588.

commercial instruments somewhat differently, focussing on issues of trust:

> I myself began with great faith in the ordinary electrical instruments; but, after taking readings sometimes with one and sometimes with another instrument, I began to lose it. Of course we are not working for ourselves, but mainly for dynamo makers, and I began to find whatever faith we had in instruments we selected, other dynamo makers had not confidence in them. They had great confidence in their own instruments, as a rule, but not much in others provided for them.[111]

Customers' reluctance to trust any instruments but their own was thus a serious problem for Willans. The initial compromise his staff adopted was to use Siemens dynamometers as there was at least some consensus that the absolute calibration of such robust devices hardly ever drifted. Yet the very null technique which made this dynamometer reliable was 'troublesome' because its time-consuming operation did not enable users to gauge momentary fluctuations in current. According to Willans, using it for this purpose was as sensible as to trying to weigh a horse and cart while they rode over the balance. The problem was especially serious as the ever-increasing power output of dynamos lead to ever greater fluctuations that needed to be monitored and properly averaged out when high-accuracy dynamo tests were undertaken.[112]

Given the publicity accorded to the unprecedented accuracy, sensitivity, and range of D'Arsonval reflecting instruments in 1888–9 and its recommendation by the company's close associate Crompton, it is not surprising that the Willans staff elected to use this instrument to meet their criteria of robustness and accuracy. Sankey and Andersen thus set up a form of this device with the calibrated proportional scale 5 ft from the mirror instrument, with adjustments for different sizes of current achieved by varying the circuit resistance. At its most sensitive setting, a change of 1/10,000,000 of an ampere was thus detectable in the motion of galvanometer's illuminated spot. Having taken 'great care' in setting up this arrangement, Sankey and Andersen claimed to be able to measure currents for ordinary work to within 0.3% of error. Nevertheless, for 'very accurate' trials they took scale readings at regular intervals to obtain a mean value, and then checked this scale reading by calibrating with a standard resistance in legal ohms and a Clark cell.[113] Presumably for reasons of commercial confidentiality the authors presented no specific evidence on particular dynamo efficiencies in their IEE paper. We do learn, ironically, of the striking honesty of the Willans and Robinson company. The upshot of using this measuring equipment with time-averaging methods was that Willans decided to decrease by 1% his claims for the maximum efficiencies that could be achieved with dynamos coupled to his company's engines.[114]

[111] Willans in discussion to ibid., p. 573.   [112] Ibid., p. 574.   [113] Ibid., pp. 519–31.
[114] William in discussion to ibid., p. 575.

Despite Willans' explicit approval, Sankey's and Anderson's innovative and contentious use of such hybrid laboratory and workshop techniques provoked warm debate among the members of the IEE present on 12 November 1891. As a regular commentator on techniques of electrical measurement, W. E. Ayrton was invited by IEE President William Crookes to present a formal response to the paper of Sankey and Anderson. Ironically, whilst Ayrton was the only auditor who openly sympathized with their use of the D'Arsonval galvanometer, he was by far the most critical of the 'accuracy' they claimed for their results. Ayrton's detailed commentary was more than twice as long as the original paper of Sankey and Andersen and offered a vast amount of analysis and an overwhelming array of detailed suggestion. Initially he complimented the authors on being 'converted' from null methods to deflectional methods (as he had long recommended), especially in adopting the D'Arsonval device that had been in daily use in his South Kensington laboratories for several years. Nevertheless Ayrton challenged their claims to have reduced errors to within 0.2% and offered characteristically didactic advice on how to achieve such a degree of accuracy by using methods developed by him and his students. Ayrton also criticized Sankey and Andersen for taking insufficient precautions against resistances that were altered by mercury amalgamation at circuit connections: Such mercury should be changed daily, as Ayrton himself had been taught to do when working for Alexander Muirhead's company in 1878–9 (see Chapter 3). More generally, Ayrton proposed that the device could achieve true proportionality as he had suggested in his 1889 paper with John Perry by rehanging the D'Arsonval coil and reshaping its magnet pole pieces. Only if that were done could the scale readings be 'trusted' to the 0.2% claimed.[115]

The next respondent, R. E. B. Crompton showed his long-standing loyalty to the Willans camp by evident irritation at Ayrton's patronizing criticisms of Sankey and Anderson. From checks made by both himself and other 'eminent' firms of dynamo manufacturers he contended it was beyond dispute that Willans' staff had obtained a 'sufficient degree of accuracy' in their electrical measurements. He accused Ayrton of having 'somewhat undervalued' the paper by Sankey and Anderson by reading it as a springboard for promoting his own preferred techniques, instead of showing concern for what they had actually accomplished for dynamo manufacturers:

So many makers of high-class dynamos have sent their engines to Messrs Willans's works to be fitted to his engines, that his works have been a common ground on which dynamo makers meet, and can judge there of the comparative merit of their dynamos from the very impartial and accurate tests that Messrs Willans have now been carrying on for some years.

[115] Ayrton in discussion to ibid., pp. 535–62.

Importantly, however, Crompton conceded that the D'Arsonval galvanometer was not 'so trustworthy' as the zero or null method by which he had found results could be reliably attained to within one quarter of a per cent – albeit at the loss of instantaneous reading.[116]

Sydney Evershed, the manager of Goolden and Trotters instrument-making factory in Westminster, raised cognate issues of trustworthiness in the material culture of measurement. Given the 'unusual' degree of accuracy claimed by Sankey and Anderson, namely, to 1 part in 1000, Evershed argued they had exercised insufficient care in considering sources of errors in their instruments. They had, for example, said nothing about the temperature coefficient of the permanent magnets and springs in their D'Arsonval galvanometer – such a possible source of error would have to be investigated quantitatively before this apparently 'excellent' device could be recommended for 'workshop practice'. To this Sankey later replied that the method they had adopted 'cut across all those kinds of errors', and that this was, in fact, one of the reasons for choosing the D'Arsonval method.[117]

Like Crompton, James Swinburne congratulated Sankey and Andersen for tests which he considered to be of great value. Yet he disagreed with their choice of measurement devices and responded by invoking his cherished dichotomy of laboratory versus workshop instruments. He held that it was not reasonable of Sankey and Andersen to dismiss commercial ammeters and voltmeters as being *of themselves* insufficiently accurate for the task. All that was needed to make them accurate, he contended, was to calibrate them properly with an appropriate (non-linear) scale. Insofar as the Willans staff had chosen the D'Arsonval instrument for measuring currents and voltages for dynamos, he was somewhat ambivalent about their decision. To the extent that he was an 'electrician' involved in the work of telegraphs and delicate experimentation, he appreciated the large range and 'fair accuracy' that could be accomplished with this instrument. Swinburne did not, however, regard a reflecting galvanometer as an engineering instrument: As an engineer he objected to spot-watching.[118] From working with power circuitry he was accustomed to being able to read a number of nearby instruments at a glance simultaneously without having to adopt any contrived bodily posture. His objection seems to have been that 'reading' the scale of a reflecting galvanometer constrained the user to watch a single dancing light-spot, and this prevented the concurrent monitoring of other important instruments in the power-station.

---

[116] Crompton in discussion to ibid., p. 564–6.
[117] Evershed in discussion to Sankey and Andersen, 'Description of the standard volt- and ampere-meter', p. 569–70. Sankey in discussion to ibid., p. 576. See Evershed's obituary in *JIEE*, 85 (1939), pp. 775–6.
[118] Swinburne in discussion to ibid., p. 567.

The Willans staff's allegedly incongruous use of a reflecting galvanometer did not, however, provoke Swinburne to *distrust* their results. Quite the contrary, for insofar as their methods had been developed in the environment of workshop engineering, Swinburne considered them to be more trustworthy than Ayrton's allegedly superior laboratory-based 'improvements'. Indeed, Swinburne typically dissented from Ayrton's assumption that techniques developed in his college laboratory could necessarily be more accurate than those developed in the commercial environment of the Willans workshop. He reminded Ayrton that the *most* accurate measurements were carried out in the environment of the trade workshop, notably the commercial analysis of weights and the use of Whitworth's machines to determine lengths (see Chapter 2). As far as he was concerned, the laboratory pedigree of an instrument was no guarantee that it could produce accurate results. Ayrton's proposed improvements to Sankey's and Andersen's techniques should thus not be taken too seriously as warrantable sources of greater accuracy.[119]

The editor of the *Electrician* and Cambridge Natural Sciences graduate A. P. Trotter[120] similarly trusted the results presented by the Willans staff and concurred with Swinburne's dichotomy of laboratory versus workshop measurements, albeit without agreeing that the latter could attain higher accuracy. In other respects, Trotter's contributions to the IEE debate furnish an interesting contrast. Unlike Swinburne, Trotter argued that Sankey and Andersen had accomplished only laboratory-style measurements but without the 'high accuracy' aimed at by laboratory workers. He thus complained that Sankey and Andersen had not shown their IEE audience how their methods could be 'converted' into electrical engineering measurements. The rationale for such a conversion, according to Trotter, was that spot-watching was a practice which the engineer would adopt only with 'considerable dislike.' Indeed, the D'Arsonval galvanometer was not an instrument which had 'come to stay for engineering purposes', notwithstanding the precedent set by the various improved forms of the Thomson reflecting galvanometer. Much more fitting for engineering work he contended was a 'needle-reading' commercial instrument that could be read to 1 part in 1000.[121]

At this point Peter Willans stepped forward to defend his employees whom he had commissioned to develop new measuring methods. He contended that averaging out a fluctuating spot-reading by Sankey's method was better than either accurate but relatively uninformative 'null' methods or Ayrton's more instantaneous but naive method of placing 'trust' in each successive

---

[119] Ibid., p. 568-9.
[120] Phillip Strange, 'Two electrical periodicals: The electrician and the electrical review 1880–1890', *IEE Proceedings* 132A (1985), pp. 574–81, esp. p. 579.
[121] Trotter did, however, allude to a third intermediate category of the 'instrument maker's class' of measurement; Trotter in discussion to Sankey and Andersen, 'Description of the standard volt- and ampere-meter', pp. 571–3.

scale deflection's always meaning 'always the same thing'. And a further advantage of the spot-watching approach was that it enabled their customers to witness and corroborate the results in ways that were not possible with a null method (and by implication more difficult with a needle method). A spot of light was at least something which 'two people can see' and they could make independent records of the results at the same time, before the final averaging of half an hour of results.[122] And in response to Evershed's challenges, Willans could not say 'too much' in praise of Andersen's scrupulousness in using the apparatus designed by Sankey. Willans himself had watched very closely and ensured that nothing was taken for granted, but 'checked again and again' at every point.[123]

When Captain Sankey replied, he took Ayrton as his chief combatant and was evidently irritated by Ayrton's suggestions on how to improve the linearity of their scale readings *apparently* to eliminate secular changes'. Such improvements were simply not necessary for them at Thames Ditton[124] for, whilst Ayrton might trust his direct-calibrated devices to accomplish high accuracies, Sankey and Andersen trusted them only for ordinary measurements. For them, accomplishing the highest accuracy was not a matter of trusting the contrived scale calibration of direct-reading instruments. Rather it involved using the standard legal ohm and Clark cell,[125] in which they had 'faith' as certifiable to within the Board of Trade requirements of accuracy to within 1 part in 10,000. And they noted sharply that whilst Ayrton's methods attained results mutually consistent to within 0.1%, this degree of accuracy did not entail that such results were true (accurate) to within 0.1%.[126] The Willans team, however, tacitly acknowledged Ayrton's support concerning the propriety of the D'Arsonval instrument. They were frankly 'surprised' that Swinburne, Trotter, and Evershed did not accept this instrument for engineering work and offered both aesthetic and quantitative arguments in its favour. Andersen considered that the spot method of reading was 'simply charming', and now that oil lamps had been replaced with electric lights in the reflecting arrangement, this method was 'far superior' to the reading of currents with pointers.[127] Once engineers had worked with a good D'Arsonval galvanometer, they would not part with it in favour of instruments with pointers as readings could be taken to 'greater accuracy' with the spot than with a pointer. And this was just one of the many advantages which they considered made their device uniquely superior to ordinary commercial instruments.[128]

---

[122] Ibid., p. 517.   [123] Willans in discussion to ibid., pp. 573–5.
[124] Sankey in discussion to ibid., pp. 575–7.
[125] Crompton in discussion to ibid., pp. 565, Sankey and Andersen discussion, p. 577 and 586.
[126] Andersen in discussion to Sankey and Andersen, 'Description of the standard volt- and ampere-meter', pp. 580–1.
[127] Ibid., p. 578.   [128] Sankey and Andersen in discussion to ibid, p. 589.

The irony here then is that, as much as Sankey and Andersen won the assent of most electrical engineers to their claims for accuracy of results, the same engineers who *trusted* their results were not readily persuaded to accept the D'Arsonval galvanometer as a proper engineering instrument. The reason they did not accept it as a universal instrument had less to do with its accuracy than with the embodied reading practices of spot-watching required of its users. Five months later at the ICE, James Swinburne reiterated his commitment to workshop instruments having 'clear' index pointers, and the only D'Arsonval device which he recognized for engineering was the non-reflecting direct-reading form.[129] The continued use of the D'Arsonval devices by the Willans staff appeared to have little effect on the reputation of Willans engines for high efficiency; nor did the premature death of Willans in a horse-riding accident in May that year. If anything, the company's advertising stressed even more assertively the unique economy of Willans engines.[130] Their longer-term fate was determined rather by the increasing success of the uncontroversially more economical Parsons turbine in the mid-1890s, which Willans and Robinson turned to making by the early twentieth century.[131]

After Swinburne's public criticisms of Ayrton in 1892, the issues at stake in these disputes faded into an unresolved pluralism about how a current-measuring device should be made, calibrated, and used. This subject was not discussed again in public until Kenelm Edgcumbe and Franklin Punga presented a paper at the IEE in March 1904, suggesting that there was still a 'certain mistrust' among engineers for the instrument-maker and 'all his works'. Unless such makers had taken 'great care' in engraving or painting scales on instruments, they advised users to develop their own card or paper scales to avoid the risk of large calibration errors.[132] Nowhere did they even bother to mention the feasibility of constructing scales with equally spaced dial markings. And their only mention of mirrors concerned those installed under the pointers of direct-reading needle instruments to help users avoid parallax errors.[133]

4.6. CONCLUSION

We have seen that the preferences of Ayrton, Swinburne et al. for different sorts of measurement instruments and practice were not simply reducible

---

[129] Swinburne, 'Electrical measuring instruments', pp. 1–2 and 9–10.
[130] See advertisements, *Engineering*, 44 (21 October 1887), supplement III; 'smaller consumption of steam than ANY other type of engine in the market', 53 (1 April 1892), p. 73. I thank Giles Hudson for tracing this advertisement.
[131] Dunsheath, *A history of electrical engineering*, pp. 196–200; Parsons, *Early days of the power station industry*, p. 113.
[132] Ibid., p. 620.
[133] Kenelm Edgcumbe and Franklin Punga, 'Direct-reading measuring instruments for switchboard use', *JIEE*, 33 (1904), pp. 620–55, 655–93, esp. p. 622.

to pragmatic or cultural 'interests'. Equally important were their practical working contexts – Ayrton's academic laboratory work, Swinburne at the lighting installation and Willans in the engine-testing room. The material hazards, temporal rhythms, and commercial constituencies[134] for these places posed different constraints on the spatial and bodily configurations into which current-measuring instruments had to be integrated. These constraints did not *determine* their choice of current-measurement technique, however. What also mattered was a practitioner's familiarity with techniques adopted in previous working contexts, such as testing telegraph cables in the case of William Ayrton and mechanical engineering in the case of James Swinburne. Thus the heterogeneous origins of electrical engineering engendered a persistent heterogeneity of instrument-reading technologies.

From this heterogeneity, we see that judgements of the numerical trustworthiness of instrumental measurement were tied in complex ways to the judgements of the instrument used. Although some doubts were raised about the trustworthiness of the D'Arsonval galvanometer, the primary issue for some was not its capability for producing the most accurate measurements. Dynamo manufacturers could accept the accuracy of its results even whilst questioning the contextual appropriateness of the light-spot reading technology that had hitherto served so well in telegraphic work. What mattered in carrying conviction in claims about accuracy of results was the trustworthiness of Willans and Robinson as the most open and honest company involved in testing the efficiency of generating equipment, which overcame distrust in the reading technology of new measuring devices. And even if other electrical experts at the IEE sought to outdo each other to identify untrustworthy elements in the reading technology of Willans' favoured instrument – or of others – many of them still bought their steam-engines from this company. Then again, most of them hardly contemplated using the supposedly universal D'Arsonval galvanometer anywhere outside the testing room.

[134] van Helden and T. L. Hankins, 'Introduction', p. 5

5

Coupled Problems of Self-Induction: The Unparalleled and the Unmeasurable in Alternating-Current Technology

The introduction of powerful alternate current machines by Siemens, Gordon, Ferranti, and others, is likely... to have a salutary effect in educating those so-called practical electricians whose ideas do not easily rise above ohms and volts. It has long been known that when the changes are sufficiently rapid, the phenomena are governed much more by induction, or electric inertia, than by mere resistance. On this principle much may be explained that would otherwise seem paradoxical.

Lord Rayleigh, Presidential Address to the BAAS, Montreal, 1884[1]

[T]he secohmmeter does not measure all that can be measured. That is true.

Ayrton and Perry, Discussion at the STEE, 1887[2]

Preceding chapters discussed how the industries of submarine telegraphy and electric lighting prompted concern for measuring electrical properties and were the main audiences for new technologies that measured electrical resistance and direct current, respectively. This chapter extends this theme to consider the emergence of self-induction as an important but troublesome parameter in the technology of ac generation developed in the late 1880s. My main focus is the fate of the 'secohmmeter', an instrument designed by William Ayrton and John Perry to measure self-induction. Hitherto practitioners had often treated self-induction as a source of major errors in measurements of electrical resistance. For example, in 1881 Lord Rayleigh diagnosed that neglect of self-inductive effects in the operation of 'whirling-coil' apparatus had lead to an error of more than 1% in the BAAS Electrical Standards Committee's determination of absolute resistance fifteen years earlier (Chapter 3).[3] Investigations of self-induction that came soonest after Rayleigh's were mostly inspired by new electrical technologies for handling

---

[1] Lord Rayleigh, 'Address', *BAAS Report* (1884), Part 1, quote on pp. 8–9.
[2] William E. Ayrton and John Perry, 'Modes of measuring the coefficients of self and mutual induction', *JSTEE*, 16 (1887), p. 391.
[3] Lord Rayleigh, 'Experiments to determine the value of the British Association unit of resistance in absolute measure', *Philosophical Transactions of the Royal Society*, 173 (1882), pp. 661–98. See discussion on whirling-coils in Simon Schaffer, 'Late Victorian metrology and its instrumentation: A manufactory of ohms', in R. Bud and S. Cozzens (eds.), *Invisible connections: Instruments, institutions and science*, Vol. IS09 of the SPIE Institute Series, Bellingham, WA: SPIE, 1992, pp. 24–55.

rapidly changing currents, namely lightning conductors and telephone lines.[4] Debate subsequently emerged about whether self-induction was an unfortunate 'evil' to be expunged from the construction of variable current technologies or a benign feature that could be optimized to accomplish technological control and stability. A later feature of these debates was a paradoxical concern about whether this most unequivocally 'real' electromagnetic parameter could ever be subjected to meaningful measurement. Participants in debates discussed in this chapter by no means subscribed to William Thomson's 1883 declaration that measurability was in all such cases a prerequisite of proper knowability.

By 1889, concerns to quantify self-induction featured prominently in discussions about whether ac power generation could be made practically feasible enough to meet the concerns of engineers, shareholders, and consumers. The vexing question – generally neglected in modern histories of ac power – was whether ac generators (alternators) could be made to run in parallel as safely and economically as dynamos in the rival dc system (Section 5.1). I consider the controversy from 1888 to 1893 over whether the magnitude of self-induction in an alternator was of critical importance in achieving parallel operation. Some denied this possibility; those who considered it to be important tended to adopt the terms of earlier debates on self-induction in lightning conductors and long-distance telephony: Should the 'choking' effects of self-induction be eliminated, or should its value be optimized for most efficient working?[5] As Jordan and Yavetz have noted, Ayrton and Perry developed the direct-reading instrument 'secohmmeter' in 1885–6 to attempt to measure the self-induction for these new technologies. In Sections 5.3–5.5, I show how they and Ayrton's students also tried in 1889 to apply the secohmmeter to solve the problem of paralleling in ac power technology. It was specifically the debate over the readings taken with this instrument that prompted public debate over the question of whether the self-induction of moving alternator components could actually be subject to trustworthy measurement.

The specific quantitative question of how much self-induction (if any) was necessary for alternators to run in parallel was linked to the question of how

---

[4] Dominic W. Jordan, 'The adoption of self-induction by telephony, 1886–89', *Annals of Science*, 39 (1982), pp. 433–61; idem, 'D. E. Hughes, self-induction and the skin effect', *Centaurus*, 26 (1982), pp. 433–61; Hunt, '"Practice vs Theory": The British electrical debate, 1888–91' *Isis* (1983), pp. 341–55; Ido Yavetz, 'Oliver Heaviside and the significance of the British electrical debate', *Annals of Science*, 50 (1993), pp. 135–73. For a comparable discussion of the role of self-induction in ac machinery in the USA see Ronald Kline, 'Science and engineering theory in the invention and development of the induction motor', *Technology and Culture*, 28 (1987), pp. 283–313.

[5] See works by Hunt, Jordan, and Yavetz, ibid. The question was whether self-induction was a deleterious choking quality to be expunged from electrical conductors or used strategically to prevent damaging current surges in lightning conductors and to minimize signal loss in long-distance telecommunication. Self-induction was later dubbed 'inductance' or 'self-inductance' by Oliver Heaviside.

Coupled Problems of Self-Induction

self-induction could be physically characterized, a topic not unequivocally resolved in the writings of James Clerk Maxwell (see Section 5.2). Following hints in Maxwell's *Treatise on Electricity and Magnetism* of 1873, Lord Rayleigh told the 1884 BAAS meeting that electricians and engineers could treat self-induction simply as the electrical analogue of mechanical inertia (see the epigraph that opens this chapter). Yet the important kindred phenomenon of mutual induction between unconnected current-carrying conductors posed a problem. Maxwell himself acknowledged that some important part of such inductive effects was located in the fields around and between such conductors. After all, why did the mere *proximity* of iron seem to make such a difference to the magnitude of a conductor's self-induction? And if there were any iron *moving* nearby, was it ever legitimate to claim that a coefficient of self-induction had a value that could meaningfully be approximated to a 'constant' and thus measured with a secohmmeter? Even when no effects of mutual induction between currents were apparent, was consideration of self-induction *sufficient* to characterize the relation between potential differences and changing currents? Or should sophisticated readers of Maxwell have concluded that the coefficient of electromagnetic self-induction was merely one of a series of coefficients needed to represent how time-varying currents generated a potential difference? Then again, if a conductor's coefficient of self-induction was dependent solely on its geometrical configuration, as Maxwell clearly indicated, why did it sometimes also seem to depend on the material composition of the conductor? These were tricky questions with no simple consensual answers.

Among all the British discussants in the late 1880s discussed in this chapter, only four figures seemed to have had clear views on these complex matters. The Cambridge Senior Wrangler-turned-dc-engineer, John Hopkinson, was convinced that the self-induction in alternator armatures could be treated as analogous to mechanical inertia and thus *approximately* a constant. This simplification was a central feature of his somewhat disputed theoretical calculations concerning alternator behaviour developed in 1883-4 (Sections 5.1 and 5.2). In 1888 the Austrian expatriate electrical engineer Gisbert Kapp claimed he had solved the paralleling problem by producing alternators that had a substantial amount of self-induction (Section 5.4). To this came the reply in 1889 by the Brush company engineer, William Mordey, that he had solved the paralleling problem by producing alternators that had effectively zero self-induction in their armatures (Section 5.5). At the height of this debate, the aristocratic James Swinburne argued that it was meaningless to use such simplistic terms as coefficients of 'self-induction' and 'mutual induction' to capture the complex behaviour of rotating ac machinery, and thus measurement results attained with a secohmmeter were entirely spurious (Sections 5.4-5.6). In my conclusion, I show how electrical engineers solved the paralleling problem, but not – *pace* Rayleigh – by reference to the subject of self-induction.

Before exploring the details of that debate, let us consider the wider context of power generation and why the problem of ac paralleling mattered so greatly.

5.1. 'WE DO NOT COUPLE MACHINES': THE TRIBULATIONS OF AC PARALLEL RUNNING

The public who through long winter evenings, and longer London fogs, sit reading by the cool and steady light of their electric lamps, but who are most indignant if by any chance it flickers or fails them, do not realize how intense the struggle has been for those pioneers of electric lighting who have toiled so hard and incessantly to surprise yet one more of Nature's secrets ... Many an engineer's wife knows how common it was four or five years ago for their husbands, who had come back late at night worn out and exhausted, to be fetched again by the message that there was 'something wrong at the works.'

Mrs J. E. H. Gordon, *Decorative Electricity*, 1891[6]

The parallel operation of generators is a curiously neglected topic in the historiography of electric lighting. In the 'battle of the systems' fought out between the protagonists of ac and dc technologies in Europe and the USA during the last two decades of the nineteenth century, the feasibility of each depended on a successful resolution of the 'paralleling' question. In his magisterial *Networks of Power*, Thomas Hughes deals with this issue only for dc lighting technology, and thus understates the broader significance of the problem for power and lighting systems. Hughes notes that Edison planned his distribution system to have all customers' lights wired in parallel so all would have the same potential difference across their lamps. That way – at least in theory – all lamps would be equally bright.[7] And as more customers switched on their lights, Edison's company station engineers could meet the rising demand for power by running extra dynamos (or batteries)[8] in parallel to supply the additional current required at the same voltage. This was crucial for coping with fluctuating consumers' demands without overloading and burning out generating equipment: That caused the effects that Mrs Gordon noted were anathema to domestic consumers (see the epigraph that opens this section). Although Edison experienced a few problems in synchronizing dynamos when he launched his New York dc supply station at Pearl Street in autumn 1882, Hughes shows how he soon solved these problems by linking together the governors on the steam engines driving them.[9]

---

[6] Mrs J. E. H. Gordon, *Decorative electricity*, London: 1891, pp. 153–4.
[7] Thomas P. Hughes, *Networks of power*, Baltimore: Johns Hopkins University Press, 1983.
[8] Ibid., p. 85.
[9] Hughes reports that this problem inspired Edison in November 1882 to replace Porter–Allen engines with Armington and Sims engines as a permanent solution to the problem. Hughes, ibid., p. 43. For discussion of the 'hunting' problem, see subsequent discussion.

Coupled Problems of Self-Induction 177

Although Hughes gives brief consideration to Edison's initial problem with parallelism in dc supply, he does not acknowledge the longer-term difficulty for ac suppliers. For ac suppliers the problems in synchronizing alternators in parallel were much more widespread and more difficult to solve – much to the gratification of rival dc suppliers.[10] The tacit inference here is that Edison's ac rival, Westinghouse, encountered no problems in this regard when his ac supply was launched in the USA during 1887. Yet an earlier historian of electricity supply, R. H. Parsons, emphasizes that the apparent impossibility of getting alternators to run in parallel was a major feature of the battle of the systems – at least until ac paralleling was widely accomplished circa 1893-4.[11] Before that, ac suppliers in Britain typically used one very large alternator for all customers on a circuit, no matter how few there were. This was because attempts to run extra, smaller alternators in parallel in order to meet rises in demand commonly led to disastrous current surges. These generally did substantial harm to generators, light bulbs, and the all-important consumer good will. Such at least was the experience of J. E. H. Gordon, who from 1884 to 1886 supervised Britain's first public ac installation in Britain at Paddington in West London.[12] During a discussion paper at the STEE, two years after completing this work, Gordon related his own company's experiences of attempts to connect alternators in parallel:

Many of us have tried . . . but they do not work together till they have run for three or four minutes; they will in that time jump, and that jumping will take months of life out of 40,000 lamps. That alone is a rather serious difficulty in coupling machines together, and I think we may take it in practice – I am not speaking about the laboratory or experiments – we do not couple machines.[13]

Gordon was not alone in finding that various chaotic effects occurred immediately after attempts were made to connect two alternators in parallel. Other engineers found some alternators could not be made to synchronize

---

[10] An editorial note in the *Electrician* in August 1889 analysed the situation thus: Those whose interests lay in the commercial development of dc working naturally held the 'very strong' opinion that attempts at parallel working in ac technology had 'never got beyond large scale workshop experiments'. See 'Alternate current working', *Electrician*, 24 (1889), pp. 324-5.

[11] Robert H. Parsons, *The early days of the power station industry*, Cambridge: Cambridge University Press, 1940, pp. 142-50.

[12] Ibid., pp. 42-51.

[13] See Gordon's contribution to the discussion of Rookes E. B. Crompton, 'Central station lighting: Transfomers vs. accumulators', *JSTEE*, 17 (1888), pp. 195-6. For a variant reporting of the discussion by the *Electrician* staff, see *Electrician*, 20 (1887-8), pp. 634-7, 655-6, discussion on pp. 656-7, 694-8, 749-52, and *Electrician*, 21 (1888), pp. 88-93. Gordon's Paddington installation was undertaken on behalf of the Telegraph Maintenance and Construction Company. When Telcon decided to discontinue its electrical-lighting projects in 1887-8, Gordon became chief engineer to the newly founded Metropolitan Supply Company; Anonymous, 'Obituary: James Edward Henry Gordon', *Electrician*, 30 (1893), pp. 417-18.

phases at all, one machine undergoing violent and potentially damaging current surges until it effectively ceased to operate. Other machines could be made to come close to synchronicity but showed the phenomenon later called 'hunting'. This was the persistent perturbation of coupled machines in and out of synchronization that led to irregular surges and drops in voltage. This was also the phenomenon that bothered Edison's dc stations in 1882 and which he had diagnosed as a problem in steam governance; ac workers seem, however, to have been unaware of both Edison's problem and this diagnosis. Because these effects tended to lead to the sorts of flickering and failing of lights which Mrs Gordon noted was so prejudicial to the interests of her husband's customers, many ac workers gave up attempts at parallel running. Dissatisfied with the many problematic elements of ac practice when working for the Telegraph Construction and Maintenance Company, even Gordon himself soon reverted to the rival technology of direct currents for the Metropolitan Supply Company. Within Hughes's quasi-military terminology for ac suppliers, the problem of paralleling was thus a 'reverse salient', a problem that was so great that for a while it severely limited incursion of ac technology into the commercial territory early occupied by dc suppliers.[14]

Nevertheless, some engineers such as Sebastian Ferranti persisted with the technology of ac supply because it offered the possibility of large-scale economic advantages in long-range transmission. The system of step-up and step-down transformers unique to ac technology meant that power could be transmitted to dispersed customers over large distances at very high voltage and very low current. This decreased the economically significant energy losses in transmission cables that varied as the square of the current. Owing to the conviction that comparable economic advantages of parallel operation could never be effectively attained with ac power, however, many early ac stations in Britain were set up without any facility for connecting alternators. Following Gordon's example of avoiding parallel operation at Paddington in 1886, ac generating stations in Cambridge (1891) and Scarborough (1893)[15] operated a system of switching some customers from one alternator circuit to a second independent machine as demand rose at peak times, such as the early evening. This was by no means a uniformly satisfactory alternative to parallel operation. In relation to the issues of instrument-reading raised in the previous chapter, it is significant that one trade journal noted in 1891 that such a system required power-station attendants to engage in 'incessant' watching of circuit ammeters to ensure that individual alternators did not

---

[14] Ibid. Ferranti's scheme for the enormous unparalleled alternators at Deptford station and its long-distance ac working across London became a prime target for the derision of the direct current lobby; John F. Wilson, *Ferranti and the British electrical industry*, Manchester: Manchester University Press, 1988, p. 38.

[15] Parsons, *Early days of the power station industry*, pp. 146–7.

get dangerously overloaded. Worse still, the switching of consumers circuits between separate alternators apparently brought a 'continual' blinking of the lights. Such station attendants would apparently 'fondly imagine' that they handled the switches with such dexterity that consumers would fail to notice such extinctions. Householders were evidently not oblivious to this, however.[16]

In formulating his radical plan for a long-distance ac supply at Deptford in 1891, Ferranti adopted the older strategy of using single huge generators for all customers to avoid the perils of paralleling. This attracted considerable derision from the dc lobby, especially when such machines broke down – not a rare occurrence – for all customers' lights would be extinguished until they could be mended or substituted. Such problems were among many that contributed to Ferranti's dismissal as chief engineer at Deptford in 1891.[17] Ironically, however, before he had even begun planning that installation, new forms of alternators were being developed that were allegedly much more amenable to parallel operation than his own machines. At the very STEE meeting in 1888 at which Gordon had declared his apostasy, the recent convert to ac supply, James Swinburne, contrasted Ferranti's alternators with those of his competitors:

The question of coupling of alternating machines in parallel cannot be discussed generally; it depends entirely on the machines. Some of these can only be coupled in parallel when doing no work, such as the Ferranti. Other types can be made which will go together perfectly with large variations of load without getting out of step. The Westinghouse is about the medium, and those run very well in parallel.[18]

By the following year, there were further reports that outside Britain certain kinds of alternators were working effectively when coupled. Zipernowski had succeeded in parallel running his alternators in Rome, Livorno, Frankfort-am-Main, and Marienbad, as had the Westinghouse company in the USA.[19] Yet even for the growing number of ac suppliers who managed to couple their alternators without mishap, there was still disagreement about what made paralleling possible for these machines and not others. When accounting for 'successful' paralleling it was unclear, for example, how much explanatory weight should be given to the role of electromagnetic factors

---

[16] [Anonymous] 'Mordey-Willans combination in parallel running', *The Electrician*, 27 (1891), pp. 298–9, quotes on p. 289.
[17] See Wilson, *Ferranti and the British electrical industry*; Parsons, *Early days of the power station industry*, p. 138.
[18] Crompton, 'Central station lighting', discussion on p. 401. Until about 1886, Swinburne was employed at Crompton's work in Chelmsford. At this meeting Swinburne admitted that it was probably he who had fuelled Crompton's scepticism about the possibility of running alternators in parallel.
[19] W. M. Mordey, 'Alternate current working', *JIEE*, 18 (1889), pp. 583–613, discussion on pp. 613–88, 584–8. Zipernowski's machines operated at a frequency of 42 cycles/s.

such as self-induction and how much to the mechanical issues of steam supply. In Section 5.3, I look at the quasi-mechanical theory of ac machines that John Hopkinson developed in 1883–4 to argue for the possibility of running alternators in parallel, and in later sections I assess his retrospective claim in 1889 that this controversial theory explained the secret of paralleling. Before that, however, I examine the pedigree of Hopkinson's claim that, in analysing ac machines, their self-induction could be treated as simply analogous to mechanical inertia.

5.2. THE PROBLEMATIC 'INERTIAL' ANALOGY: MAXWELL'S ACCOUNT OF SELF-INDUCTION

It is difficult, however, for the mind which has once recognised the analogy between the phenomena of self-induction and those of the motion of material bodies, to abandon altogether the help of this analogy, or to admit that it is entirely superficial and misleading.
James Clerk Maxwell, *Treatise on Electricity and Magnetism*, Vol. 2, 1873[20]

For mid-nineteenth-century physicists and electricians it was by no means a self-evidently appropriate strategy to treat electromagnetic phenomena as analogous to those familiar from mechanics. In the researches of Michael Faraday and Joseph Henry in the early 1830s, 'electromagnetic induction' emerged as having spatial characteristics hitherto unseen in the interactions of ordinary material bodies. In his first series of experimental researches on electricity published in 1831, Faraday noted that electrical currents could be induced differentially by the *change* of current in, or physical movement of, other current-carrying wires. Such was the peculiarity of mutual-induction that instantaneous changes of current could be achieved without any direct contact between interacting conductors. In 1834, Faraday followed up these observations with a study of circuits containing electromagnets in his article 'On the Influence by Induction of an Electric Current on Itself'. His starting point was the well-known phenomenon that a spark would be observed – or a shock felt – when a physical break was made in a circuit containing an electromagnet. Initially, he considered this to be due to the way in which 'electricity circulates with something like momentum or inertia in the wire'. The force of the analogy was this: Trying to force a current to stop suddenly by breaking its path was not unlike trying to stop the movement of an object very suddenly. The dramatic shocks or sparking effects in wires were analogous to the jolting reaction experienced in the latter case. At first Faraday considered this inertia of current to be a simple matter of physical quantity of a conductor carrying a larger quantity of current. The longer the wire,

---

[20] James Clerk Maxwell, *Treatise on electricity and magnetism* 2 Vols., Oxford: Clarendon, 1873, Vol. 2, p. 181; 3rd edition, 1891, p. 196.

the greater the apparent sparking effect, and indeed long wires produced sparking effects that a short wire could not.

Yet the importance of spatial factors led him to reject the notion that such effects simply were a result of the current's 'momentum'. Such explanations, wrote Faraday, were 'at once set aside' by the fact that a length of wire was not the sole determinant of the inductive effects. A given length of wire would produce greater sparking effects if it were wound up into a helix, and much greater still if this helix were made into an electromagnet by the insertion of an iron core. Thus the inductive effect of a current on itself was not a fundamental characteristic of the current; rather this effect was governed by both the proximity of iron and the spatial relation of different parts of the conductor to each other. Accordingly, post-Faradayan commentators attributed the characteristic of self-induction not to the currents but to the *configuration* of the conductors. In this construal, the disanalogy between electromagnetic induction and physical inertia was obvious: The inertial properties of mechanical bodies did not change if their constituent parts were rearranged into a different configuration, or if other bodies came into close (non-contiguous) proximity.[21]

For the four decades after work by Faraday and Henry, however, interest in the curious phenomenon of self-induction was largely confined to mathematical practitioners of natural philosophy. Apart from Robert Kirchhoff's theoretical discussion of the effects of self-induction in 1857,[22] few dedicated much time to analysing its significance. William Thomson largely discounted it in his contemporaneous theory of submarine telegraph signalling. In analysing what governed the movement of a signal pulse along a long submarine cable submerged in the dielectric medium of seawater, Thomson argued that the tiny self-induction of the cable was of far less significance than its resistance and capacitance. And as Dominic Jordan has shown, this well-vindicated Thomsonian position served as a useful precedent for later figures to downplay the significance of self-induction in new electrical technologies. Indeed, it was one of the arguments used, for example, by William Preece in 1886–8 against Oliver Heaviside, S. P. Thompson, and others, who argued that an optimum value self-induction of telephone cables was crucial for preserving signal form in long-distance telephonic communication. Following Thomson's telegraphic exemplar, Preece argued that the self-induction was a deleterious quality that should, as far as possible, be eliminated from such technologies.[23] Advocates of the importance of self-induction thus had to

---

[21] M. Faraday, *Experimental researches in electricity*, London: 1839, Vol. 1, articles 1–139, pp. 1–41 (1st series, 1831), articles 1048–1118, pp. 322–43 (9th series, 1834). Quotation from article 1077, p. 330.

[22] Jordan, 'The adoption of self-induction by telephony', pp. 436–8. On Kirchhoff, see C. Jungnickel and R. McCormmach, *The intellectual mastery of nature: Theoretical physics from Ohm to Einstein*, 2 Vols., Chicago/London: University of Chicago Press, Vol. 1, 1986.

[23] Jordan, 'The adoption of self-induction by telephony'.

look outside the Thomsonian corpus to gain guidance on how to analyse this subject.

Anyone wishing to learn about the theory and measurement of self-induction and mutual-induction before circa 1880 had to turn instead to the works of James Clerk Maxwell, specifically his *Treatise on Electricity and Magnetism* of 1873. This work was not as little read as some have supposed. Several academic electrical engineers were quite familiar with Maxwell's treatment of this subject and recommended it to both their students and readers of their textbooks. Silvanus Thompson alluded to it in the appendices of his 1884 textbook on dynamos, and, in introducing the secohmmeter to the STEE in 1887, William Ayrton described the *Treatise* as the electrician's 'standby'. Indeed, Ayrton noted that Maxwell's book contained a complete exposition of the principles of self-induction and several – albeit rather impractical – means of measuring it. John Ambrose Fleming, Maxwell's last student at Cambridge before he died, was also a devotee of this work, as was Oliver Heaviside – whose contemporary work on resistance optimization had been cited in Maxwell's *Treatise*.[24]

Nevertheless, readers of Maxwell's *Treatise* would have had to work hard to extract a coherent account of self-induction and mutual-induction from the dispersed and rather heterogeneous discussions it covered. They would have found a few scattered references to it under the appellation 'coefficient of self-induction' or the more cumbrous 'coefficient of electromagnetic capacity of self-induction'. However, Maxwell offered them no explicit cross referencing to help them link the separate discussions together.[25] They would,

---

[24] Silvanus P. Thompson, *Dynamo-electric machinery: A manual for students of electrotechnics*, London: 1884, esp. p. 387, but compare discussion on mutual induction on p. 388; Ayrton and Perry, 'Modes of measuring the coefficients of self and mutual induction', p. 293–4. See Ayrton's remarks at the STEE in 1886 in discussing a paper on dc dynamos, viz., 'Those present who are skilled mathematicians, and have their Clerk-Maxwells at their fingertips. Gisbert Kapp, 'The predetermination of the characteristics of dynamos', *JSTEE*, 15 (1886), pp. 518–30, discussion, on p. 530. Some of John Ambrose Fleming's notes on the comparison of coefficients of self-induction in Maxwell's very last lecture were interpolated in the third edition of his *Treatise*: Maxwell, *Treatise*, 3rd edition, 1891, Vol. 2, p. 396. In collaboration with his telegraph engineer brother Arthur, Oliver Heaviside also referred to Maxwell's book in preparing some highly technical analyses for the *Philosophical Magazine* and the *Journal of the Society of Telegraph Engineers* – see Bruce Hunt, *The Maxwellians*, Ithaca, NY/London: Cornell University Press, 1991. For Heaviside's work, see discussion in Yavetz, 'Oliver Heaviside and the significance of the British electrical debate', pp. 135–73. Maxwell cites the youthful Oliver Heaviside's analysis of the 'best' resistances to use in a Wheatstone Bridge in the *Philosophical Magazine*, February 1873, in Maxwell, *Treatise*, 1891, 3rd edition, Vol. 1, p. 482. For readership of Maxwell, see Hunt, *The Maxwellians*, pp. 13–14, and Andrew Warwick, *Masters of theory: Cambridge and the rise of mathematical physics, 1760–1930*, Chicago: University of Chicago Press, 2003, pp. 286–356.

[25] Maxwell, *Treatise*, 1st edition, Vol. 2, pp. 354–7, 377–82; 1891, 3rd edition, p. 395, pp. 425–30. Many other parameters go under the rubric of L in this book, and correlatively the symbol for current is variously $x, y, C, \dot{y}$, or γ. The use of the term L for inductance or self-inductance in the mid-1880s was largely due to Heaviside.

moreover, have found a considerable ambiguity about whether they should interpret self-induction as a form of electromagnetic 'inertia'. Self-induction is first introduced in the *Treatise* under the quasi-Faradayan heading 'On the Induction of a Current on Itself', and Maxwell faithfully summarized Faraday's observations on sparks in broken circuits, as previously discussed. However, he also expanded Faraday's preliminary discussion of the inertia of currents by using a hydrodynamic analogy: The shocking effect of trying to stop a current was akin to the jolt experienced in trying to bring the flow of water in a pipe to an instantaneous halt.[26]

Yet Maxwell then also followed Faraday in arguing for the disanalogy between electromagnetic and mechanical-inertial phenomena. He reiterated the point that the inertial effects of electrical currents were not confined to the conductor in the same way that the inertia of water was confined to the material liquid in the pipe. The inertia of water in a pipe was unaffected by either its shape or adjacent objects, whereas the self-inductive quality of a wire was strongly dependent on its geometrical form and the proximity of iron. Thus factors *outside* the wire were at least as important as those within. Alluding to his highly mechanized 'dynamical theory of the electromagnetic field',[27] Maxwell hinted that there might be 'some motion' going on in the space around the wire in which the electromagnetic effects of current were manifested. It was this phenomenon of the electromagnetic field that manifested the 'retarding' effects apparent in both the mutual-induction between two separate circuit currents and the self-inductive action between different elements of the same current.[28] Later in the *Treatise*, Maxwell thus showed how the self-induction and mutual-induction of conductors could be calculated from their geometrical properties of shape and separation.[29] By implicit contrast to the phenomenon of resistance, these attributes of conductors were *not* affected by temperature or impurities or dependent on the particular material used; only in the third edition of the *Treatise* did the editor J. J. Thomson identify iron as an exceptional case in this regard.[30]

For the point of view of electrical engineers reading Maxwell's *Treatise* in the 1880s, it is notable that the book offered formulae only for the limited cases of static laboratory apparatus in which the self-induction or mutual-induction was a well-defined constant. Maxwell gave no equations for cases involving changing coefficients of self-induction such as, for example, the inductive interactions between conducting circuits and iron that were in *relative motion* to each other. Presumably Maxwell had no great interest in these because they had no obvious practical bearing for his technological

---

[26] Faraday, *Experimental researches*, article 1077, p. 330.
[27] See Peter Harman, *The natural philosophy of James Clerk Maxwell*, Cambridge: Cambridge University Press, 1998, pp. 113-24.
[28] Maxwell, *Treatise*, Vol. 2, p. 181; 3rd edition, p. 196.
[29] Maxwell, *Treatise*, Vol. 2, pp. 278-98; 3rd edition, pp. 306-330.
[30] John Joseph Thomson's editorial footnote in Maxwell, *Treatise*, Vol. 2, pp. 295-6; 1891, 3rd edition, pp. 323-5

interests in telegraphy, and the equations were probably too complex to solve neatly by his Cambridge analytical methods anyway. But in the years immediately after Maxwell's death in 1879, the case of variable self-induction was entirely germane to the problems facing ac engineers trying to understand the behaviour of alternators. As we shall see, one strategy adopted by ac theorist John Hopkinson was simply to deem the complexities of such cases to be unimportant and to treat the self-induction of moving machinery as if it were as well-defined a quantity as the inertia of a mechanical body. Indeed, Maxwell's *Treatise* could on a liberal reading be interpreted as licensing just such a move. Having recognized the analogy between self-induction and mechanical inertia, Maxwell observed that it was 'difficult' to abandon altogether the 'help' it offered or admit that it was 'entirely superficial and misleading'.

For Maxwell, the fundamental idea of matter in motion as the bearer of energy and momentum was 'so interwoven with our forms of thinking', that a glimpse of it in any subject gave one the sense that it furnished the path to 'complete understanding'.[31] Thus although Maxwell did not unequivocally subscribe to the full force of the analogy between mechanical inertia and self-induction, he did use it selectively as a heuristic to help his readers understand the development of his dynamical theory. For a pair of interacting circuits with currents $\dot{y}_1$ and $\dot{y}_2$ and self-inductions $L_1 + L_2$, respectively, with M representing the coefficient of mutual induction, the electrokinetic energy would be $T = \frac{1}{2} L_1 \dot{y}_1^2 + \frac{1}{2} L_2 \dot{y}_2^2 + M \dot{y}_1 \dot{y}_2$. Whilst terms like $\frac{1}{2} L \dot{y}^2$ were explicitly likened to that of the kinetic energy of a material body ($\frac{1}{2} mv^2$), the mixed term $M \dot{y}_1 \dot{y}_2$ indicated significantly how much the energy of the circuits was locked up in their mutual-inductive interactions outside the wire – a telling divergence from the analogy with mechanical inertia.[32] Then again, elsewhere in Maxwell's account it transpired that, even where there were no effects of mutual induction, the coefficients of self-induction and resistance might not always be sufficient to capture the complexity of inductive effects. Maxwell noted that such complexity might arise from the non-uniform distribution of currents even in the most uniform cylindrical wire and higher- order effects of the time variation of the current. Thus he formulated the following general equation for the potential difference E between the ends of a wire of length and resistance R would, in general, be dependent on the current C as given in the following infinite series:

$$E = RC + l(A + \tfrac{1}{2})\frac{dC}{dt} - \frac{l^2}{12R}\frac{d^2C}{dt^2} + \frac{l^3}{48R^2}\frac{d^3C}{dt^3} - \frac{l^4}{180R^3}\frac{d^4C}{dt^4} + \&c.$$

Whilst the term RC represented the conventional term for resistive potential difference, in Maxwell's interpretation the term $l(A + \tfrac{1}{2})$ represented the

[31] Ibid., p. 181, 3rd edition, p. 196.   [32] Ibid., pp. 206–7; 3rd edition, p. 223.

Coupled Problems of Self-Induction 185

self-induction L and thus the term $l(A + \frac{1}{2})dC/dt$ represented the part of the electromotive force which 'would be employed in increasing the electrokinetic momentum' of the circuit. The higher-order terms, not usually discussed in conventional accounts, were interpreted by Maxwell to constitute the 'correction' that would ideally be implemented to allow for the nonuniformity of the current distribution, most of the current being crowded near to the surface of the wire.[33] Maxwell devoted no further discussion to these non-linear terms and certainly did not consider them of sufficient general importance to develop methods of measuring them.[34] Indeed, the sections of the *Treatise* that Maxwell devoted to measuring inductive phenomena dealt with only first-order effects of self-induction and mutual induction. Nevertheless, as one of Maxwell's readers in the 1880s, William Ayrton noted that effects representing the higher-order terms in Maxwell's equation could impinge importantly on attempts to measure coefficients of self-induction (see subsequent discussion).

Given the subtlety and selectivity with which Maxwell used the inertial analogy as heuristic guide in the *Treatise* of 1873, it is rather ironic that Maxwell's most widely known representation of self-induction appears to have been a purely *mechanical* model. (Figure 8). He used this in illustrative teaching work at the Cavendish Laboratory in Cambridge, and it appeared only in the third posthumous edition of the *Treatise* edited by his later successor, J. J. Thomson. This device reified the analogy previously discussed, showing how the *differential* action of self-induction could be represented as a form of mechanical inertia. If the user turned the main handle on the rotating pulley – disc at a constant speed (representing a constant current) – nothing of great interest would occur. Only if the user tried to *change* the rotational speed of the disc (representing a change in current) would another linked disc start to turn in the opposite direction by means of a differential gearing. The rotation of this latter disc represented the reverse current 'induced' by the changing current in the first and would cease only once the first disc attained a constant speed again. Crucial to the inertial representation of self-induction was the operation of an adjustable weighted flywheel placed between the discs. When a differential motion was set up representing a changing current, this flywheel would rotate and its angular momentum would represent the effect of self-induction on this process, seeming to act *against* any attempts to change its rotational motion.[35] By adjustment of the position of weights on the wheel, different magnitudes of self-induction could be represented.

The analogical lesson of this device was that only when currents changed would self-induction become manifest as a form of electromagnetic resistance

---

[33] Ibid., pp. 292–3, 3rd edition, p. 322.
[34] See Jordan, 'D. E. Hughes, self-induction and the skin effect'.
[35] See the reproduction of this model interpolated in the third edition of Maxwell, *Treatise*, Vol. 2, 1891, p. 228 – probably identical to that used by Hopkinson.

to changing currents. It hardly need be said that this model aimed neither to capture the way in which rearranging the spatial form of a conductor would change its self-induction nor the comparable effect of the proximity of iron. Nevertheless, it was this very device that John Hopkinson used to explain the phenomenon of self-induction to the ICE in April 1883[36].

5.3. SELF-INDUCTION AS MOMENTUM: JOHN HOPKINSON'S THEORY OF AC PARALLELING

There are really two ways of looking at self-induction, and they both come to the same thing and give precisely the same results. One way is that in which I think Faraday looked at it. Faraday considered that the circuit itself produced a magnetic field, the variations of which had an effect as electromotive force upon that circuit. Another way of looking at it is that which Clerk Maxwell introduced, of treating the current as if possessing the property of momentum . . . When it is a case in which there are iron magnets in the neighbourhood of the circuit, I am inclined to think that the old method of Faraday is more convenient, but I really doubt whether really the wisest plan is not to be ready to adopt either mode of expressing the facts, as may be most convenient as the cases arise.

       John Hopkinson, Discussion of 'On the Theory of Alternating Currents' at the STEE, 1884[37]

The first general theory of alternator operation in Britain was developed by John Hopkinson and presented in two papers in 1883–4. Hopkinson's lecture on electric lighting delivered to the ICE in 1883 was perhaps the earliest public challenge to the prevailing wisdom among engineers that ac paralleling was not generally possible. As Senior Wrangler in the Cambridge Mathematics Tripos of 1871, Hopkinson's views on this were not easily dismissed, particularly as he noted with a judicious commercial observation that a 'great deal' turned on the possibility of ac paralleling. Nevertheless, his audience might still have been rather surprised to hear him offering advice on that subject. At the lecture Hopkinson gave no evidence that he had any practical experience of working with ac generators, and much of the knowledge he displayed at this lecture could easily have been gleaned by reading published documents on ac technology. Moreover, as a recently appointed consultant to the English Edison company, his specialized expertise in dc supply techniques might have raised doubts about the trustworthiness of his judgements about ac technologies. Hopkinson's strategy, nevertheless,

---

[36] John Hopkinson, 'On some points in electric lighting' [1883], in Bernard Hopkinson (ed.), *Original papers by the late John Hopkinson, D.Sc, F.R.S.*, 2 Vols., Cambridge, 1901, Vol. 1, Technical Papers, pp. 57–83, quote on p. 59.

[37] John Hopkinson, 'The theory of alternating currents, particularly in reference to two alternate current machines connected to the same circuit', JSTEE, 13 (1884), pp. 496–515, discussion on pp. 524–58, quote on p. 554–5.

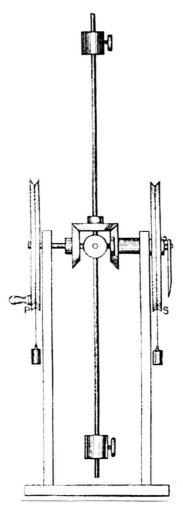

Figure 8. James Clerk Maxwell's pulley and flywheel model for illustrating how self-induction was analogous to mechanical inertia (see main text for explanation). Maxwell used a version of this device in the Cavendish Laboratory at Cambridge, and four years after Maxwell's death John Hopkinson used it to illustrate the principles of self-induction in his lecture 'On Some Points in Electric Lighting' to the ICE in 1883. [*Source*: John Hopkinson, 'On Some Points in Electric Lighting', (1883), in B. Hopkinson (ed.), *Original Papers by the late John Hopkinson, D.Sc, F.R.S.*, 2 Vols., Cambridge, 1901. Vol. 1, Technical Papers, p. 60.]

was to show how theoretical reasoning could illustrate what ought to be *possible* with ac machines. He did so by adopting an entirely mechanical model for alternator operation that made few concessions to the possible disanalogy between mechanical and electromagnetic phenomena.[38]

Out of deference to the familiarity of his engineering audience with mechanical illustrations, Hopkinson used Maxwell's analogue flywheel device to explain how an electrical circuit behaved as though it had inertia.[39] He admitted that the magnetic aspects of self-induction were not 'completely represented' by this model, hinting that the self-induction of the armature circuit might be significantly affected by its motion past adjacent iron pole pieces. Nevertheless, he characterized self-induction as a form of 'momentum' which enabled him to use the graphical methods of mechanics to analyse the performance of a Siemens alternator presented to the ICE three weeks previously.[40] Presuming for the sake of exegetical convenience that the waveform of an ac generator was sinusoidal in form, Hopkinson used standard engineering graphical methods to illustrate the effects of self-induction in such an ac machine. First, he showed the effect in which self-induction made the cycle of current variation lag behind the cycle of potential difference; hence the mean power output could not accurately be calculated by simple multiplication of maximum current and maximum potential difference. More significantly, the effect of self-induction was to *decrease* the maximum current available from an alternator. This implied that a large self-induction was economically undesirable, as it would decrease the mean power available from the machine.[41] Hopkinson's main purpose, of course, was to analyse the effects of connecting two alternators together. He thus moved from the analysis of one machine to study the combined waveforms of a pair of machines in the two possible connections. Having first shown that alternators could not usefully be connected in series because one machine would absorb all energy generated by the other,[42] he found no *theoretical* reason why power could not be extracted from a pair of alternators in parallel. So long as their voltage cycles were exactly synchronized, he reasoned that two alternators could with 'confidence' be run in parallel. Hopkinson's only proviso was that power-station attendants would have to exercise a 'little care' in

---

[38] I am grateful to Andrew Warwick for pointing out to me that Hopkinson's methods of seeking only analytical solutions to physical problems in 'On some points in electric lighting' were akin to those in which Cambridge Mathematics Tripos students were coached to use. See Warwick, *Masters of theory*. For evidence of Hopkinson's work on alternators in relation to his optical work for lighthouse installations ca. 1880–1, see Hopkinson, 'The theory of alternating currents, p. 514.

[39] Hopkinson, 'On some points in electric lighting', p. 60.

[40] A paper on Siemens alternators had evidently been given recently at the ICE.

[41] These invoked the periodic time of the machine T, I was the number of lines of force 'embraced' by the coils of the armature, R was the resistance of the circuit.

[42] Hopkinson, 'On some points in electric lighting' pp. 59, 67–8.

ensuring that the second machine was connected only when it attained the appropriate frequency of rotation.[43]

Eighteen months later, in the same room of the ICE building, Hopkinson presented a detailed algebraic analysis of linked alternators to the STEE. (see subsequent discussion). He explicitly revealed that to arrive at a neat analytic equation for a linked pair of alternators he had to assume that the self-induction of an armature was a constant. He admitted that this was 'not exactly' the case, but this made his mathematical analysis much simpler, as did his decision to ignore all mutual-induction effects on the copper wire in the alternator. Thus we see the force of Hopkinson's comment in the following discussion that what mattered most to him in analysing self-induction was adopting the 'most convenient' mode of 'expressing the facts'. Having made such assumptions as simplified his mathematical task, Hopkinson produced some 'equations of motion' for alternators – the very terms of which illustrate his commitment to dynamical models. The first of these showed that the leading machine of a paralleled pair generated most of the useful power, thus showing that parallel operations with a lighting circuit as the load were in principle feasible (see Figure 9).

Hopkinson's claim to have secured 'experimental verification' of his theory was problematic, however.[44] Since July 1884 Hopkinson had collaborated with the President of the STEE, William Grylls Adams, in testing three large De Meritens alternators supplying the arc lights at the South Foreland Lighthouse at Plymouth. Adams' candid account at this meeting shows that they managed to accomplish parallel running only if a third alternator were connected in series to govern them: Without this, arc lamps burned with 'irregularity', and in some instances the leading machine would lose speed until it stopped completely. Adams suggested that a high coefficient of self-induction in the alternators served to reduce the danger of mishap through damaging current surges when two were connected together – an empirical suggestion not matched anywhere in Hopkinson's theoretical treatment.[45]

---

[43] Ibid., p. 68. For Hopkinson's life and work, see James Greig, *John Hopkinson: Electrical Engineer*, London: Her Majesty's, Stationery Office, 1970. In the following year, Hopkinson publicly gave priority to the notoriously litigious Henry Wilde for having achieved parallel operation of alternators as long ago as 1868. John Hopkinson, 'The theory of alternating currents', p. 496; Greig, *John Hopkinson*, pp. 15–17. Henry Wilde, 'On a property of the magneto-electric current to control and render synchronous the rotations of the armatures of a number of electromagnetic induction machines', *Philosophical Magazine*, 4th series, 37 (1869), pp. 54–62.

[44] Hopkinson, 'The theory of alternating currents', p. 501.

[45] William Grylls Adams, 'The alternate current machine as motor', *JSTEE*, 13 (1884), pp. 515–28, esp. pp. 515–18. For Adams' career as Professor of Natural Philosophy at Kings College, London, from 1865, see Gooday, *Precision measurement and the genesis of physics teaching laboratories in Victorian Britain*, University of Kent at Canterbury, unpublished PhD thesis, 1989, Chapter 5. See Greig, *John Hopkinson*, p. 16, for an illustration of the De Meritens alternators.

II. *Two machines are coupled parallel and connected to an external circuit resistance R.*

Let x_1, x_2 be currents in the two machines. The external current will be $x_1 + x_2$, and consequently the difference of potential at the junction, $R(x_1 + x_2)$.

Let the electromotive forces of the two machines regarded in this case as connected parallel be $E \sin \frac{2\pi(t \pm \tau)}{T}$ and let the self-induction and resistance of each be 2γ and $2r$.

The equations of motion then are

$$2\gamma x_1' + 2rx_1 = E \sin \frac{2\pi(t+\tau)}{T} - R(x_1 + x_2),$$

$$2\gamma x_2' + 2rx_2 = E \sin \frac{2\pi(t-\tau)}{T} - R(x_1 + x_2);$$

whence

$$\gamma(x_1' + x_2') + (R+r)(x_1 + x_2) = E \sin \frac{2\pi t}{T} \cdot \cos \frac{2\pi \tau}{T},$$

and

$$\gamma(x_1' - x_2') + r(x_1 - x_2) = E \cos \frac{2\pi t}{T} \cdot \sin \frac{2\pi \tau}{T}.$$

Solving these

$$x_1 + x_2 = \frac{E \cos \frac{2\pi \tau}{T}}{(r+R)^2 + \left(\frac{2\pi \gamma}{T}\right)^2} \left\{(r+R) \sin \frac{2\pi t}{T} - \frac{2\pi \gamma}{T} \cos \frac{2\pi t}{T}\right\}$$

$$x_1 - x_2 = \frac{E \sin \frac{2\pi \tau}{T}}{r^2 + \left(\frac{2\pi \gamma}{T}\right)^2} \left\{r \cos \frac{2\pi t}{T} + \frac{2\pi \gamma}{T} \sin \frac{2\pi t}{T}\right\}.$$

Electrical work done by the leading machine

$$= \tfrac{1}{2} E \sin \frac{2\pi(t+\tau)}{T} \{x_1 + x_2 + (x_1 - x_2)\}$$

$$= \tfrac{1}{2} \frac{E^2}{(r+R)^2 + \left(\frac{2\pi \gamma}{T}\right)^2} \left\{(r+R) \cos^2 \frac{2\pi \tau}{T} - \frac{2\pi \gamma}{T} \sin \frac{2\pi \tau}{T} \cos \frac{2\pi \tau}{T}\right\}$$

$$+ \tfrac{1}{2} \frac{E^2}{r^2 + \left(\frac{2\pi \gamma}{T}\right)^2} \left\{r \sin^2 \frac{2\pi \tau}{T} + \frac{2\pi \gamma}{T} \sin \frac{2\pi \tau}{T} \cos \frac{2\pi \tau}{T}\right\}.$$

This expression shows that *the leading machine does most work in all cases.* Suppose r is small compared with R and $\frac{2\pi \gamma}{T}$, also that $R = \frac{2\pi \gamma}{T}$, we have the work done per second

$$= \frac{E^2}{8R} \left\{\cos^2 \frac{2\pi \tau}{T} + \sin \frac{2\pi \tau}{T} \cos \frac{2\pi \tau}{T}\right\}.$$

Figure 9. John Hopkinson's account of the parallel coupling of alternators: In characteristic Cambridge Wrangler fashion he treats the phenomenon as a problem in dynamics governed by 'equations of motions'. He equates the sum of the inductive and resistive potential differences inside each alternator (left-hand side) to the total electromotive force impressed on the external circuit, viz., the sinusoidally time varying component less the potential drop across *R*, the external resistance. From this he infers that because one machine will do more work than the other then useful work can be got out of the parallel coupling (in contrast to series connection in which one alternator exactly absorbs the power of the other). From this possibility Hopkinson moved to the conclusion that parallelling alternators should not be impossible. [*Source*: John Hopkinson, 'The Theory of Alternating Currents, Particularly in Reference to Two Alternate Current Machines Connected to the Same Circuit', *JSTEE*, 13 (1884), pp. 496–515, esp. pp. 503–4]

The electrical engineers present at the 1884 STEE meeting would not have been surprised to hear of such practical problems experienced by Adams and Hopkinson. Llewellyn Atkinson, an apprentice engineer and later President of the IEE, recollected that 'nobody' at the time believed Hopkinson's claim that parallel running was possible. It flew in the face of common knowledge among electricians that the Siemens alternators most widely used at the time were 'practically impossible' to run in that configuration. And Atkinson specifically rejected the mechanical model used by Hopkinson, treating the problem 'simply as the equation of motion of two moving bodies' with 'certain' forces between them. Such was the remoteness of his model from every experience that it might 'almost as well have been the sun and the moon' that he was modelling rather than the linking of two pieces of electromagnetic machinery.[46] The professorial lobby too expressed its doubts about some of Hopkinson's assumptions, especially his oversimplification of inductive effects in alternators. W. E. Ayrton, for example, insisted that Hopkinson should introduce non-linear multiplying factors to allow for the variable magnetic saturation of alternator iron cores, hinting thereby at a possible inconstancy in the armature's self-induction.[47] S. P. Thompson argued that Hopkinson should allow for the variability of the 'coefficient of mutual-induction' between the field magnets and the armature as the latter rotated past the former, as discussed in an appendix to his recent textbook *Dynamo-Electric Machinery*.[48] In reply, Hopkinson simply denied that the conditions of practice vitiated his idealized assumptions: Although he admitted the self-induction of the alternator did depend to some extent on the relative position of coil and magnets, he contended this was 'very nearly' a constant for practical purposes.[49] Tellingly, however, he gave no analysis of the quantitative error likely to be introduced by the adoption of this convenient simplifying assumption.

In a more reflective vein, Hopkinson did acknowledge that, when iron magnets were in proximity, using Faraday's interpretation of self-induction as a product of a circuit's magnetic field might be more appropriate than the alternative 'momentum' model that he attributed to Maxwell (see the epigraph that opens this section). Indeed, in his contemporaneous textbook, S. P. Thompson treated the role of self-induction in ac machinery in terms of the Faradayan model of lines of inductive force. In this view the coefficient of self-induction represented 'the number of lines of force which the circuit would possess or induce on itself' for a unit of absolute current flowing. The

---

[46] Llewellyn B. Atkinson, Untitled speech at IEE fiftieth anniversary commemoration, *JIEE*, 60 (1922), pp. 441–3, quote on p. 443. As a novice engineer Atkinson had been one of W. E. Ayrton's students at Finsbury Technical College.
[47] Hopkinson, 'The theory of alternating currents', p. 531.
[48] Ibid., pp. 538–9. See Thompson, *Dynamo-electric machinery*, 1884, Appendix IV, 'On the general equations of dynamo electric machines and the theory of Joubert', pp. 392–4.
[49] Hopkinson, 'The theory of alternating currents', p. 554.

fertility of Thompson's approach was manifested in two important results that were taken up by later analysts of ac machines: First, an alternator produced maximum work when its ratio of resistance to self-induction was 2π times the rotational periodicity. The second was his introduction of the concept of 'armature reaction' based on the interactions of lines of force between armature and field magnets in relative motion. This seems to capture the complexity of an alternator's electromagnetic characteristics – notably the lag of current behind potential difference – more comprehensively than an account based simply on inertial notions of mutual- or self-induction.[50]

When Thompson's more nuanced and widely researched dynamo theory is compared with Hopkinson's facile algebraic manipulation, it is not surprising the STEE discussion of Hopkinson's 1884 paper revealed undertones of polite scepticism. His manifestly simplistic theory of ac machinery could offer neither explanation nor remedy for the industrial problems of ac paralleling. Both Ayrton and Thompson aired suspicions that something might have been lost in Hopkinson's reductive treatment of alternator self-induction as an approximately constant coefficient. This is all the more important in the light of Nancy Cartwright's demonstration that the adoption of expedient approximations has often had non-trivial and unpredictable consequences for deductive predictions from theories.[51] Given the apparent falsity of Hopkinson's approximating assumptions that alternator self-induction was a constant and mutual-induction negligible, contemporaries might well have regarded Hopkinson's prediction of the possibility of parallel running as being entirely ill-founded. Indeed, there was little further discussion of his theory of ac machines until controversy erupted five years later about the accomplishment of paralleling.

5.4. FROM CURRENT BALANCE TO SECOHMMETER: MEASURING SELF-INDUCTION AT THE STEE

The only thing gained by using Maxwell's method was that it formed a pleasant way of spending an afternoon.
    W. E. Ayrton and J. Perry, 'Modes of Measuring the Coefficients
    of Self and Mutual Induction', STEE, 1887[52]

It has been for some a time a matter of necessity to have an accurate and simple method of determining coefficients of self-induction.
    J. A. Fleming, comment on Ayrton's and Perry's STEE paper, 1887[53]

---

[50] Thompson, *Dynamo-electric machinery*, p. 70–80, 260–6.
[51] For a discussion of how the process of approximation generally has non-trivial analytical consequences, see Nancy Cartwright, *How the laws of physics lie*, Oxford: Clarendon, 1983, pp. 100–127, and the discussion of this in Warwick, 'The laboratory of theory' in Wise (ed.), *The values of precision*, Princeton, NJ: Princeton University Press, 1995, pp. 311–51.
[52] Ayrton and Perry, 'Modes of measuring', pp. 293–6, 394.
[53] Discussion in ibid., p. 341.

Efforts to find new ways of measuring self-induction and mutual-induction in 1886 were not initially stimulated by problems in the theory and practice of paralleling alternators. Such efforts emerged instead from David Hughes' controversial attempts in 1886 to challenge Maxwell's treatment of self-induction by means of a quantitative instrument. Acquiring fame in 1878 for his invention of the microphone, the Welsh-born Hughes developed an induction balance in the following year to investigate the relations between sound perception and electrical induction. This 'balance' was modelled on the Wheatstone bridge for resistance, determining an unknown self-induction by comparison with a (variable) known value. It used a rheotome to generate an oscillating current, the null condition (zero current flowing) in the induction balance being attained when the induction elements were as appropriately balanced as the resistance elements. Hughes showed how a telephone (see Chapter 2) could sensitively detect changes in inductive balance that were due to the proximity of iron objects, and this technique was used to locate the bullet in the body of assassinated US President Garfield in 1881. With his sonometer, a device for registering sound intensity, Hughes was able to differentiate quantitatively between the metals introduced to the coils of the induction balance. Hughes claimed this could detect 1/10,000 part impurity in a metal, and a version of it was later reportedly ordered for the Royal Mint. Such was the impact of his technological demonstrations that Hughes was elected a Fellow of the Royal Society in 1880 and received a Royal Medal in 1885.[54]

As Dominic Jordan has shown, however, Hughes' work on self-induction became controversial in January 1886 when he broke with institutional protocol in his Presidential Address to the STEE. Instead of the more customary irenic synopsis of the previous year's electrical research and engineering, Hughes presented a survey of his own electromagnetic researches. His paper purported to show that the self-induction of a conductor depended on the *material* of its conductor as well as on its geometry: For the same shape of conductor, he claimed that iron showed a significantly greater self-induction than copper, especially at higher frequencies. He also claimed that the non-inductive resistance of a circuit element to an oscillating current was almost always greater than its resistance to a steady current – a radical claim because it was conventional to consider only *inductive* resistance as increasing with frequency.[55] Hughes' presentation received an unusual amount of criticism for a presidential lecture when discussed at the STEE two weeks later. Lord Rayleigh argued that Hughes' instrument did not strictly measure

---

[54] George Burniston Brown, 'David Edward Hughes, F.R.S., 1831–1900', *Notes and Records of the Royal Society*, 34 (1979–80), pp. 227–39, esp. p. 231.; Joe Marsh and J. G. Roberts, 'David Edward Hughes: Inventor and scientist', *Proceedings of the IEE*, A 126 (1979), pp. 163–76.

[55] David E. Hughes, 'The self-induction of an electric current in relation to the nature and form of its conductor', *JSTEE*, 15 (1886), pp. 7–25.

self-induction because its balancing operation could not adequately differentiate between the effects of resistance and self-induction. He noted that Hughes had not only neglected to allow for the frequency dependence of the balance of his circuit, but also that, to restore the circuit to null balance in each experiment, Hughes should have adjusted both resistance and inductive elements – as Maxwell's techniques required. His failure to do so, judged Rayleigh, accounted for Hughes' rather anomalous results.[56]

W. E. Ayrton and S. P. Thompson, by contrast, attempted to lessen the embarrassment at presidential ineptitude. They redeemed some of Hughes' observations of changing self-induction by reinterpreting them as practical evidence to *support* rather than challenge a Maxwellian view of high-frequency currents. They noted that Hughes' experiments fulfilled Oliver Heaviside's prediction, following Maxwell, that, as the frequency of electrical signals increased, currents would tend to travel in only the outermost periphery – the 'skin' – of a conductor where the effects of self-induction were lowest.[57] After further experimentation Hughes withdrew or reformulated some of his contentious claims during the following year, but became marginalized when he refused to accept the canonical 'skin-effect' interpretation of his Presidential lecture.[58] Jordan thus suggests that this episode marked a 'decisive' victory of Maxwellian theoreticians over untutored practitioners such as Hughes. Yet the aftermath of the STEE debate was not actually so clear-cut. One aspect of Hughes' challenge to Maxwell was soon vindicated. In an editorial interpolation to the third edition of Maxwell's *Treatise* in 1891, J. J. Thomson showed how Maxwell's analysis of variable currents might be reconciled with different values of self-induction per unit length for magnetic and non-magnetic metals.[59] Also, as Jordan himself notes, efforts by Rayleigh, Ayrton, and Heaviside to further theorize the role of self-induction in variable currents inspired derision rather than capitulation from some ordinary practitioners.[60]

More importantly for our purposes, Jordan and Yavetz note that one indirect outcome of Hughes' experimental efforts to 'measure' self-induction was the development of Ayrton's and Perry's secohmmeter.[61] This appears to have followed a sympathetic suggestion by the *Electrician* on 13 February

---

[56] Jordan, 'D. E. Hughes, self-induction and the skin effect', pp. 126–29.
[57] Oliver Heaviside, 'Electromagnetic induction and its propagation. Section 2: On the transmission of energy through wires by the electric current', reprinted from the *Electrician*, January 1885, in Oliver Heaviside, *Electrical papers*, 2 Vols., London: 1892, Vol. 1, pp. 434–41. See Jordan, 'Skin effect', p. 130, and p. 151, n. 17.
[58] Ibid., n. 33.   [59] Maxwell, *Treatise*, Vol. 2, 3rd edition, pp. 323–5.
[60] Jordan, 'D. E. Hughes, self-induction and the skin effect', esp. pp. 142–49. Yavetz, 'Oliver Heaviside and the significance of the British electrical debate', pp. 148, 153. See scathing letter from 'Coil' to editor of the *Electrical Review*, cited in Jordan, 'Skin effect', n. 44.
[61] Yavetz, 'Oliver Heaviside', pp. 154–5.

1886 that someone should revise Hughes's induction balance – a 'bold and ingenious' adaptation of the Wheatstone bridge principle – to overcome the 'uncertainty' that had long hung over the character of induced currents.[62] Both Ayrton and Perry in Britain and Ledeboer and Maneuvrier in France soon developed techniques for 'measuring' self-induction. Only the former pair, however, attempted to make an instrument of the direct-reading form discussed in Chapter 4.[63] When Ayrton and Perry presented their new secohmmeter to the STEE in April 1887 they not only represented their work as an extension of Hughes' experimental efforts but also diplomatically construed these as an extension of Maxwell's *Treatise*. Citing Maxwell's equation (discussed in Section 5.2) relating the electromotive force produced by a current of varying intensity over a cylindrical conductor,[64] Ayrton charitably interpreted Hughes as having investigated the higher-order terms 'neglected' by previous scholars beyond the simple self-inductive coefficient $(A + \frac{1}{2})$ usually known as L. According to Ayrton, Hughes' rather 'delicate' measurements had shown these higher-order terms to be far from negligible for certain cases; this effectively vindicated Hughes' controversial claim that the effective resistance of a conductor varied in response to an intermittent current.[65] The strategic purpose of such comments is evident from the fact that Ayrton's and Perry's instrument was not designed to measure any of these higher-order coefficients. As they admitted later, 'the secohmmeter does not measure all that can be measured.'[66] For this device, in fact, they borrowed heavily on Hughes' representation of self-induction as having effects akin to altered circuit resistance. The secohmmeter measured self-induction (or mutual-induction) by using a rotating commutator to convert its intermittent effects into a form of effectively permanent resistance to be balanced comparatively in a Wheatstone bridge.

Ayrton's and Perry's ostensible rationale for creating the instrument was that electrical practitioners could come to have as clear and 'instinctive' an understanding of self-induction as they now had of resistance only if they had a reliable technology for making repeated measurements of it (see Chapter 2). This was especially important as the mathematical experts argued that self-induction was at least as important as resistance in governing the

---

[62] 'Professor Hughes's discoveries', *Electrician*, 16 (1886), p. 270, quoted in Jordan, 'D. E. Hughes, self-induction and the skin effect', p. 146.

[63] For contemporary French approaches to measuring self-induction, specifically for the case of a Siemens dynamometer, see Pierre Ledeboer and Georges Maneuvrier, 'Sur la détermination du coefficient de self-induction', *Comptes Rendues d'Academie Francaise*, 104 (1887), pp. 900–2, translated as 'On the determination of the co-efficient of self-induction' in *Telegraphic Journal and Electrical Review*, 20 (1887), pp. 255–6.

[64] Maxwell, *Treatise*, Vol. 2, 1873, pp. 292–3; 3rd edition, 1891, p. 322.

[65] Ayrton and Perry, 'Modes of measuring', pp. 292–3, 390–1. Compare Jordan, 'D. E. Hughes', p. 145.

[66] Ayrton and Perry, 'Modes of measuring', p. 391.

operation and efficiency of long-distance telephony, high-speed telegraphy, and ac generation. To make an instrument that rendered self-induction easily and quickly quantifiable for the ordinary electrical worker, Ayrton and Perry drew on the kinds of techniques that they had used in developing the portable direct-reading ammeter since 1881 (see Chapter 4). The device that Ayrton began working on with his South Kensington student W. E. Sumpner and John Perry at Finsbury Technical College from spring 1886 was known as the secohmmeter in reference to their adoption of the 'secohm' as the unit of self-induction. Because self-induction had the dimensions of time × resistance, Ayrton and Perry drew on the internationally recognizable etymology of 'second' and 'ohm' to label the unit of self-induction as the 'secohm'. This was chosen in preference to the alternative that would have sounded too risibly like the London Cockney dialect pronunciation for 'homesick'.[67]

When they began designing the secohmmeter Ayrton and Perry looked to Maxwell's canonical methods for measuring self-induction but found them inappropriate to the pressing temporal constraints of practical engineering work. Maxwell presented several means of measuring these quantities in static systems by complicated two-stage processes in which augmented forms of the Wheatstone bridge were used. These were null methods that enabled an unknown value to be compared with a precalibrated or calculated standard of self-induction, mutual-induction, or capacitance. However, they required the tricky and time-consuming iterative business of balancing the bridge for both zero permanent currents as for a conventional Wheatstone bridge and also for zero transient currents when the circuit was completed.[68] Ironically, in the *Treatise*, Maxwell did not include the deflectional 'electric-balance' method which he had so cogently employed to measure the self-induction of a coil in 1865. This was later employed by Lord Rayleigh in his 1881 determination of the error caused by the hitherto undetermined self-induction of the BAAS 1864 resistance measurement apparatus.[69] In this latter method, the crucial skill was to gauge the maximum transient

---

[67] Editorial note, 'The unit of self-induction', *Telegraphic Journal and Electrical Review*, 20 (1887), p. 429. The 'Secohm' was one of several candidate units for self-induction supplanted by the 'henry' at the Chicago Congress of 1893 in deference to the historical claims made on behalf of Joseph Henry by the US hosts. See Graeme Gooday, 'Faraday re-invented: Moral imagery and institutional icons in Victorian electrical engineering', *History of Technology*, 15 (1993), pp. 190–205. The aim of Ayrton and Perry was to render explicit the relationship between resistance and self-induction by incorporating the unit of resistance (ohm) and that of time (second) in their name for the unit of self-induction.

[68] Maxwell, *Treatise*, Vol. 2, pp. 353–7; pp. 393–8 with appendix in 3rd edition on pp. 399–401. Hughes' induction balance was of a similar form and thus comparably time consuming to use.

[69] Rayleigh, 'Experiments to determine the value of the British Association unit of resistance in absolute measure', *Philosophical Transactions of the Royal Society*, 173 (1882), pp. 661–98, 677–84. This was abstracted by Ayrton in *JSTEE*, 11 (1882), p. 152; Ayrton and Perry, 'Modes of measuring', p. 295.

Coupled Problems of Self-Induction 197

deflection of a galvanometer in two different states of resistive balance. From the size of these angular deflections and of (absolute) values of resistance in the balance, the value of the self-induction could be calculated with a highly complicated set of equations.[70]

Maxwell's techniques for measuring self-induction were impracticable, however, for engineering work.[71] Ayrton and Perry described his double-adjustment bridge methods as so time consuming and awkward as to be 'nearly hopeless', and his 1865 method of taking two readings of a galvanometer's maximum deflection as merely a 'pleasant way of spending an afternoon'.[72] Instead they used a mechanized surrogate technique that manipulated the temporal attributes of self-induction in a bridge circuit. When a sufficiently rapid regular ac was used, the repeated 'kicks' of current change produced by the unknown self-induction would be registered cumulatively as equivalent to a steady increase in circuit resistance. To simulate the effect of an ac Ayrton and Perry used a complex commutator device which rotated at a measurable frequency/speed. The action of the commutator was to make and break the galvanometer and battery circuits independently in such a rapid and regular way that the effect of self-induction on the current flow was to create the 'apparent steady definite increase' of resistance that could be measured by a Wheatstone bridge. In the simplest approximation for this set-up, they calculated the self-induction L by multiplying the time T for which the rotating commutator connected *both* the galvanometer and battery in circuit by the apparent increase in circuit resistance σ created by insertion of the unknown self-induction into the bridge. The linear equation $L = T\sigma$, on which the secohmmeter was based, thus met the proportionality condition on which Ayrton and Perry somewhat contentiously made all of their measuring instruments for the purposes of quick and easy reading (Chapter 4). It was, however, only an *approximate* equation insofar as its derivation invoked simplifying assumptions about the linearity of the induced changing current accomplished by making and breaking the circuit connections with the rotating commutator and by excluding more complicated self-induction and resistance effects of other circuit elements.[73] (see Figure 10).

---

[70] The original source for this was James Clerk Maxwell, 'On a dynamical theory of the electromagnetic field', *Philosophical Transactions of the Royal Society*, 155 (1865), pp. 475–7, and was republished as an editorial interpolation in the third edition of the *Treatise*, Vol. 2, pp. 399–401.

[71] Thompson, *Dynamo-electric machinery*, p. 387; but compare discussion on mutual induction on p. 388; Ayrton and Perry, 'Modes of measuring', pp. 293–4. Compare Ayrton's remarks at the STEE in 1886 with Gisbert Kapp's paper on dc dynamos.

[72] Ayrton and Perry, 'Modes of measuring', pp. 293–6, 394.

[73] See ibid., pp. 303–5 for the full equation: $L = \sigma(T + \lambda/g - B/A)$, where λ is the self-induction of the galvanometer, g is its resistance, and B and A refer to the bridge circuit resistances.

Figure 10. The secohmmeter – Ayrton's and Perry's revised (1887) device for measuring self-induction; this version incorporated a mercury column to record that speed of handle rotation at which the internal induction balance attained its null position. It thus enabled one person alone to be able to use the device and determine all relevant readings. [*Source*: William Ayrton and John Perry, 'Modes of Measuring', *Telegraphic Journal and Electrical Review*, 20 (1887), p. 479.]

Early versions of the secohmmeter embodying this principle required two persons to operate it. T was determined as the speed at which the hand-driven spinning double commutator accomplished the relevant circuit contacts and interruptions for a constant reading. Yet the same person could not easily determine the speed at which a constant galvanometer deflection was obtained at the same time as reading the actual value of the deflection itself. So, for the commercial version to be manufactured by Nalder Brothers,[74] Ayrton and his collaborators produced a 'zero' version of the secohmmeter for a single operator to use. After the variable resistance in the bridge had been suitably increased to allow balance to be attained at a reasonable speed, the speed of the rotating commutator would be increased until a zero-balance reading was attained. The speed of rotation was registered by a column of mercury connected to a centrifugally expanding box reservoir, and at the point a zero reading was attained, the solo user could press a trigger which locked the length of mercury so that the speed at zero reading could be read off thereafter. Yet the sense in which Ayrton and Perry's instrument was direct-reading was not the same as that in which their 1884 ammeter had been direct reading (see Chapter 4). To secure a reading of self-induction, the secohmmeter user had to make the calculation of dividing the resistance increase by the speed of rotation at balance. This instrument thus did not give instantaneous readings nor could it register fluctuations of self-induction – precisely the objections that Ayrton had made about the Siemens dynamometer as a current-measuring device (Chapter 4). More ironic still was the matter of scalar calibration. Although self-induction had the dimensions of length in the electromagnetic system of units (resistance having the dimensions of velocity), the makers of the secohmmeter did not consider it practicable for this device to have a single linear scale with a length marking calibrated in units of secohms.

Ayrton and Perry did not discuss such finer matters when they presented their device to the STEE in April 1887. Instead they demonstrated the purported accuracy that could be achieved with their new direct-reading device which produced measurements of a solenoid to within 0.5% of its calculated self-induction of 0.0215 'second-ohms'. The credibility of such refined values depended, of course, on the perceived reasonableness of the approximations made in arriving at the instrument's theoretical law in operation. Ayrton's and Perry's general claims for the accuracy of the instrument's operations were supported in a detailed analysis by the student Sumpner[75] and confirmed by two pieces in the *Telegraphic Journal and Electrical Review*

---

[74] Ibid., p. 323.
[75] William E. Sumpner, 'The measurement of self-induction, mutual induction and capacity', *JSTEE*, 16 (1887), pp. 344–79.

in subsequent weeks.[76] Sumpner also illustrated the broader utility of the secohmmeter by presenting data showing how the armature self-induction in Ferranti alternators was not a constant but varied according to the degrees of magnetic saturation in the iron cores.

Given that such useful data emerged from use of the instrument, the responses from people at the STEE meeting in April 1887 were positive, Hughes describing the paper as a 'distinct advance' on all previous methods of measuring self-induction. S. P. Thompson considered it to be a 'really very important addition' to practical knowledge of self-induction, the secohm meter providing engineers with a 'real' means of acquainting them with the 'actual working values' of coefficients of self-induction.[77] Just two weeks later E. C. Rimington, of the School of Electrical Engineering in Regent Street, read a paper at the Physical Society of London on a 'A Modification of a Method of Maxwell's for Measuring the Coefficient of Self-Induction.' This incorporated Ayrton's and Perry's 'admirable' secohmmeter as a refinement of Maxwell's technique.[78] Their new secohmmeter was in fact so well received that Ayrton's and Perry's paper was abstracted in *Nature* and reproduced in full in both the *Electrician* and the *Telegraphic Journal and Electrical Review*.[79] An editorial note in the latter suggested that readers who were still 'hazy' about self-induction should attentively peruse and master this paper, for its importance could not be 'overestimated'.[80]

Not all readers of the *Telegraphic Journal and Electrical Review* were so positive about the secohmmeter, however. Rankin Kennedy soon wrote in to complain about the potentially misleading principle on which the instrument operated. Representing self-induction as if it had a cumulative effect equivalent to an increase in a circuit's resistance would only confuse the 'practical' man about the distinction between these parameters.[81] And, as we shall see in the next sections, James Swinburne was particularly scornful of this new instrument and the claims of those who tried to apply it to the contexts of ac power generation.

---

[76] E. C. Rimington. '[The Secohmmeter]', *Telegraphic Journal and Electrical Review*, 20 (1887), pp. 373–5; John Cooper, 'Simple proof of the law of the secohmmeter', *Telegraphic Journal and Electrical Review*, 21 (1887–8), p. 577.
[77] Ayrton and Perry, 'Modes of measuring', pp. 383–4.
[78] E. C. Rimington 'A modification of a method of Maxwell's for measuring the coefficient of self-induction', *Telegraphic Journal and Electrical Review*, 21 (1887–8), pp. 62–4, quote on p. 63.
[79] See William Ayrton and John Perry, 'Modes of measuring', pp. 431–4, 457–60, 478–82, and idem, 'The secohmmeter', *Nature* (London), 36 (1887), pp. 129–32.
[80] Editorial note, Self and mutual induction', *Telegraphic Journal and Electrical Review*, 20 (1887), p. 452.
[81] Rankin Kennedy, 'The second-ohm unit', *Telegraphic Journal and Electrical Review*, 20 (1887), p. 471.

5.5. THE SECOHMMETER IN ACTION: GISBERT KAPP AND THE PARALLELING OF ALTERNATORS

[T]hose who prefer to make the data fit their mathematics are sorely tempted to ... deal with the coefficients of self- and mutual induction where such things cannot be said to exist.

James Swinburne, 'Practical Electrical Measurement', *Electrical Review*, 1887[82]

In a series of articles on 'Practical Electrical Measurement' he produced for the *Telegraphic Journal and Electrical Review* in 1887–8, James Swinburne omitted any mention of the secohmmeter. This snub was repeated when Swinburne re-edited these articles into the monograph *Practical Electrical Measurement* in 1888. His aversion to the secohmmeter should not, however, be interpreted simply as a practical man's disregard or ignorance of self-induction. Recognition of its complicating effects was a recurrent feature of Swinburne's analysis of alternate currents. He noted, for example, that a Siemens dynamometer could be the most error-free instrument for measuring ac currents, for if wound with very few turns of wire it had no 'appreciable' self-induction at all.[83] Swinburne's rejection of the secohmmeter was rather one of the more extreme manifestations among contemporaries of the growing recognition of the complex nature of self-induction as a parameter not amenable to simple inertial interpretation.

By 1887 it was conceded, even by John Hopkinson, that self-induction was not in general a constant in ac machines. Having developed a magnetic-circuit theory to analyse the dc dynamo in collaboration with his brother Edward,[84] Hopkinson used this theory to argue in a paper to the Royal Society that the self-induction of an alternator armature generally varied with simple and compound sinusoidal dependence on the phase of the alternator cycle.[85] To make this claim, however, Hopkinson had assumed that all ac machines generated a sinusoidal current. Whilst this simplified the mathematics of Hopkinson's analysis, even this assumption was directly at odds with what ac engineers knew of actual alternator outputs. In his series on electrical measurement for the *Telegraphic Journal*, Swinburne pointed out that his own patented alternator was like Westinghouse's machine in producing a rounded 'sawtooth' waveform. And it was well known that such machines performed unlike those with a (nearly) sinusoidal output, such as the older

---

[82] James Swinburne, 'Practical electrical measurement', *Telegraphic Journal and Electrical Review*, 21 (1887–8), esp. pp. 603–5.
[83] Ibid., esp. pp. 603–5. James Swinburne, *Practical electrical measurement*, London: 1888.
[84] John Hopkinson and Edward Hopkinson, 'Dynamo electric machinery', *Philosophical Transactions of the Royal Society*, 177 (1887), pp. 331–58; Greig, *John Hopkinson*, pp. 17–22.
[85] John Hopkinson, 'Note on the theory of the alternate current dynamo', *Proceedings of the Royal Society*, 42 (1887), pp. 167–70.

form of Siemens alternator.[86] Swinburne nevertheless had his own reasons for agreeing with Hopkinson that the self-induction of an alternator was not a constant in an operational alternator.

Swinburne accepted that some alternators, such as the older Siemens types, had armature coils far enough from the iron of the field magnets that they actually possessed a well-defined coefficient of self-induction. Nevertheless, it was clear 'at a glance', wrote Swinburne, how complicated the action of an alternator was when unsaturated (i.e., incompletely magnetized) iron was in closer proximity. Indeed, for many recent alternators the armature iron was sometimes so close to the coils that it nearly touched them; thus the self-induction could not possibly be described as a constant property of the armature. Even in devices in which there was still a significant air-gap, such as the Ferranti and Westinghouse machines, Swinburne argued that Foucault currents developed in the iron armature cores which opposed changes in magnetization. Although the net effect of such processes was to oppose any change of the current – as in the cases of self- and mutual-induction – Swinburne considered it 'simply incorrect' to call all this self-induction and treat it as if it had a constant coefficient. Where iron cores were present, it was not even possible to speak of separately varying components of self-induction and mutual-induction: One could only speak in Faradayan terms of varying electromotive forces produced in the circuits. Accordingly he was scathing about those theorists – like Hopkinson – who preferred to make the data 'fit their mathematics' and assume that they were dealing with the coefficients of self- and mutual-induction in situations in which such things 'cannot well be said to exist'. Not only were the laws they produced 'more or less wrong', but the curious respect of electricians for mathematics left them believing that writers who introduced 'formidable-looking' expressions were authoritative, whilst the 'real' empirical laws were not investigated.[87]

Two years later a candidate for a 'real' practical law of parallel alternator operation relating to self-induction was put forward. It was proposed by the expatriate Austrian Gisbert Kapp, who in 1885 had been co-discoverer with Hopkinson of the theoretical law of dc dynamo operation.[88] On 19 February 1889 Kapp presented a long paper to the ICE which set out what he believed to be the electrical and mechanical requirements for an ac system

---

[86] Swinburne, 'Practical electrical measurement', pp. 628–9. Swinburne particularly criticized those who, purely for the sake of mathematical convenience, assumed that all alternators had a sinusoidal waveform. Swinburne, *Practical electrical measurement*, 1888, pp. 133–4.

[87] Swinburne, 'Practical electrical measurement', pp. 603–4; *Practical electrical measurement*, pp. 134–6.

[88] David G. Tucker, *Gisbert Kapp 1852–1922: First professor of electrical engineering at the University of Birmingham*, Birmingham: University of Birmingham, 1973, pp. 15–17.

to work in parallel.[89] He sought to generalize from the particular success of Westinghouse's parallelling to a more universal account. Kapp's analysis focussed particularly on the current surges that commonly occurred at the moment of coupling two alternators: This could be very damaging to generating machinery and thus could seriously inconvenience both supply engineers and consumers. The problem was that the 'skill and attention' necessary to prevent such surges were not widely possessed by ordinary generating-station staff using conventional ac machines designed with low resistance and low self-induction (to minimize power loss). Echoing Adams' position of 1884, Kapp argued that, to prevent current surges but not 'seriously prejudice' their efficiency, alternators should have a 'sensible' – i.e., substantial – amount of self-induction in the armature circuit. This was to be achieved by the inclusion of an iron core in the alternator armature. A further condition was that an external circuit should not contain any appreciable amount of self-induction, a condition he suggested would be 'easily fulfilled' if the alternators were connected to a bank of transformers.[90]

As was often the case during engineers' debates there was a vigorous parading of diverse views at the ICE after Kapp's paper. Swinburne unsurprisingly attacked Kapp for speaking of armature self-induction as if it were a constant when in motion.[91] Ayrton passed over this issue in his typically expansive monologue, and instead took the opportunity to present data to criticize the second part of Kapp's argument. Using the secohmmeter, an instrument he expected would be 'familiar' to at least some members of the ICE, Ayrton and a corps of his South Kensington students had established that the majority of transformers in commercial use actually had a very substantial self-induction. Accordingly, they could not meet the requirement for safe paralleling as specified by Kapp.[92] Ayrton's claims that the secohmmeter gave 'accurate' and 'sensitive' measurements on such matters[93] were reiterated when he and Perry were invited to give an impromptu paper at the IEE a few weeks later. They produced further results to challenge Kapp's claim that transformers had negligible self-induction and to enjoin all engineers to avoid such mistakes by using the secohmmeter.[94] Yet, in the next stage of the controversy over the role of self-induction in the parallel operation of alternators, their secohmmeter was not to prove as decisive as Ayrton and Perry had perhaps hoped.

---

[89] Gisbert Kapp, 'Alternate current machinery', *PICE*, 97 (1889), pp. 1–42, discussion 43ff.
[90] Ibid., pp. 1–21.   [91] Ibid., discussion on p. 62.   [92] Ibid., discussion on pp. 68–72.
[93] Ibid., discussion on pp. 72–4. Ayrton also challenged Swinburne's claim that there was no self-induction in dc armatures, and to support this claim introduced, for the first time, experiments from the Central Institution labs on the determination of waveform, ibid., pp. 75–7.
[94] William Ayrton and John Perry, 'Laboratory notes on alternating current circuits', *JIEE*, 18 (1889), pp. 284–8.

5.6. DIAGNOSING SELF-INDUCTION: MORDEY'S CONTESTED ANALYSIS OF PARALLEL WORKING

The practical dynamo-builder does not care two straws whether the mystery of parallel running of alternators has or has not been packed up in somebody's mathematical equations to be extricated therefrom subsequently as from a conjuror's inexhaustible bag, nor does he derive much guidance from the differences of experts, some of whom consider the secret depends upon abundance of self-induction in the armature, and some of whom think it does not.

'Alternate Current Working', editorial note in *The Electrician*, 1889[95]

By spring 1889 news was afoot among London's electrical engineers that the Anglo–American Brush Company in Lambeth was manufacturing a radically new design of alternator that was fully capable of parallel operation. This was built to the designs of the company's technical manager William Mordey to have a stationery and completely *ironless* copper armature. An editorial note in *The Electrician* announcing Mordey's imminent paper on this machine at the IEE predicted that it would prove to be of the 'highest' interest.'[96] Indeed Hopkinson, Adams Ayrton, Thompson, Kapp and Swinburne were all present on 23 May to hear Mordey launch his paper with an attack on Kapp's thesis on ac working. Mordey announced that he had seen enough of the use of iron in armatures to 'wish to do without it', and if iron were really necessary, he maintained that the armature was the 'very worst' place to put it'. Examining the evidence as to whether self-inductive iron really was the secret of successful parallel working, Mordey found that as many examples showed success as showed failure. Whilst Zipernowski and Lowrie–Parker alternators were successful in this regard, Westinghouse iron-cored alternators worked effectively in parallel only at loads at half their maximum, and an anonymous but well-known company admitted that they could achieve parallel running only when all machines and arrangements were 'identical in every way'. After further investigation of various other possible combinations of resistance and self-induction in the alternator or external circuit,[97] Mordey concluded iconoclastically that parallel working required an alternator to have *neither* resistance *nor* self-induction. In so saying he admitted he was in the 'unenviable' position of being out of accord with the theories and practice of other 'very able' workers in the field.'[98]

To support his case, Mordey presented the results of eight experiments with alternators carried out on his own design of a copper armature. These were carried out under 'exacting and onerous' conditions: Being taken up to full speed and then switched into parallel configuration with zero load or an

---

[95] Editorial note, 'Alternate current working', *Electrician*, 24 (1889), p. 325.
[96] Notes, *Electrician*, 24 (1889), p. 23.
[97] William M. Mordey, 'Alternate current working', *JIEE*, 18 (1889), pp. 583– 613, discussion on pp. 613–88.
[98] Ibid., pp. 590–1.

inductionless load and then running drastically varying loads under them. Other tests included acts as uncongenial as connecting the alternators while in opposite phases or whilst at radically different potentials. Despite diverging from Kapp by not employing self-inductive choking coils or resistances to check the current surges between the alternators, Mordey reported that there was 'never a single case when they got out of step, even momentarily'. Indeed, he attributed the very success of his trials to the 'negligible' self-induction and resistance possessed by the alternators – these conditions ensuring that there was 'instantaneity of action' when the two were connected.[99] And the day after his IEE paper (24 May), Mordey won what appeared to be a major strategic victory by arranging for the entire IEE council, its President, and journalists from the technical press to visit the Brush works to witness a replication of his results.

The correspondent of the *Electrician* recounted that each experiment was 'attended with perfect success' and that nothing further was needed to demonstrate the practicability of ac parallel working.[100] On the following day an editorial in that journal reported Mordey's 'liveliest' of papers. Seldom, it opined, had the IEE heard results that so thoroughly overturned so many widely accepted beliefs. Accordingly the *Electrician* reproduced Mordey's paper in full over the next four weeks so readers could judge for themselves. What readers would have seen unfolding in following issues, however, was a remarkable phenomenon: widespread acceptance of Mordey's accomplishment accompanied by equally widespread disagreement with his explanation of it. Although S. P. Thompson praised him for demonstrating that 'good design' of alternators was the 'first essential' of parallel operation,[101] only the *Electrician* reviewer agreed with Mordey's account of his success as lying in the 'smallness' of his alternator's self-induction.[102]

Following the reconvening of the IEE discussion of Mordey's paper a week later, there were months of disagreement about the magnitude and relevance of armature self-induction in explaining Mordey's results, and about the significance of secohmmeter measurements undertaken to attempt to resolve such disputes. A further recurrent question in this debate was whether Mordey's accomplishment had been predicted by Hopkinson's 1884 theory. Although not present at the IEE discussion on 30 May, Hopkinson arranged for his somewhat lofty commentary to be relayed by freelance electrical engineer George Forbes (see Chapters 2 and 6). Hopkinson disagreed with both Mordey's and Kapp's diagnosis of the conditions for parallel running. He contended that it was an 'obvious immediate consequence' of the fourth paragraph of his 1884 paper (dealing with the conditions for achieving

[99] Ibid., pp. 591–3. [100] Notes, *Electrician*, 24 (1889), p. 79.
[101] Mordey, 'Alternate current working', discussion on p. 661.
[102] Notes, *Electrician*, 24 (1889), p. 51.

maximum work from coupled alternators) that this would be accomplished when the ratio of armature resistance to self-induction was 2π times the frequency of rotation. In his 1884 textbook, Thompson had identified this to be the condition at which maximum work could be extracted from a single alternator. Yet it is far from obvious from Hopkinson's 1884 paper that this would be the same as the necessary or sufficient condition for parallel operation: Hopkinson merely claimed that his original theory was 'sufficient' to predict all Mordey's results. Tellingly he conceded that it was not 'exactly true', as he had claimed five years earlier that armature self-induction was constant. Yet to treat the self-induction as a variable – as he had admitted was necessary at the Royal Society in 1887 – would render the equations 'unmanageable', that is, unsolvable by his Cambridge methods.[103] Nonetheless, he still would not concede that this theoretically problematic approximation compromised the credibility of his theory, especially insofar as it was a putative account of what made parallel operation possible.

Mordey emphatically disagreed with Hopkinson's claims.[104] Mordey had in fact been present when Hopkinson had presented his 1884 theory to the STEE, and now denied five years later that this theory was 'sufficient' to predict the Brush company's new results. Mordey now had cogent evidence against Hopkinson's privileging of the resistance/self-induction ratio as the definitive criterion of ideal parallel working conditions. This evidence was a set of secohmmeter measurements of armature self-induction produced by Ayrton and a phalanx of his assistants. From these, Mordey concluded that this ratio had a value that was 'about the same' for both Mordey's machine and another type of alternator that definitely could *not* be run in parallel. Given this apparently fatal empirical confutation of Hopkinson's L/R thesis, Mordey asked Hopkinson 'kindly' to look into the matter again. He also noted archly that if Hopkinson had in fact known the secret of paralleling as long ago as 1884, his silence about it since then was somewhat surprising.[105]

Ayrton placed an alternative interpretation of the secohmmeter measurements. Praising Mordey's paper for demonstrating the practical conditions required for parallel running, he added (with no apparent irony) that electrical engineers owed Mordey an 'enormous debt' for showing the empirical correctness of Hopkinson's mathematical conclusions. Ayrton in fact agreed with Hopkinson that the only incomplete point in the latter's 'admirable' theory had been the neglect of the complex reaction between the armature currents and magnetic fields. Having anticipated questions about the variability of self-induction in the Mordey armature, though, Ayrton had taken

---

[103] See the preceding discussion on Hopkinson, 'Note on the theory of the alternate current dynamo'.
[104] Although Mordey's reply to all the discussants was delivered at the end of the meeting, I have merged the two components of the debate for maximum exegetical coherence.
[105] Mordey, 'Alternate current working', discussion on pp. 675–6.

secohmmeter measurements with Sumpner and his fellow students to find out the effect of changing the relative position of armature and iron pole pieces. They visited an installation of a Mordey alternator *in situ* and took measurements on the *static* armature in different relative positions to the iron pole pieces, using the secohmmeter with a commutator rotating at the normal operating frequency of a Mordey alternator. Their results indicated that, during the course of one rotation, the armature's self-induction varied by roughly 12% from a mean of around 0.034 secohms.[106] In reply to this Mordey claimed these measurements showed that the self-induction was not 'sensibly altered' during its rotational cycle – interpreting the 12% variation as effectively negligible![107]

Further participants in the debate gave rather less credence to Mordey's and Ayrton's interpretation of the secohmmeter measurements. Kapp's response was unsurprisingly among the most critical as he obviously had most to lose from any credence Mordey won for his revisionist claims. Kapp effectively dismissed the secohmmeter evidence by asserting that one of Mordey's experiments at the Brush works could not have been performed if the alternators were 'devoid of self-induction' – as the secohmmeter readings seemed to indicate. When two alternators at 2000 and 1000 V were connected and quickly adjusted to a common potential of 1500 V, the orderly behaviour of the former could be explained, Kapp argued, only by the presence of a significant amount of self-induction. This was not commensurate with the 'small' amount of self-induction purportedly revealed by the secohmmeter.[108] James Swinburne accorded even less importance to the secohmmeter's ability to capture the essential aspects of self-inductive action. He concurred with Kapp that, if there had been no self-induction in the armatures, the alternators would not have been able to work synchronously. On the contrary, it was 'obvious' to Swinburne that the Mordey alternators had 'considerable' self-induction because otherwise much damage would have been caused by the 'enormous' current through their armatures.[109] As the IEE President chairing the meeting, Sir William Thomson challenged Kapp's and Swinburne's explanation of Mordey's 'exceedingly beautiful experiment.' Thomson argued that the effect cited by Kapp was due solely to the 'reaction' of armature currents

---

[106] Ibid., 661–4. See table on p. 662. For the sake of comparison the self-induction of a Morse telegraph receiver was approximately 0.1–0.5 secohms, and for a Thomson reflecting galvanometer, 2–70 secohms. These figures taken in May–June 1889 were published in Ayrton and Perry, 'Laboratory notes on alternating current circuits', p. 194.

[107] Mordey, 'Alternate current working', discussion on p. 690.   [108] Ibid., pp. 652–5.

[109] Because all the experiments were in fact 'perfectly in accord' with the existing theories that were due to Kapp and Hopkinson, Swinburne claimed that Mordey was completely unjustified in asserting that these theories were fallacious. Swinburne in discussion in ibid. pp. 655–7.

on the fields; an *Electrician* reporter noted that this argument was 'naturally' heard with great attention and carried 'conviction' to all present.[110]

In the wake of this debate at the IEE, the *Electrician* was conspicuously indulgent towards Mordey's and Ayrton's secohmmeter measurements and uncharitable towards their critics. Its house writer noted, for example, that Kapp's claims about the substantial and variable self-induction in the Mordey armature were not borne out by the secohmmeter measurements. However, it did concede that the reaction between the armature and field-magnets was 'evidently extremely complex' and probably needed a full mathematical investigation that did not take self-induction as a 'starting point' for the analysis of parallel running as Hopkinson and other theorists had done.[111] In the ensuing years Swinburne himself significantly worked on his own form of 'armature reaction' theory as an alternative approach that did not rely on assumptions about constancy of self-induction. Yet, to try to resolve the debate, the *Electrician* still promoted Ayrton's and Perry's programme of secohmmeter measurements on static armatures, thereby sparking a colourful altercation with Swinburne – and considerable indifference from Ayrton and Perry.

5.7. COUNTING 'AYRTON'S AND PERRY'S THINGS': THE SECOHMMETER FURTHER CONTESTED

Ayrton and Perry were less sanguine than the *Electrician* about the possibility that secohmmeter measurements could ever decisively resolve questions about the (variable) quantitative value of armature self-induction. Indeed, they never explicitly declared that their instrument could fulfil such a role in the Mordey debate. When they were asked to fill in an unexpected vacancy in the schedule of IEE meetings in June 1889, they did not mention this debate. This was despite the fact that the largest part of their impromptu paper, 'Laboratory Notes on Alternating Current Circuits', was devoted to reporting students' secohmmeter measurements of the self-induction of galvanometers, telegraph equipment, voltmeters, transformers, and alternator armatures. Yet Ayrton's and Perry's reports of these measurements offered no theoretical interpretation of their significance. This was especially notable with regard to secohmmeter measurements on the static armatures of alternators made by Ferranti and Mather and Platt. These showed that, when stationery, the Ferranti armature had a self-induction of 0.0011–0.0013 secohms and the Mather and Platt armature had a self-induction of 0.005 secohms. Ayrton and Perry remained silent, though, on the seemingly obvious

[110] 'Alternate current working', *Electrician*, 23 (1889), p. 325. This editorial was a response to the recently published account of the IEE debate that had taken place two months before.

[111] *Electrician*, 23 (1889), p. 108. Emphasis added – note that Hopkinson's claim is not given as a 'deduction' or 'consequence' of this primary theory.

fact that the Ferranti alternator could not be run in parallel, despite having a self-induction on an order of magnitude smaller than that of the Mordey armature (about 0.034 of a secohm). Most strikingly of all, Ayrton and Perry made no further comment on this issue in future publications.[112] Neither of them separately undertook any researches with the secohmmeter from this point onwards – and indeed the pair never collaborated on research again, deciding that the distance between Perry at Finsbury and Ayrton at South Kensington made joint work impracticable.[113] Ayrton moved instead to study D'Arsonval galvanometers with his assistants Sumpner and Mather, as discussed in Chapter 4.

Despite Ayrton's and Perry's reluctance to be further drawn on the matter, the *Electrician* was determined to promote secohmmeter measurements as the key to resolving the debate on paralleling. On 28 June 1889 it drew readers' attention to Ayrton's and Perry's recent survey of secohmmeter measurements at the IEE and reproduced their 'Laboratory Notes' in full in the same issue.[114] On 5 July a further editorial noted that although the secohmmeter was a rather expensive piece of apparatus, it should be regarded as 'indispensable' for all workers in alternating currents,[115] and hoped that either Ayrton himself or 'others interested' would add considerably to the number of such measurements. Yet the persistent attempts by the *Electrician* to enforce interpretive closure on the Mordey controversy by reference to secohmmeter measurements did not succeed.

Just after Mordey's paper was published in the JIEE, a rather tetchy leading article in the *Electrician* on 2 August argued that engineers still lacked a comprehensive set of rules for accomplishing successful ac paralleling. They derived little guidance from the 'differences among the experts' as to whether the solution depended on 'abundance' or otherwise of self-induction in the armature. Although the writer admitted that both the self-induction and the resistance of Mordey's alternator were small in magnitude, it allowed that his critics might well disagree that these factors alone could explain his success in paralleling.[116] The journal's editorial staff were now prepared to entertain Hopkinson's post hoc stipulation that the crucial consideration in alternator paralleling was the frequency of rotation; that the ratio of armature self-induction to resistance should be optimized to a value of about $1/6$ (i.e., $1/2\pi$) of the machine's rotational period. There had thus far been no explicit calculation of this ratio for Mordey's alternators to see if they

---

[112] Although a former student of Ayrton's, R. W. Weekes, listed numerical values for self-induction and resistance of a Ferranti alternator, no link was made to the earlier debates. See *Electrician*, 25 (1890), p. 187.

[113] See obituary by John Perry, 'William Edward Ayrton', *Proceedings of the Royal Society*, 85 A (1911), pp. i–viii; Graeme Gooday, 'William Edward Ayrton', in *New dictionary of national biography*, Oxford: Oxford University Press, forthcoming, 2004.

[114] *Electrician*, 24 (1889), p. 186.   [115] *Electrician*, 24 (1889), pp. 211–12.

[116] 'Alternate current working', *Electrician*, 24 (1889), pp. 324–5.

met this condition at paralleling. And if Hopkinson were right and Mordey wrong, alternators with iron-cored armatures and hence large self-induction would not be at a disadvantage so long as their internal resistance were accordingly higher too.[117] The journal thus challenged those engineers who considered that some specific amount of armature self-induction (small or great) was essential for paralleling to now produce measurements to prove this. If only Ayrton and his students would use the secohmmeter in a 'large campaign against all sorts and conditions of alternators', the quantitative evidence produced was likely to promote 'real progress'.[118]

As is clear from a letter published in the *Electrician* on 9 August 1889, James Swinburne strongly disagreed with this prognosis. Notwithstanding the favourable opinions generally accorded to Ayrton's and Perry's secohmmeter, Swinburne contended that everybody except he had been confused 'hopelessly' on the issues of self-induction in Mordey's experiments. The average 'paper man' still appeared to think, he complained, of self-induction as having a constant coefficient whereas a great many 'dynamo men' knew better. He bluntly denied that the behaviour of moving armatures could possibly be captured by reference to only self-induction,[119] especially if it were measured on static alternators, as Ayrton's and Perry's measurements had been:

I must say I do not think we will get much further if self-induction and armature reactions are sometimes mixed-up and sometimes separated, and I really doubt whether an army of third-year students, glittering with secohmmeters telling us how many of Ayrton and Perry's things there are in each stationary armature, will get information fit for anything better than a paper.

Swinburne concluded that at least one of the participants at the IEE discussion must have talked a 'great deal of nonsense', and effectively agreed with Hopkinson's account that an optimum ratio of self-induction to resistance was the solution to the paralleling problem.[120] A week later an editorial note of 16 August in the *Electrician* now reverted to attacking Hopkinson's position and supporting Mordey. The reason for this was that Mordey had produced compelling evidence that Brush alternators could run in parallel at *any* frequency. Its hostility to Swinbure was undiminished, however. Noting sarcastically that Swinburne was 'fond of assuring us' that self-induction was a variable, it declared that, unless he could show that the

---

[117] Ibid., p. 326.
[118] Ibid., p. 325. The literary allusion is to Walter Besant, *All Sorts and Conditions of Men*, London, 1882, on the philanthropic founding of an educational institution for London's diverse East End.
[119] Hopkinson had explicitly argued the exact opposite against Ayrton and Thompson in 1884: see precedng discussion.
[120] *Electrician*, 24 (1889), p. 360.

value of L in Mordey's machine varied inversely with the speed (thus keeping R/L equal to $2\pi \times$ freqency of rotation), they were 'at a loss to imagine' how Swinburne could possibly continue to maintain his support for Hopkinson.[121]

The continuation of Swinburne's vituperative correspondence with the *Electrician* elicited a variety of responses from its readers.[122] One of Ayrton's former students who had been present at Hopkinson's 1884 lecture (see Section 5.3) took against the *Electrician*'s polemical line. Llewellyn Atkinson wrote in on 23 August to question whether Hopkinson's thesis was necessarily incompatible with Mordey's results, and whether indeed there really was a single optimal solution to parallel running:

> [Mordey] has not, I maintain, proved that [resistance and self-induction] should both be absent, or that the ratio between them ought to have some value very different to that pointed out by Dr Hopkinson. Neither has he proved that an alternator with iron in the armature cannot be made to satisfy the conditions of running in a parallel as well as his own, whilst experiment shows that with properly designed iron cored armatures all the experiments he described... can be repeated with precisely similar results. [I] will conclude by asking what experimental evidence or what theoretical reason is there, to suppose that two alternators, in whose combined armature circuit there is no resistance and no self-induction, will run in parallel?[123]

Yet after a further burst of invective against Swinburne on 30 August, the combative editorials came to a sudden halt. As everybody concerned was so 'little agreed' as to the data on which the discussion should be based, it seemed 'useless' to pursue the argument further, although it hinted that Mordey had granted the journal confidential access to 'certain experimental data' that might yet resolve things. The journal's staff indeed had little time for such a dispute at that juncture as they were busily preparing for the imminent Electrical Congress in Paris.[124] Despite occasional mention of successful parallel installations of Mordey alternators hereafter,[125] this unresolved debate disappeared from the pages of the technical press, displaced in part by new discussions such as the name and magnitude to be given to an international unit of self-induction to be discussed at the Paris Congress.[126] In 1890, Swinburne himself pursued his own line on the subject in the journal *Industries* that he owned and edited. He reiterated 'somewhat dogmatically' that the key to understanding the conditions of paralleling lay in an

---

[121] Ibid., 24 (1889), pp. 367–68.
[122] William B. Sayer considered it 'strange' that at least some of the contested points had not been settled. See his letter to the editor, *Electrician*, 24 (1889), pp. 387–88.
[123] Ibid., p. 413. [124] Ibid., pp. 417–8.
[125] See, for example, a letter from George Hooker to the editor in *Electrician*, 25 (1890), p. 690.
[126] William E. Ayrton, 'The practical unit of self-induction', *Electrician*, 23 (1889), p. 569.

understanding of how a substantial 'armature reaction' served to govern the performance of alternating machines when they were coupled together.[127]

By the time the *Electrician* debate ceased in the summer of 1889, it was becoming harder to doubt that alternators of radically different design patterns could all be be run commercially in parallel. Westinghouse's 'iron-type' machines had been paralleled in the USA since 1888; from late 1889 Mordey's iron-free armature machines had been supplying Bournemouth, and similar Elwell–Parker-designed alternators were used at Brompton and Kensington. Such pluralism bore out Atkinson's suspicion that minimal alternator self-induction was neither a necessary nor sufficient precondition of paralleling. Not surprisingly then, the debate over ac paralleling which continued for several years thereafter no longer focussed on self-induction and made no further reference to secohmmeter measurements. R. H. Parsons notes that disputes over the relative merits of Mordey 'copper-type' armatures and Westinghouse 'iron-type' armatures were only one part of the paralleling debate, and attention was soon focussed on other aspects of ac practice as the possible differentiating cause of success or failure in paralleling. Some commentators maintained that paralleling had been possible at Bournemouth and Brompton because the elasticity of the belts or ropes used to connect the alternators to steam engines helped to absorb and correct any lack of synchronization between them. Accordingly they maintained that alternators *directly* driven by steam engines could not be run in parallel. However, in July 1891, Mordey demonstrated at the Willans and Robinson plant in Thames Ditton that his alternators most definitely could be paralleled when directly driven by Willans high-speed engines (see Chapter 4). In response to that a commentator in *Industries* suggested somewhat fawningly that under Mordey's personal supervision 'any' large alternators could be made to run in parallel.[128] Although other commentators speculated that parallel running depended largely on the use of alternators with identical waveforms, specifically of a sinusoidal form,[129] much of the debate thereafter moved away from the alternators and their connections to the mechanical arrangements for driving them.

By the time Mordey made his next public pronouncement on the subject at the IEE in 1893, his treatment of parallel working acknowledged that it was as much a question of 'prime motors' as of electrical apparatus. As Edison had privately observed of dc supply in 1882, this was especially important as regards the regulation of steam to the engine for each alternator (see

---

[127] James Swinburne, 'The coupling and control of dynamos in central stations', *Industries*, 9 (1890), pp. 290–1.

[128] Parsons, *Early days*, pp. 143–4. See editorial, 'Mordey–Willans combination running in parallel' in *Electrician*, 27 (1891), pp. 298–9 and compare with editorial, 'The coupling of alternators', *Electrician* 26 (1891), pp. 394–5 and Anonymous, 'Running alternate machines in parallel', *Industries*, 11 (1891), p. 86.

[129] On the issue of waveforms, see Brian Bowers, *A history of electric light & power*, Stevenage, England: IEE/Peregrinus, 1981, p. 67.

Figure 11. Alternators running in parallel at the Engine House at the Amberley Road Electric Lighting Station, West London, ca. 1893, showing layout of belt-linked alternators and steam generators, all superintended by the station foreman. Note the array of direct-reading instruments on the switchboard on the rear wall, and the activity of the switchboard attendant in monitoring them. [*Source*: 'Amberley Road Electric Lighting Station – The Engine House', *Engineer*, 75 (1893), p. 523.]

Figure 11). Mordey's recommendations no longer concerned the resistance or self-induction of armatures, but now included the use of a separate engine for each alternator, each supplying steam at the same pressure.[130] As a trained mechanical engineer, Swinburne disagreed, arguing that to secure parallel running the generating station had to be considered as a whole, with a common governing device for all the steam-engines to keep them synchronized. It was 'foolish', he maintained, to try to run a number of alternators in parallel off independently governed engines.[131] By 1894, when paralleling was relatively common in the ac industry, the further significance of prime movers

[130] William M. Mordey, 'On testing and working alternators', *JIEE*, 22 (1893), pp. 117–34, esp. pp. 131–4 and following discussion.

[131] Ibid., discussion, p. 180. At that 1893 IEE meeting Hopkinson was present, but at a meeting the following year when Hopkinson was absent, Mordey reverted to his former position: 'For four years past I have maintained that the practical solution to this problem is quite different from that deduced by Dr Hopkinson from his differential equations. That is to say without disputing the correctness of his deduction, which made self-induction the controlling factor in the operation, I have argued on the basis of experience, that the best and most perfect synchronizing as to be obtained by ... the absence rather than the presence of

was apparent when the Scottish House-to-House Electric Lighting Company found it could not run Mordey alternators in parallel when driven by gas-engines. This was final proof, if any were needed, for contemporaries that Mordey's 1889 specification of alternator design was insufficient to accomplish paralleling: Only when the gas-engines were replaced with a steam plant in the following year could this company run his alternators in parallel.[132] Indeed, the historian of early electrical supply, R. H. Parsons himself, considered that the whole issue of synchronizing alternators was a matter of using identically constructed alternators with the same waveform and the consistent matching of their speeds through well-regulated mechanical driving.[133]

Given the shift to mechanical engineering concerns in the 1890s, the electromagnetic principles of subsequent alternator design practices evinced considerable pluralism. For example, in the fifth (1896) edition of his best-selling *Dynamo-Electric Machinery*, S. P. Thompson offered a long list of recommended mechanical conditions for the construction and operation of alternators to be run in parallel. Not especially prominent among these was the early Mordey specification that in 'well-designed' machines the resistance and self-induction should be as small as possible.' This was immediately qualified by the recommendation that the ratio of resistance to self-induction should be related to the working frequency to the order of magnitude recommended by Hopkinson in 1889.[134] Yet it is clear from the prevalent diversity of design specifications listed by Thompson that neither Mordey's nor Hopkinson's electromagnetic stipulation was sufficient or necessary for effective engineering practice. Thompson included not only a lengthy discussion of Mordey's famously low self-induction model[135] but also a brief account of Hopkinson alternators. '[I]n spite of' their high self-induction, these were suitable for working in parallel, and had 'so much armature reaction' that they could be operated with safety, that is, without current surging.[136] Evidently Hopkinson the erstwhile theoretician had learnt more from the practical stipulations of Kapp and Swinburne than from those of Mordey, but with no less success. And it is pertinent to note that when W. E. Ayrton reviewed the late Hopkinson's career in 1904,[137] he noted that it was now

self-induction.' W. M. Mordey, 'On parallel working, with special reference to long lines', *JIEE*, 23 (1894), pp. 260–80, discussion on pp. 280–314.

[132] Parsons, *Early days*, p. 145.
[133] William E. Ayrton, 'The life of a scientific engineer', *Nature* (London): 70 (1904), pp. 169–72, quote on pp. 169–70.
[134] Sylvanus P. Thompson, *Dynamo-electric machinery*, 5th ed., 1896, p. 600.
[135] Ibid., pp. 619–24.   [136] Ibid., p. 613.
[137] Hopkinson and three of his children died in a catastrophic alpine mountaineering accident in 1898; Greig, *John Hopkinson*, pp. 39–40. At the time, Ayrton told the BAAS 'It has often been said that he who leads in mathematics cannot follow the engineering life of Westminster. But a striking disproof of the generality of this statement was furnished by

acknowledged that Hopkinson's theoretical account of 1884 did not actually apply to all machines. Like many other engineers who struggled to deal with the technological manifestations of self-induction, Hopkinson had been 'more influenced' by his experimental trials on commercial alternators than by his 'theoretical reasoning'.

5.8. EPILOGUE: THE LINGERING MARGINAL CAREER OF THE SECOHMMETER

The obituary of neither William Ayrton nor John Perry mentioned the secohmmeter as one of their more significant accomplishments, and no account of the development of ac supply grants it any significant status in the resolution of the problems encountered in parallel operation. Indeed, the secohmmeter never featured in any heroic success story of electrical engineering, nor was it granted a substantial niche in the everyday work of electrical engineers. The apparent subsequent marginality of the secohmmeter is partly because both Ayrton and Perry ceased to show much interest in this device in 1889 and did not seek opportunities to present the secohmmeter as a source of decisive measurements of self-induction in technological problem-solving. Significantly, Ayrton's and Perry's unit of self-induction as the secohm was displaced by the 'henry' at the Chicago International Congress of 1893, so the name of the instrument would thereafter have appeared somewhat idiosyncratic and local to London. Although Ayrton's students continued to use this instrument (and the unit name) on their course at the City & Guilds College in South Kensington, they also learned non-secohmmeter methods to ascertain the self-induction of ac machinery in normal working conditions. One trio used such alternative methods in 1891 to show that the self-induction of a Ferranti armature did not vary with external load and that the 'armature reaction' could be fully accounted for in terms of armature self-induction.[138] Thus even those taught by Ayrton did not treat secohmmeter techniques as a universal solution to the problem of measuring self-induction. Unlike the ammeter it could not be used on moving machinery and could not be used to take instantaneous readings of changes in self-induction – a critical matter, given Swinburne's relentless assertion that coefficients of self-induction were not constant in working ac machinery.

When Ayrton's counterpart at the University College of London, J. A. Fleming, published a standard set of advanced laboratory exercises for electrical engineering students in 1898, nowhere did measurements of

---

the brilliant work in the domains of theory and practice which was accomplished by him for whom we mourn.' William E. Ayrton, Presidential Address to Section A, *BAAS Report* (1998), Part 2, p. 767.

[138] See R. W. Weekes in discussion to Mordey, 'On testing and working alternators', pp. 186–8.

self-induction by the secohmmeter or any other technique appear.[139] This perhaps reflected the ways in which self-induction never became as pressing a subject for measurement in commercial practice as current, resistance, or potential difference. Even so we know that secohmmeters were evidently used by some practitioners outside of Ayrton's laboratory. In 1895 Nalder Brothers of London were still developing the commercial form of this device and the variable inductance standard that often accompanied it, and production was continued by Pye into the 1920s.[140] A commentary in the *Electrician* for March 1895 described the secohmmeter as having 'increased in simplicity and reliability since the day of its birth' in the previous decade.[141] Nevertheless it existed only as a relatively minor part of the electrician's repertoire and even then not necessarily in the role that Ayrton and Perry had made for it. The article just mentioned, clearly written by a sympathizer, recommended that it could be used to measure not only 'inductances' but also capacitances and 'polarizing resistances'.[142]

Ironically, when the secohmmeter was used to measure self-induction, it was not as an autonomous instrument but in an auxiliary role to other rival techniques that were developed in the 1880s and 1890s – but not generally as direct-reading instruments. In his *Handbook for the Electrical Laboratory* of 1902–3, Fleming noted that, although methods for using double-circuit interruptors to measure self-induction were still known by the name of secohmmeter, this was just one of many techniques available for this purpose. Fleming regarded the secohmmeter primarily as a useful *adjunct* to enhance the sensitivity of other techniques of measuring self-induction. Yet he had found he encountered a 'great deal' of problems with the moving contacts in this device, and these inspired him and his assistant W. C. Clinton to devise their own motorized form for their own laboratory usage.[143] And it was in the role of an auxiliary aid to bring greater sensitivity to induction balances that the secohmmeter featured in successive editions of B. Hague's *Alternate Current Bridge Methods* into the middle of the twentieth century.[144]

5.9. CONCLUSION

Self-induction did become interesting and important to electrical engineers, as Rayleigh had predicted in 1884. Yet it never become as ubiquitous a

---

[139] John Ambrose Fleming, *Laboratory notes and forms*, London: 1898.
[140] See Kenneth Lyall, *Electrical and magnetic instruments*, Cambridge: Whipple Museum, 1991, (unpaginated), item 459.
[141] Anonymous, 'The measurement of inductances, capacities and polarizing resistances', *Electrician*, 34 (1895), pp. 546–7.
[142] S. R. Milner, 'The use of the secohmmeter for the measurement of the combined resistances and capacities', *Philosophical Magazine*, 6th series, 12 (1906), pp. 297–317.
[143] John Ambrose Fleming, *Handbook for the electrical laboratory*, 2 Vols., London: 1901–3, Vol. 2, pp. 170–207.
[144] Brian Hague, *Alternate current bridge methods*, 5th edition, London: revised 1957, pp. 3–6.

Coupled Problems of Self-Induction 217

preoccupation of ac lighting engineers as resistance, current, and potential difference, and (instantaneous) measurement of it was not commercially embodied in a direct-reading apparatus in quite the same way as it was for those other parameters. Insofar as self-induction mattered to them in the operation of technologies for communication and lighting, its magnitude could be predicted and constrained by the use of design theories. These drew to a greater or lesser extent on the work of Faraday, Maxwell, and others, so the need for measurements of self-induction was not pressing. And insofar as self-induction mattered in the operation of ac generators it was latterly held to be only as a constitutive part of the highly complex phenomenon of 'armature reactance' on which purely Maxwellian theories of self- and mutual-induction had little to say. Later Victorian electrical engineers, such as James Swinburne, did not generally attempt to measure this phenomenon of armature reactance but rather to use this phenomenon to their advantage in stabilizing the operation of alternators, finding discussion of self-induction to be singularly futile.

As the arguments among Swinburne, Ayrton, and the *Electrician* in 1889 showed, self-induction per se was not an uncontroversially measurable property anyway. In situations in which self-induction was measurable with a secohmmeter, the utility of the values so obtained was far from clear. If these measurements were taken of a tractable piece of stationary equipment, these did not obviously represent useful values of self-induction for when the machine was moving, but then the secohmmeter simply was not made to register variations that were suspected of occurring in a commercially operating machine. The complexities of self-induction in ac machinery challenged Ayrton's and Perry's expertise in electrical-measurement techniques well beyond what they considered usefully achievable: They did not try to produce a direct-reading instrument that could *instantaneously* represent the changes in so circumstantially contingent a first-order variable attribute of conductors as self-induction. They had more useful things to measure and enhance their reputations thereby.

With regard to the Mordey–Hopkinson debate in 1889, it is easy to see how a conviction emerged that self-induction was a crucial component in establishing parallel alternator running. This would have been a rather obvious inference from the insistence of Rayleigh, Thompson, and others that self-induction was the uniquely new feature of ac generation that explained why ac machinery behaved so perplexingly differently to the more familiar 'commonsense' dc machinery. Given also that in recent years self-induction had been identified as the crucial technological parameter for other new developments in lightning conductors and long-distance telephony, it is not surprising that a similar identification was made in the case of ac power. Yet the often-understated significance of mechanical engineering issues in electrical technology displaced the debate away from the expertise of professors and practitioners in self-induction: The problem of paralleling was solved by drawing on a broader spectrum of engineering resources that were much more to do with the operation of steam technology and the mathematical analysis of mechanical stability than with the abstruse electromagnetic presence of self-induction.[145] Insofar as that is a legitimate conclusion of this chapter, the preceding discussion has been an account of the *non-victory* of Maxwellian 'theory' over untutored engineering 'practice'.

Whilst my account has re-evaluated the 'battle of the systems' in terms of a 'reverse salient' for ac generation that profoundly affected the consumer's experience of electric lighting, the next chapter will examine how the challenges of securing reliable and trustworthy domestic meters was a critical problem for both suppliers and customers and for both technologies of ac and dc power.

---

[145] Stuart Bennett, *A history of control engineering*, Stevenage, England: IEE/Peregrinus, 1979, pp. 170–1. I thank Colin Hempstead for drawing my attention to this source. See also Bennett's more recent piece, '"The industrial instrument – master of industry, servant of management": Automatic Control in the Process Industries, 1900–1940', *Technology and Culture*, 32 (1991), pp. 69–81.

6

Measurement at a Distance: Fairness, Trustworthiness, and Gender in Reading the Domestic Electrical Meter

Happy will be the man who succeeds in inventing a meter combining simplicity with exactness.
            Willoughby Smith, Discussion at STEE, 1883[1]

By far the most important instrument of all is the meter... The enormous difference in the revenue due to inaccurate meters does not seem to be fully realised. A meter that reads 2 or 3 per cent wrong, may make all the difference between working at a loss or at a profit. It may be urged that inaccuracy does not matter, because it tends to average about right. This is, however, very doubtful... Meters which are corrected when adjusted at the same temperature, may vary very largely if one is placed in a cold cellar and the other in a warm front hall, or in a kitchen.
            James Swinburne, 'Electrical Measuring Instruments', ICE, 1892[2]

Of all the measurement devices used in early electric-lighting projects, the domestic meter was the most commercially significant. More of them were manufactured for the constituency of domestic consumers than the ammeters produced for practising electrical engineers. Yet, as the only electrical instrument expected to operate reliably for long periods in inclement conditions far away from company surveillance, it was also the most problematic to construct and operate within the commercially useful degrees of accuracy cited by James Swinburne in the second epigraph in the opening of this chapter. Several historians have noted the way that trust in physical measurements is a problem when undertaken out of view of powerful expert tribunals and at some remove from centralized means of calibration.[3] In the early 1880s it was hard enough to get an ordinary electrical instrument such as a tangent galvanometer to read consistently to within about 1% in the

---

[1] Willoughby Smith, Presidential remarks in chairing discussion of James Shoolbred, 'The measurement of electricity for commercial purposes', *JSTEE*, 12 (1883), pp. 84–107, discussion on pp. 107–69, quote on p. 169.
[2] James Swinburne, 'Electrical measuring instruments', *Proceedings of the Institution of Civil Engineers*, 110 (1892), pp. 1–32, quote on p. 19–20.
[3] Rob Iliffe, '"Aplatisseur du monde et de Cassini": Maupertuis, precision measurement, and the shape of the earth in the 1730s', *History of Science*, 31 (1993), pp. 335–75; Joseph O'Connell, 'Metrology: the creation of universality by the circulation of particulars', *Social Studies of Science*, 23 (1993), pp. 129–73.

well-controlled laboratory environment. And even that could only normally be achieved with the care of expert hands and with regular daily or even hourly recalibration for changes in magnetic field (see Chapter 4). So what kind of trustworthiness could be expected in the readings of domestic meters that had to be left unattended for weeks on end and open to the meddlings of consumers?

The supply meter was one of several possible solutions in the mid to late 1880s to the problem of how to measure the consumption of customers connected to newly emerging central supply systems.[4] This quantification had to be undertaken in a manner that both householders and suppliers could trust, or at least find it hard to make successful imputations of untrustworthiness against it. In the financially risky new enterprise of electric-lighting supply, companies needed a reliable billing method to avoid operating at a loss, but equally importantly had to know how to pre-empt customer suspicions of overcharging – suspicions that were by no means always unjustified. This latter issue was a crucial strategy for handling fickle consumers who might revert back to gas lighting if dissatisfied with the expensive and often unreliable new electric alternative. Those suppliers familiar with the levels of customer complaints in the gas-supply industry clearly could not take it for granted that householders would all be passively uncritical consumers. Not only did some complain persistently about overcharging or the irregular performance of meters, but others adopted the more practical approach of fraudulently tampering with meters to reduce their readings to a level more congenial for household budgets.[5] As we shall see, the pragmatic demand between both electricity suppliers and customers for a metering technique that both could trust in principle was tempered by suspicions of untrustworthy conduct on both sides.

As discussed in Chapter 2, the initial problem for potential customers of Edison in the early 1880s was that it was difficult for them to see how such an intangible and ineffable a commodity as electricity *could* be meaningfully metered like piped water or acrid coal-gas.[6] Such a problem added to householders' difficulties of trusting and transacting with remote and unseen individuals by the technological mediation of meters to which the

---

[4] Meters were not generally used for the stand-alone domestic installations in the first half of the 1880s for which an overall periodic rental fee was common. Some rich and famous households, however, were offered free installations for the purposes of publicity and marketing. See, for example, the discussion of Lord and Lady Randolph Churchill's installation in their London home in 1883 in Rookes E. B. Crompton, *Reminiscences*, London, 1928, pp. 108–9.

[5] The reading of early electricity meters was arguably as prone to fraud and mishap as Carolyn Marvin has shown to be the case for the contemporary novelty of the domestic telephone. Carolyn Marvin, *When old technologies were new: Thinking about electric communication in the late nineteenth century*, Oxford: Oxford University Press, 1988, pp. 92–7.

[6] Frank Lewis Dyer and T. M. Morton, *Edison: His life and inventions*, 2 Vols, New York/London: 1910, Vol. 1, p. 406.

former were prohibited from gaining access. Domestic meters were probably already familiar to many householders from gas and water installation, and the operation of these in registering bulk flow through a supply pipe was probably understood if not entirely trusted (see subsequent Sections 6.2). A further problem arose for the new technology: How could customers trust the testimony of an electric meter if they could not understand the arcane electrical mechanism by which it worked? Further questions arose specific to electrical meters that were not precedented in the case of gas meters. Should the customer be entitled to read the meter anyway or should this be the unique prerogative of the supplier, such as the Edison company, in whose honesty the customer was obliged to place confidence? And then again, what actually was it that a given meter registered, and was it the appropriate or fair quantity to meter? More specifically, was it fair or practicable to bill householders for consuming 'electricity', energy, or lighting, and with what guarantees about the constancy of terms and settings at which these were supplied? These were the sorts of questions central to the first two decades of metering, and to which this chapter is devoted to exploring.

Whereas previous discussions in this book focussed on the electrical engineers who designed measurement instruments and who constituted the primary constituency of users, this chapter engages more with the agency and discretion of the electrical consumer. The constructivist approach of Bijker and Pinch is a useful pointer to considering how domestic householders were a particularly relevant social group – 'the domestic audience' in the construction of various kinds of domestic meters. Notwithstanding Willoughby Smith's Presidential response to the first major discussion of electrical meters at the STEE in 1883 (see the epigraph that opens the chapter), it was neither necessary nor sufficient for meters to be 'simple' and 'exact' to bring happiness to designer or consumer. Indeed, in Sections 6.2–6.4, we shall see that, by the 1890s, customer preferences had effectively forced engineers, most notably those of the Edison company, to build meters modelled on contemporary gas meters, albeit with the complexity and awkwardness of their counter-rotating dials.

In Section 6.5, though, I will show there was some dispute among electrical experts about whether the construction of many meters to register the quantity of electricity passing rather than the quantity of energy or lighting supplied was in fact in accordance with consumer interests. Finally, in Section 6.6 I examine the identity of the domestic meter reader from the perspective of gender issues, noting that even those women who were well informed about the costs of early domestic electrical lighting represented the responsibility for reading the meter as if it were a uniquely masculine prerogative. My conclusion will draw out the problems in the ambiguity of consumer interest and of consumer identity faced by a social-constructivist account in attempting to document the development of the domestic electrical meter.

6.1. THE HISTORIOGRAPHY OF THE DOMESTIC ELECTRICAL METER

At present, while electric light is, to some extent, a novelty, the housing of the meter receives some little extra attention; but it will soon share the fate of the gas-meter, and be put in any out-of-the-way cellar, where it will be exposed to dust and damp, and perhaps traces of corrosive gases. In addition to this, most meters are slightly warm, and this attracts all sorts of insects... it is foreseeing such contingencies and providing for them, that constitutes the principal difficulty in developing a new industry.

James Swinburne, 'Electrical Measuring Instruments', ICE, 1892[7]

Although absent from most previous histories of electrical lighting written by socio–cultural historians or historians of engineering, the domestic meter has at least been discussed in the curatorial and economic accounts of early electrical supply. It is these bodies of literature that most usefully set the scene for my consumer-oriented account. Neil Brown's survey of dc and ac meters between 1880 and the early 1890s, based on holdings at the Science Museum in London, highlights the diversity and varied development of mechanisms deployed. Just as the account by Stock and Vaughan (see Chapter 4) focussed on the internal mechanisms used in galvanometers and ammeters for producing instantaneous registration of current, Brown's account focuses on the way that designers sought a continuous means of registering the temporal passage of current by harnessing various secondary effects. First developed was the electrolytic meter, Edison's 1882 form of which hinged on the deposition of a metal by electrodes in a current-carrying electrolyte, with Lowrie-Hall adapting a form for ac use in 1887. In the double-clock mechanism produced by Ayrton and Perry (1882), Aron (1884), and by Albert Einstein's father and uncle (1890 – see subsequent discussion), the passage of a current provided an electromagnetic braking effect on one clock relative to another. Latterly predominant in Britain was the motor meter that appeared in two major forms. From 1884 Ferranti produced a dc 'mercury-motor' meter in which the current-carrying liquid metal both rotated a vane and provided the braking friction that gave it a constant speed. The 'induction-motor' meter produced by Shallenberger in 1889 for Westinghouse's ac supply relied on mutual-induction effects (see Chapter 5) to rotate an iron disc, with braking provided by attached air-vanes.

Brown's account hints that the Shallenberger ac meter and the Ferranti dc meter were ultimately the most widely used in Britain because these were the first to receive formal endorsements from the Board of Trade – in 1892 and 1896, respectively – uniquely recommending each as the best for their respective system. In contrast to a standard constructivist account, Brown's implied emphasis is thus on authoritative state approval as the rationale for the widespread acceptance of these meters rather than Pinch's and Bijker's

[7] Swinburne, 'Electrical measuring instruments', pp. 21–2.

emphasis on stake-holder consensus. Brown's approach here is thus more readily reconciled with Mackenzie's notion of ordinary technology users' reliance on authority to authenticate the properties of technology.[8] Other aspects of Brown's analysis do more easily fit a constructivist analysis because he emphasizes the many other shorter-lived forms of meter which were patented but not widely taken up in the market place. These included George Forbes' thermal-convection meter, Hopkinson's and Hooker's commutator motor meters, and a wide variety of intermittent integrating meters made in France, Britain, and the USA which were highly popular circa 1890 – a form of the latter even being patented by Sir William Thomson. In terms of cultural comparisons, Brown usefully notes that the lattermost of these were most popular in France, their merits being established in characteristic fashion by open prize competition.[9]

Brown suggests that the widespread use of meters in connection with electrical supply was a phenomenon still fairly new in 1889–90. This is not surprising, given that this period saw a large growth in the centralized supply system begun in Britain in response to the more supplier-friendly legislation of 1888, which rescinded the somewhat restrictive terms of the 1882 Electricity Supply Act.[10] Nor is it surprising, given the expense and questionable reliability of meters available before then, that early small-scale electricity suppliers elected to use other methods of charging customers. Brown discusses two other methods used in Britain and France during the 1880s. One was the fixed-price contract in which customers were charged an annual fee based on the number of lamps installed; for example, the Metropolitan supply company's dc installation launched at Whitehall Court, London, in 1888 charged customers 30 shillings per annum for each 8 candlepower (CP) lamp, although it did use meters when it opened its ac installation at Sardinia Street in 1889. Another less widely used system was that of charging per unit of time governed by clockwork devices attached to each lamp fitting. Only one instance of this in use has thus far been identified in the Colchester installation of the South Eastern Brush Electric Light Company from 1884 to 1886; customers were charged half a penny per lamp hour, with an alternative tariff of 2d per day in summer and 3d in winter.[11]

---

[8] Donald Mackenzie, 'How do we know the properties of artefacts? Applying the sociology of knowledge to technology', in Robert Fox (ed.), *Technological change: Methods and themes in the history of technology*, Amsterdam: Harwood Academic, 1996, pp. 247–63.

[9] C. Neil Brown, 'Charging for electricity in the early years of electricity supply', *PIEE*, 132 A (1985), pp. 513–24.

[10] See Thomas P. Hughes, *Networks of power: Electrification in western society, 1880–1930*, Baltimore: Johns Hopkins University Press, 1983, and Leslie Hannah, *Electricity before nationalization: A study of the development of the electricity supply industry in Britain to 1948*, London: Macmillan, 1979.

[11] Brown, 'Charging for electricity', pp. 513–4.

Brown suggests that the decision to abandon such systems and move to metering was specifically a supplier's decision taken on the grounds that supply companies that charged by a metering system were more profitable than those which did not. It is indeed rather unlikely that this transition could have been a result of customer pressure as the fixed-price contract system in particular favoured the interests of customers. They could leave their lamps on all day and all night and still be charged the same as a less illuminated householder – much to the detriment of the supplier's profits! The plausibility of Brown's supplier-centred economic argument for the adoption of metering gains further strength if one considers that, throughout the 1880s, electric lighting was still a highly risky field of entrepreneurial speculation, and there were various financial disasters that repelled many potential investors. Among the most notorious of these were the bursting of the £7 million Brush company 'bubble' in 1882–3, the commercial failure of Edison's first lighting projects in Pearl Street in New York and Holborn Viaduct in London in 1884, and the collapse of the Ferranti Hammond company in 1885. It did not help either that in the last two decades of the nineteenth century, the average cost of electricity supply in the UK was roughly twice that of gas: This price differential was nearer 3:2 in the USA.[12] Hence, unsurprisingly, the take-up of electricity in the UK was minimal until the mid-1890s – see subsequent discussion in Section 6.6.[13]

In his study of Sebastian Ziani de Ferranti's career, J. F. Wilson points out that the young engineer's agent and lawyer, Francis Ince, rejoined him as a partner in a renewed ac venture in 1885 only because Ferranti had developed commercially feasible meters. He quickly sold these to suppliers in Britain, France, and Belgium, presenting them in his first advertising copy of July 1885 as 'the only commercial meter up to the present time'; this was presumably an arch remark directed at the rather problematic Edison dc meter. Wilson also observes that Ferranti shrewdly made meters for both ac *and* dc systems, and it is (ironically) the latter that were initially the source of his main profits, and these kept the company afloat during the difficult years following the problems at his ac installation at Deptford in 1891.[14] Whilst some early suppliers were content to buy meters from Ferranti, others designed their own as integral parts of their supply system. In 1890 Hermann and Jakob Einstein installed the patented Einstein–Kornprobst double-clock meter to customers of their Munich-based dc supply business until their failure to win crucial local contracts led to bankruptcy four years later.[15]

---

[12] Hughes, *Networks of power*, pp. 63–4.
[13] Ian C. R. Byatt, *The British electrical industry, 1875–1914 : The economic returns to a new technology*, Oxford: Clarendon, 1979.
[14] John F. Wilson, *Ferranti and the British electrical industry, 1864–1930*, Manchester: Manchester University Press, 1988, pp. 13–23, 26–7, 59–63.
[15] Hermann and Jakob were, respectively, Albert Einstein's father and uncle. Lewis Pyenson notes that the ill-fortune of J. Einstein & Cie was a significant feature of Albert's difficult

The comprehensive centralized planning of electricity supply systems was, as is well known, developed by Thomas Edison on the model of the gas system from 1879. One historian of the Edison business and two biographical studies have noted the role of the Edison electrolytic meter in emulating the role of the domestic gas meter, and noted the problems encountered by Edison in getting the meter to work reliably and to be accepted by his customers.[16] Nevertheless, such topics are not mentioned at all in Hughes' account of Edison's work in *Networks of Power*.[17] This work focusses instead on the large-scale engineering questions of economy and reliability and the comparative international politics of civic planning. Within the systems-building historiography of electric light espoused by Hughes, there is no significant place allowed for the contingent responses of customers, particularly their concerns about the reliability of meters. Instead, Hughes appears to assume that customers passively accepted the supply given them by companies in the master plans of the great system builders Edison and Westinghouse and does not consider that economic issues of payment by them to be as significant as the financial issues of machinery and transmission. I argue, however, that problems with his electrolytic meter were a major 'reverse salient' for Edison in the first decade of his supply. Indeed, by looking more carefully at the debate over meters, I explore the extent to which domestic consumers had at least some power to resist the Edison company's imposition of its metering system on them.

6.2. THE GAS-METER PARADIGM OF MEASURING AT A DISTANCE

A meter should be direct-reading; preferably it should have an index like that of a gas meter. This is not a matter of real importance; it is a concession to prejudice. When gas came into use gas chandeliers were made, and it was pretended that candles were

---

teenage years and prompted him to retreat into studying mathematics. Pyenson dismisses the possibility that the dual clock meter influenced Einstein's formulation of the 1905 'special theory of relativity': The use of two independent clocks to represent two frames of reference in relative motion was, he argues, merely an artefact of later popularizations. Lewis Pyenson, *The young Einstein: The advent of relativity*, Bristol: Hilger, 1985, pp. 35–57, esp. pp. 40–4.

[16] Francis A. Jones, *Thomas Alva Edison: Sixty years of an inventor's life*, London: 1907, p. 123; Frederick Jehl, *Menlo Park reminiscences*, 3 Vols. Dearborn, MI: 1936, Vol. 2, pp. 637–69; Frank L. Dyer and Thomas C. Martin, *Edison: His life and inventions*, 2 Vols., New York/London: 1910.

[17] The only allusion to Edison's meter in this work is in an illustration of the Menlo Park laboratory reproduced from *Lesley's Journal* of 1880 in which a rudimentary form of 'metre' [sic] is pictured next to other major artefacts of the Edison supply business; see Hughes, *Networks of power*, p. 35. Hughes notes that Edison's laboratory was strewn with such standard current-measuring equipment as Thomson reflecting galvanometers, astatic galvanometers, and a Helmholtz–Gaugain tangent galvanometer, but he does not discuss the purpose of these instruments in Edison's project; ibid., p. 24.

being burnt. Now that electric light has been introduced, the fittings are altered a little more, and it is pretended that gas is being used. The chief advantage of direct reading is that it allows the consumer to see how much he is using; and this gives him confidence.[18]

James Swinburne, 'Electrical Measuring Instruments', ICE 1892

During the 1880s both electricity and gas-supply industries were expanding in a process of direct mutual competition. Gas lighting was much the cheaper and better established of the two, with its price falling throughout the decade, although still affordable only by middle-class households.[19] Nevertheless, the historian can trace several continuities between electric and coal-gas lighting in both patterns of supply and consumption. In his systems-centred account Hughes notes that Edison's model of central station supply for electricity borrowed heavily on the gas paradigm. Edison indeed shrewdly undertook a detailed study of the system of gas distribution in New York before setting up a similar pattern of electricity supply in 1878. By contrast with Hughes' focus on engineering considerations, a more inclusive social-constructivist standpoint would recognize the significance of consumer demands in shaping the take-up of electricity supply. Most significant in this regard is the way in which consumers' expectations of, and demands on, electricity suppliers were conditioned by their contemporary experience of gas consumption. The technological and aesthetic conservatism of householders is striking in this regard. Wolfgang Schivelbusch, for example, argues that many consumers preferred newer forms of domestic light consumption to be modelled closely on older forms.[20] This is borne out by James Swinburne's address to the ICE in 1892 in which he noted with typical condescension that consumers preferred to have both electric lamps and electric meters modelled on pre-existing forms of gas technology (see the epigraph that opens this section).

---

[18] Swinburne, 'Electrical measuring instruments', p. 21.

[19] Carol Jones, 'Coal, gas and electricity', in Rex Pope (ed.), *Atlas of British economic and social history from c. 1700*, London: Routledge, 1989, pp. 68–95, especially pp. 81–6 on gas. The gas appliance manufacturer William Sugg reckoned in 1884 that consumption in a winter period for 'a fair sized dwelling house' in which all lighting and cooking was done by gas, based on a rate of 3s per thousand cubic feet, would be £8.20 – more than could easily be afforded by the poorest households. Willliam Sugg, *The domestic uses of coal gas as applied to lighting cooking and heating, ventilation: With suggestions to consumers of gas as to the best mode of fitting up houses and using gas to the best advantage*, London: 1884, p. 36–7. Occasional press scares about the dangers of electricity, though, might in part be attributable to fears among the gas lobby that the price differential would not indefinitely stay in their favour. See the chapter on fire risks by James E. H. Gordon in Mrs J. E. H. Gordon, *Decorative electricity (with a chapter on fire risks by J. E. H. Gordon)*, London: 1891, pp. 17–28. James Gordon points out that newspapers regularly reported accidents involving electrical installations, but remained virtually silent about frequent explosions involving gas equipment.

[20] Wolfgang Schivelbusch, (trans. A. Davies) *Disenchanted night: The industrialization of light in the nineteenth century*, Oxford/New York: Berg, 1988, pp. 161–2.

Schivelbusch plausibly speculates that users of candles and oil lamps cherished the *visible* consumption of the older fuel types as if they were miniaturized hearth fires. Accordingly they were not so easily satisfied with or comforted by domestic lighting linked to centralized supply systems in which the consumption of 'piped' fuel from an external source was not directly visible.[21] This can be extended to consider the ways in which consumers were used to being able to judge how *much* lighting fuel they were consuming in a given time. Whilst with candles and paraffin lamps it was easy to gauge daily or weekly consumption at a glance, it was a more daunting technological challenge for suppliers of gas lighting and, later, electric lighting to facilitate comparable consumer surveillance of their consumption.[22] And once gas consumers had become accustomed to being metered for gas supply, there was a particular kind of user interface to which they had become strongly familiarized by the time electric lighting arrived.

The standard form of gas meter common by the 1880s (and probably much earlier) had a set of three counter-rotating dials marked respectively in units 0–9, the first and the third being marked clockwise and the middle being marked anticlockwise (see Figure 12). Moving from right to left, the dials recorded successively higher orders of magnitudes of gas consumption. Typically these represented thousands, ten thousands, and hundred thousands of cubic feet of gas (see subsequent discussion of this 'quantity' unit of consumption). A smaller, fourth, faster-moving dial was not read for the purposes of billing householders for their monthly consumption. This could be used either by the supplier to check the meter's calibration or by customers to identify the hourly consumption of individual household appliances. There was one important feature about this meter design, however, which did cause recurrent problems in taking 'readings' (see Chapter 4): Adjacent dials were marked in mutually contrary senses, and thus had to be read in physically opposite directions – one clockwise, the other counterclockwise. This kind of counter-rotational mechanism for gearing the indexical pointers was presumably chosen for reasons of efficiency and accuracy: It was the simplest means of arranging for a complete 10-unit rotation of one dial to register one unit of rotation on the adjacent higher-order dial. It was thus presumptively the least error prone and cheapest direct-reading mechanism to produce, not

---

[21] Ibid., pp. 161–2. I thank Sophie Forgan for drawing my attention to this point.
[22] Peter Kjaergaard has pointed out that, for householders accustomed to private monitoring of their own usage of candles and oil lamps, it was a radical shift of practice in the first decade of the nineteenth century to gauge their consumption by gas meters inspected in regular visits from supply company officials. In 1858, *Punch* magazine passed a satirical comment on how this prospect might have been seen in 1809: 'We...hear that contrary to all notions that an Englishman's house is his castle, the minions of the gas associations are to have ingress to our dwellings to adjust the juggling machinery for measuring this previous humbug.' Anonymous, 'Three chapters from the book of Cant', *Punch*, 17 (1858), p. 1; I thank Richard Noakes for drawing this piece to my attention.

Figure 12. Gas meter ca. 1884 with three conter-rotating dials reading 540,000 ft³, plus auxiliary upper dial for calibration purposes only. (*Source*: William Sugg, *The Domestic Uses of Coal Gas*, London, 1884, p. 35).

to mention the form least liable to need regular maintenance or mending after breakdowns. Not all consumers found this sort of gas-meter interface easy to read, however. As we shall see, some users evidently had difficulty in understanding the need to read the middle dial in contrary fashion.

Nevertheless, by the early 1890s the incorporation of such a set of indexical dials was common not only in gas meters in Britain but also for electricity meters. Each dial was usually calibrated in the relevant order of Board of Trade 'thermal units' of energy. And as James Swinburne noted in his ICE lecture to his fellow engineers, the adoption of this sort of reading technology in electricity meters was not determined by engineering necessity. It was merely a matter of deference to a consumer 'prejudice' for preferring novel electrical technologies to be modelled on pre-existing forms.[23] Notwithstanding Swinburne's derision of consumer foibles, it is worth noting that the fates of the many different kinds of electricity meters produced in the 1880s were closely linked to designers' recognition of this prejudice. If was simply not *enough* that meters should be 'accurate' in recording electrical consumption: They also had to be presented in a congenial and *well-trusted* format.

In Ayrton's and Perry's two 1882 models of electric-lighting meters, they explicitly engaged with the gas-meter paradigm whilst also attempting to break out of it. Their first patent specification proposed a form of meter in which a motor rotated at a speed proportional to the current through it, and

---

[23] See epigraph that opens this section.

the recording of revolutions was determined in 'much the same way' as in gas meters by indexical dials.[24] They did not produce a commercial form of this motor meter as they were unable to overcome the problems of friction in the dial-reading mechanism; from the point of view of suppliers, that caused it to underregister consumption rather undesirably.[25] Ayrton and Perry instead developed their clock meter, commercially produced and probably marketed to customers circa 1883–4. Having learned from their previous efforts, they deployed a pendulum-driven reading mechanism that was as frictionless as contemporary clock-makers could make it. In this meter, a pendulum was fitted with an electromagnetic brake that slowed down its readings exactly (or so its makers suggested) in proportion to the amount of energy supplied to the consumer. Consumers and suppliers could read such a meter by comparing its 'slow' time with that of an ordinary household clock and multiplying the time difference by a calibration constant to determine the value of energy consumed. When explaining this to the STEE in 1883, Perry noted that a high-quality (30s) clock was required for obtaining a meter that registered 'as accurately as a gas-meter' – that is, within a few per cent. He and Ayrton at Finsbury Technical College were indeed fortunate to be able to draw on 150 years of experience from clock-makers in nearby Clerkenwell.[26]

Ayrton's and Perry's clock meter was in theory quite straightforward to use, requiring no new skills at reading moving dials nor even to read dials in contrary motion. Users needed only to have a commonplace ability to compare the reading of two clocks and a facility for multiplication. It was an unstated assumption, however, that electric-meter users were sufficiently affluent to possess a clock of the same quality as the meter clock so that their discrepancy could be registered in a meaningful way. Nevertheless, neither the simplicity of principle nor putative accuracy of construction was sufficient to win this meter the confidence of many early electricity suppliers or consumers. After hearing Swinburne's comments on lighting consumers at the ICE nearly a decade later, Ayrton's somewhat melancholic remarks about the Finsbury clock meter were reported in the ICE journal as follows:

---

[24] William E. Ayrton and John Perry, 'Registering the amount of work given electrically to any part of an electric circuit in a given time', British patent number 2642, 1882 (5 June and 5 December).

[25] See Ayrton's comments in discussion of George Forbes, 'Electricity meters for central stations', *The Electrician*, 22 (1888–9), pp. 371–5, 393–6, and discussion pp. 436–8, Ayrton's remark on p. 437. Original version published in *Journal of the Society of Arts*, 37 (1888–9), pp. 148–59. All subsequent citations are from the *Electrician* version. For Forbes' career, see John Ambrose Fleming and D'Arcy W. Thompson, 'George Forbes, 1849–1936', *Obituary Notices of Fellows of the Royal Society*, 2 (1936–9), pp. 283–6.

[26] Apart from the challenge of matching the accuracy of gas meters, there was also the difficulty of ambiguous readings once the meter had reached the 12-hour pointer, but Perry considered that could easily be dealt with by installing a secondary recording hand. See James Shoolbred, 'On the measurement of electricity for commercial purposes', *JSTEE*, 12 (1883), pp. 84–107, discussion on pp. 107–22, Perry's comments on pp. 109–13.

The Author had made an interesting remark with reference to conservatism requiring that the scale on an electricity-meter should look like the scale on a gas-meter. Professor Perry and he had seriously erred in that respect when they devised their gaining-clock meter. They had fancied that when people had to measure something totally different from what they had been accustomed to – viz. electrical energy instead of gas – they would be satisfied to use a totally different meter, but they were wrong.[27]

The fate of the carefully designed yet still off-putting Ayrton and Perry single-clock meter contrasts with that of the double-clock meter. In 1884 the German engineer Hermann Aron patented a form in which both clocks were installed within the instrument, and this won a prize in Paris six years later. The differential reading between the two clocks was registered internally in Aron's meter and was directly converted by a cog mechanism into a registration on conventional indexical dials. This ease of reading and deference to conventional consumer habits made all the difference. Not only was the Aron meter adopted by most electrical suppliers in Berlin, but also in London by the Chelsea Electric Supply Company from its opening in 1889 and the St Pancras municipal station when it began supplying electricity in 1891. In his ICE address Swinburne was impressed by the fact that the pairs of clocks used in the Aron meter agreed to within one minute in a week, a remarkable accomplishment. It thus avoided a more than 2% cumulative error against supplier or customer between monthly checks.[28] And yet, according to Ayrton, this was not the sole reason for the success of the Aron meter. He ruefully noted that Dr Aron had been 'wiser' through acting on the principle of making a 'new thing to look very like something else that people had been accustomed to use'.[29]

There was nevertheless a certain irony in electricity customers' preferences for the adoption of the gas-meter pattern as recurrent problems arose with the readability of counter-rotating dials. The counter-rotating movement of the indices was evidently a source of confusion to consumers who erroneously applied their clock-reading habits when attempting to read the middle dial that registered anticlockwise. In his promotional treatise *The Domestic Uses of Coal-Gas*, the manufacturer of domestic gas equipment William Sugg tellingly noted that the reading of the meter 'index' was indeed popularly supposed to be a 'very mysterious' operation.[30] One of the purposes of this

---

[27] Ayrton in discussion on Swinburne, 'Electrical measuring instruments', p. 36.
[28] Brown 'Charging for electricity', pp. 516–7; Swinburne, 'Electrical measuring instruments', pp. 29–30.
[29] Ayrton in discussion on Swinburne, 'Electrical measuring instruments', p. 36. On Aron's meter, see Forbes, 'Electricity meters for central stations', pp. 373–4.
[30] Sugg, *The domestic uses of coal gas*, p. 35. For examples of Suggs' patented equipment, see the advertising in a work by his daughter, Marie Jenny Sugg, *The art of cooking by gas*, London: 1890; in that year Sugg sold gas burners, pressure governors, water heaters, updraft ventilators, and ventilated gas lights from his company headquarters at the Vincent Works in Westminster.

1884 book was to overcome this supposition. He thus offered a paradigmatic example – of a consumption of 540,000 ft^3 of gas, shown in Figure 12 – to show how reading the dial configuration was the 'simplest thing possible' if users obeyed the rule that no figure should be recorded unless the needle pointer had passed it. If such a carefully taken reading were checked against the information left on the card by a visiting Gas Company officer there would be 'no possibility' of a mistake.[31]

Yet from the results of introducing this pattern of counter-rotating dials to electricity meters it is evident that mistakes were continually made. A guidebook produced in 1899 by the Thomson–Houston Company for prospective agents and customers in the USA and UK noted that the 'inaccurate reading' of electricity meters had in the past been a 'prolific' source of error and dispute. Although somewhat reduced in recent years, it was still quite a prevalent difficulty, and the company represented the cause of the problem in terms strikingly more sympathetic to customers than Sugg's account. The problem Thomson–Houston diagnosed was a lack of *care* on the part of meter-reading officers. For obvious economic reasons these were low-paid workers, and such 'cheap men' could seldom be relied on to practice 'due care' in meter work, especially if they were entrusted with the 'duty' of translating the showing on the dial. To reassure potential customers it had properly recognized and solved the problem, the company issued this account:

The reading of a meter dial, whether it be a gas, water or electric dial, is always somewhat puzzling, and in some positions difficult even to the initiated. The actual translation of the dial indication should not be intrusted [sic] to the men who visit the meters to take their state. These men should go no further than to mark with a pencil the position of each hand upon a printed diagram of the dial. This the most unintelligent can do with sufficient accuracy.

This process of graphical rather than numerical meter reading apparently enabled the 'average' worker to process 75% more meters per day; the task of translating the recorded diagrams could then be devolved to a specialist skilled clerk at the company office. Even if a dispute subsequently did arise, the central station management claimed at least to have the 'best' documentary evidence to determine the actual state of the meter.[32] As we shall see in Section 6.5, though, consumers and suppliers might disagree about the 'true' reading of a gas or electricity meter for reasons relating to the quality of supply, the integrity of meters, and the trustworthiness of both suppliers and users. In relation to the discussion in Chapter 4, we can see that, for domestic meters, the business of accurate instrumental reading was far more

---

[31] Sugg, *The domestic uses of coal gas*, pp. 35–8.
[32] The British Thomson–Houston Company, *The management of central station meter systems*, London, 1899, pp. 15–16, in the S. P. Thompson collection at the IEE Archives.

than a matter of spatial judgement. It could involve judgement of moral and material virtue as well.

The most telling evidence of the power of the gas paradigm in the construction and reception of electricity meters can be gleaned from a comparison of Edison's and Ferranti's efforts to produce dc meters in the 1880s.

6.3. THE DIAL-LESS METER: EDISON'S TECHNIQUE FOR MEASURING AT A DISTANCE

Now, you may say that this is not exactly the kind of meter we require, because customers cannot check the weighings and are absolutely dependent on the good faith of the company supplying the electricity... I believe that there is a very general impression in this country that the Edison meter is not a satisfactory meter, and is not actually used in practice.

George Forbes, 'Electric Meters for Central Stations', Address at the Society of Arts, January 1889[33]

For those engineers and instrument makers who were designing electric meters in the early 1880s, it did not at first seem self-evidently necessary to follow the gas paradigm of the user-readable index. Making dials that the domestic user could read – even the simplest and most challenging to read – added complexity to the electric meter that also increased its expense, fallibility, and susceptibility to error. There were other priorities to which designers could choose to devote more attention such as accuracy, security, reliability, and cheapness. Yet because within the commercial constraints of meter production and sales these could not all be equally pursued, some trade-off was required between such qualities. This was the sort of decision faced by Edison from 1878 when grappling with the many problematic features of his as yet untried dc lighting system and by Sebastian Ziani Ferranti when seeking to support the finance of his ac system by the sale of meters to dc rivals in 1882. Although the development of their respective meters has separately been discussed in early biographies and histories of Menlo Park and in studies of Ferranti's career, no direct comparison has hitherto been made of their strategies and their commercial success in the meter venture.[34]

At the same time as Edison first accomplished useful results with his new filament light in 1878, a newspaper journalist had apparently asked him whether he would actually be able to measure what his future lighting customers would be consuming. Edison reportedly made this reply: 'It can be done.'[35] He and his team at Menlo Park experimented with a variety of mechanisms in ensuing years. As explained in Chapter 2, they focussed on

---

[33] Forbes, 'Electric meters for central stations', p. 373.
[34] Jones, *Thomas Alva Edison*, p. 123; Jehl, *Menlo Park reminiscences*, Vol. 2, pp. 637–69; Dyer and Martin, *Edison*, Vol. 1, p. 31.
[35] Jehl, *Menlo Park reminiscences*, Vol. 2, p. 638.

Faraday's well-established electrolytic techniques to transform the measurement of electricity into the measurement of weight. The total quantity of electric charge that had flowed in a circuit in a given time could reliably be measured by passing (a known fraction of) the current through a metal salt solution and weighing the quantity of metal that this current took from one electrode and deposited on another. Edison and his team tried to design meters of this type that offered users a direct visual indication of quantitative consumption.

For example, they attempted a 'rocking–beam' meter in which a critical weight differential of the two electrodes would cause the mechanism to reverse and record the switching of the beam on a counter. Although such a meter was apparently displayed at the Paris Electrical Exhibition in 1881, Edison and his team were dissatisfied by a number of its features: George Forbes later reported that this device suffered from very temperature-dependent performances and troublesome contacts for the electrical reversals.[36] James Shoolbred reported in 1883, though, that the major problem with this device was the comparatively large proportion of the current used that was required for operating the reading mechanism: This device thus significantly overestimated householders' actual consumption. This Shoolbred expected to be a particular problem if, as was anticipated, typical domestic usage was at a fairly low level of a few household lights used for a short period each evening.[37] Presumably Edison did not expect customers to tolerate meters that led to their being overbilled, and this offers at least one reason why he chose not to dedicate his primary efforts towards such designs.[38]

After a number of other ventures in meter design in the 1880s, Edison abandoned the effort to have directly readable meters.[39] Instead, his approach privileged fidelity of registration over user readability. To accomplish this goal, Edison got his assistant Francis Jehl to develop a metering system that translated the entire process of monitoring consumption out of the user's home to the company laboratory. The premise was that the monthly weighing of customers' meter electrodes would be undertaken most *accurately* in the controlled conditions of the Edison company laboratory in which

---

[36] Ibid., pp. 658–60; Forbes, 'Electric meters for central stations', p. 373.

[37] Shoolbred, 'On the measurement of electricity for commercial purposes', p. 96.

[38] This was not a *forced* move, however: Edison was familiar with the popular 'sonorous voltameter' by which the strength of an electric current could be gauged aurally by the noise of electrolytically generated gas bubbles. Evidently he did not give sustained consideration to implementing this as a 'real-time' monitoring of consumption to his own meters, although in 1879 Edison mischievously persuaded a *Scientific American* reporter that it was the Menlo Park laboratory's most recent innovation; Jehl, *Menlo Park reminiscences*, Vol. 1, pp. 286–7.

[39] For a contemporary overview of Edison's meter making by one of his assistants, see William J. Jenks, 'Six year's practical experience with the Edison chemical meter', *Journal of the American Institution of Electrical Engineers*, 6 (1888–89), pp. 27–57, discussion on pp. 58–69.

no consumers could observe or interfere. Company officials were thus dispatched once a month to unlock the meters and remove (and replace) the meter electrodes, taking them back to the company laboratory where customers' electrical consumption would be calculated from the weighing data and a billing subsequently mailed out. As a precaution against tampering by either consumer or Edison staff, a 'quarterly' cell was installed in a separately secured part of the meter against which Edison staff (not customers) could check every third month for signs of longer-term discrepancies. By December 1878, Edison claimed to a reporter for the New York *Herald* that this meter was so well developed that it would 'infallibly' register domestic consumption to within 1 part in 1000 (0.1%).[40]

On 21 December 1879, Edison explained to readers of the *Herald* in terms calculated to reassure potential customers that his meter was significantly similar to a gas meter and that no problems would arise in billing their consumption by his unprecedented supplier-centred method. His apparatus for 'measuring the amount of electricity used' by each householder was a 'simple' contrivance about half the size of an ordinary gas meter and which, like a gas meter, could be located anywhere in the home. Moreover, the 'amount of electricity consumed' was also determined by a 'simple' calculation on the weight of copper.[41] During three years of development work on prototype meters, however, Jehl replaced the copper with zinc that was less prone to error-inducing polarization effects. After Edison was granted a patent for the zinc/zinc sulphate meter on 27 December 1881,[42] Edison turned his attention to the constituencies of stockholders, creditors, and agents. To maintain their support for his Pearl Street system in New York he reassured them in his company's promotional *Bulletin* of January 1882 that this meter was now completely 'satisfactory' in its operations. Ironically, more than three years after he had first told the *Herald* the device was perfected, he now declared that it registered currents only with 'almost absolute exactness'. Rather defensively he noted that his meters at least certainly compared favourably with gas 'meters', the reliability of which he alleged to be a matter of 'looseness and uncertainty' (see subsequent discussion).[43] Much accompanying publicity material was issued that showed the meter opened up with its inner working rendered visible and its operations 'transparent' (Figure 13). It is notable that very few of these illustrations issued showed the meter with its doors closed and thus diverted the attention of potential customers away from the lack of any directly readable user interface on the Edison meter.

---

[40] Jehl, *Menlo Park reminiscences*, Vol. 1, p. 287.   [41] Reproduced in Ibid., Vol. 1, p. 289.
[42] The US patent number was 251,545. Jehl, *Menlo Park reminiscences*, Vol. 2, p. 641.
[43] Cited in Robert Friedel, Paul Israel, and Bernard Finn, *Edison's electric light: Biography of an invention*, New Brunswick, NJ: Rutgers University Press, 1986, p. 218.

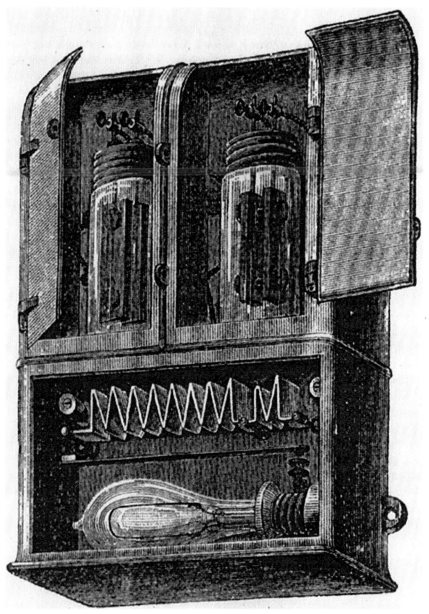

Figure 13. An Edison meter opened to show the electrolytic cells from which a company official removed electrodes for monthly weighing and billing. Most representations of this meter produced by the Edison company were of this 'open' mode – presumably to reveal the 'trustworthy' simplicity of operation and divert viewers' attention away from the complete absence of a user-readable dial. (*Source*: William J. Jenks, 'Six year's practical experience with the Edison chemical meter', *Journal of the American Institution of Electrical Engineers*, 6 (1888–9), pp. 27–57, p. 36).

To promote this new meter in London, Paris, and Milan, Jehl was dispatched over the Atlantic in early 1882 and secured commendations for it from the company's two British experts, John Ambrose Fleming and John Hopkinson. Next Jehl composed a pamphlet, *The Edison Electric Light Meter*, that explained its operations in detail as well as the careful precautions taken when the device was being calibrated and the weights of electrodes were read. This was evidently a form of 'virtual witnessing' of the form described by Schaffer and Shapin.[44] Given that Jehl's document was printed for 'private circulation only' with the prefatory aim of aiding the 'incipient scholar' whose 'duty' it was to 'manipulate and manage' the meter, it presumed its readership to be fellow electrical engineers rather than consumers.[45] Not all such engineers were impressed, however. In his *Practical Treatise on Electric Lighting* published early in 1884, J. E. H. Gordon considered that the three patented forms of meter available in a 'practical' commercial form were 'crude' devices which had not yet been made to work 'satisfactorily'. Among these Gordon counted not only Edison's electrolytic device but also even his own clockwork meter; hence his testimony cannot be dismissed as a mere partisan attack on his American rival.[46]

Notwithstanding his own press pronouncements, Edison himself took some time to be persuaded that this meter worked reliably. His company installed meters for their customers in Pearl Street in New York and the Holborn Viaduct in London in 1882 and, according to one estimate, 300 such meters were installed in New York in February 1883.[47] Friedel and Israel observe, nevertheless, that Edison did not in fact charge customers according to the meter readings until late 1883, resorting instead to the economically risky business of charging customers per lamp (see Section 6.1). Given that Edison's company was persistently making 'improvements' in the meter throughout this period, it seems plausible to follow Millard's suggestion that these meters did not perform as reliably when installed in consumers' homes as they had done in trials in the Menlo Park laboratory.[48] It was, for example, during this development period that the meters were first exposed to the chilly New York winters. Thereafter the meters were made with a thermostat that switched on an incorporated Edison light bulb

---

[44] Simon Schaffer and Steven Shapin, *Leviathan and the air-pump*, Princeton, NJ: Princeton University Press, 1985.

[45] Jehl, *Menlo Park reminiscences*, Vol. 2, pp. 655–6. Frederick Jehl, *The Edison electric light meter*, ca. 1882, privately published (copy in S. P. Thompson collection at IEE Archives in London). See review by *Electrician* on 3 June 1882.

[46] James E. H. Gordon, *Practical treatise on electric lighting*, London: 1884, p. 202.

[47] Shoolbred, 'The measurement of electricity', p. 96. Evidence on London customers is harder to find, but Jehl does cite an installation for a Mr Jones at 57 Holborn Viaduct on 20 March 1882, first checked on 20 April that year; Jehl, *The Edison electric light meter*, p. 22.

[48] Friedel and Israel, *Edison's electric light*, p. 219, André Millard, *Edison and the business of innovation*, Baltimore: Johns Hopkins University Press, 1990, p. 92.

when the temperature dropped below 40° F, thereby preventing the freezing of the electrolyte that typically brought the meter mechanism to a complete halt – giving customers a free supply.[49] Although Edison remained publicly silent on the troublesomely load-dependent and temperature-dependent performance of the meters, nearly three decades later Edison himself conceded that the 'old chemical meters' gave his company a 'lot of trouble'.[50] By 1889, the meter's defenders claimed only that it was 'accurate' to within 3% – not the 0.1% projected by Edison in 1878 – and typically only over a fivefold to tenfold range of currents.[51]

It was not just the technical problems of making meters reliable away from the congenial environment of the laboratory that bothered Edison, however. Once his company began to charge customers according to the weighing of the electrodes, householders began to object to the obscurity of the metering process. The problem was that they were not able to participate by monitoring the rate of their consumption, as they had been accustomed to doing with gas meters. As Jehl reported in his *Menlo Park Reminiscences*, New York's Pearl Street station in the period 1882–4 saw the first disputes between electricity companies and their customers.[52] Suspicions about the trustworthiness of Edison and his company were rampant, some apparently suspecting that Edison officials simply made up meter readings that would make their business profitable.[53] George Forbes told the Society of Arts in February 1889 that he had heard many criticisms of the Edison meter. There was a 'general belief among a great number' in stations in the USA where the Edison meter was used that the members of the metering staff 'never weigh at all, they simply guess at the bills and put them down.' Whilst Forbes admitted that Edison himself had said that his was not an 'ideal' meter, he nevertheless denied that Edison's staff were fraudulent. Instead he cited the somewhat less

---

[49] The introduction of heating light bulbs inside the meter confused some customers even more: Every five minutes in cold weather a customer would ring the company telephone to say 'Our meter's red hot. Is that all right?' or 'Our meter's on fire inside, and we poured water on it. Did that hurt it?' See Edison's testimony to the American *Electrical Review*, of 12 January 1909, reproduced in Jones, *Thomas Alva Edison*, p. 123.

[50] Ibid.

[51] See comments on load dependence in Shoolbred, 'The measurement of electricity', p. 96, and Forbes, 'Electric meters for central stations', pp. 373–4. A few months before Forbes' paper, Edison's assistant, William Jenks, had used extensive testimony from Edison central station managers to try to persuade members of the American Institution of Electrical Engineers that the Edison meter was reliable, notwithstanding critical comments in UK engineering journals. See Jenks, 'Six year's practical experience with the Edison chemical meter'.

[52] Jehl, *Menlo Park reminiscences*, Vol. 2, pp. 1085–6.

[53] Brown suggests that in France and the UK there was strong antipathy to any venture associated with Edison's name, probably as a matter of financial anxiety about threatened gas investments and to his unfulfilled early prophesies about the cheapness of electrical lighting. Brown 'Charging for electricity', p. 515 and Anonymous, 'Obstacles to the success of Edison's lighting system', *Electrician*, 4 (1880), p. 173, reproducing a criticism of Edison published by the *New York Times* on 7 February 1880.

than disinterested testimony of US managers of Edison generating stations that their customers were 'quite content' with the manner of reading the meter.[54]

Edison was clearly aware, though, that his customers were not content with his meters. In the late 1880s his company staff returned to working on new patents for meters that displayed user-readable interfaces. One concerned a revived form of the early rocking-beam meter that harnessed the gas evolved in electrolysis to register ongoing consumption; a model of this was shown at the Royal Institution by William Ayrton in 1892.[55] It could be argued, of course, that consumer pressure was not the only factor that prompted Edison to make this move. Millard has noted that the system of removing meter electrodes and weighing them at the company lab was expensively labour intensive – in effect an economic reverse salient.[56] In his 1892 ICE lecture Swinburne noted the 'great prejudice' that existed against the standard Edison electrolytic meter and suggested that it was by then no longer used in Britain. Swinburne speculated, moreover, that the cost of inspecting it would inevitably have outweighed any financial advantage in avoiding the inaccuracy of a dial-reading system.[57] It is significant, then, that the Edison company later adopted the Thomson–Houston recording wattmeter with gas-meter-like dials rather than its own new patented efforts at direct-reading meters.[58] Edison's original biographers, Dyer and Martin, significantly do not challenge the fate of the Edison meter discussed in Edwin Houston's *Electricity in Everyday Life* of 1905:

> The Edison chemical meter is capable of giving fair measurements of the amount of current passing. By reason, however, of dissatisfaction caused from the inability of customers to read the indications of the meter, it has in later years, to a great extent, been replaced by registering meters that can be read by the customer.[59]

This Edison case supports Ted Porter's argument discussed in Chapter 1 in that professionals could be forced by public distrust to be more accountable in their financial dealings by displaying their quantitative activities in more

---

[54] Forbes, 'Electric meters for central stations', p. 373.

[55] William Ayrton, ' Meters, motors and money matters', *The Electrician* 28 (1892), pp. 640–1. For Edison's meter, see G. S. Bryan, *Edison: The man and his work*, London: 1926, p. 147, Millard, *Edison and the business of innovation*, pp. 92, 345–6, n.16; D. Woodberry, *Beloved scientist: Elihu Thomson*, New York: McGraw-Hill, 1944, pp. 198–9. Developments in the 1890s of the electrolytic meter used the gauging of volume of gas or liquid metal generated to give a direct reading of consumption. See discussion in Rankin Kennedy, *The book of electrical installations*, 3 Vols., London, ca. 1902, Vol. 1, pp. 187–94.

[56] In Boston alone Edison reputedly spent as much as $2500 on the annual reading and servicing of 800 customer meters; Millard, *Edison and the business of innovation*, pp. 90–3.

[57] Swinburne, 'Electrical measuring instruments', 1892, p. 22.

[58] Brown, 'Charging for electricity supply', p. 515.

[59] Edwin J. Houston, *Electricity in every-day life*, 3 Vols., New York: 1904, cited in Dyer and Martin, *Edison: his life and inventions*, p. 431.

transparent ways.[60] The case of Sebastian Ferranti's meters, by contrast, shows how it was possible to pre-empt public distrust in a new technology by deploying a strategic understanding of consumer's conservative foibles.

6.4. EMULATING THE GAS PARADIGM: FERRANTI'S MERCURY-MOTOR METER

The next meter which [sic] I wish to describe is that of Mr Ferranti... This meter is so beautifully simply in its principle as to please on at first sight.
George Forbes, Lecture at Society of Arts, 1889.[61]

Whilst constructing plans for ac supply in 1882, the 21-year-old Sebastian Ferranti had an astute business strategy. He began to develop dc meters that would enable him to profit from any financial success of his rivals in the 'battle of the systems'. Perhaps aware of Edison's problems in developing his dc meter, he quickly abandoned experiments on electrolytic mechanism[62] and from 1883 to 1884 worked on a device that was premissed on the importance of a consumer-readable dial. Early drawings show how Ferranti planned the user interface with gas-index-like dials, and only *then* developed a mechanism that could conveniently produce readings in a user-friendly fashion.[63] To produce a rotational motion of the form used in gas meters, Ferranti drew on his reading on the magnetic interactions of currents in Ganot's *Physics* to devise a motor form of instrument. He made full strategic use of the unique electromagnetic properties of mercury as a viscous conducting metal liquid, designing his meter such that the passage of a current through the mercury would move the body of the fluid in a circular fashion around its containing vessel. This flow dragged a paddle wheel, and this in turn was attached to a gearing mechanism that turned a set of customer-readable dials as the paddle rotated. Like Ayrton's and Perry's contemporary efforts in designing reliable ammeters, Ferranti's aim was to develop a meter that had a reliable cumulative estimate of electricity flow by a mechanism that responded *proportionately* for all values of current. To do this he had to contrive the meter's

---

[60] Theodore Porter, *Trust in numbers*, Princeton, NJ: Princeton University Press, 1995.
[61] Forbes, 'Electric meters for central stations', p. 374.
[62] In the late 1880s, Ferranti was one of Edison's main rivals in the UK: For their controversial encounter in 1889 see *Punch* poem reproduced in Gertrude Ferranti and Richard Ince, *Life and letters of Sebastian Ziani de Ferranti*, London: 1934, p. 64.
[63] See drawings reproduced in Ferranti and Richard Ince, Life and letters, pp. 134-5 and Arthur Ridding, *S.Z. de Ferranti: A brief account of some aspects of his work*, London: Her Majesty's Stationery Office, 1964, p. 7. Many versions of Ganot's non-mathematical illustrated treatment of experimental physics were available to Ferranti, e.g., Adolphe Ganot, (trans. ed. Edmund Atkinson), *Elementary treatise on physics: experimental and applied*, 8th edition, London: 1877. Ganot's original *Cours de physique purement expérimentale* was published in Paris, 1859.

components so that that friction between mercury and paddle ensured that the speed of rotation was a linear function of the current – no easy matter.

However, as Ferranti later admitted, his mercury-motor meters were at first 'very incorrect'; unlike Edison's meter they gave far from proportionate readings and were commercially of 'no use at all'.[64] As I emphasized in Chapter 4, accomplishing proportional calibration in the 1880s was a phenomenally resource-intensive problem. Accordingly Ferranti spent the best part of two decades using trial-and-error methods of counterbalancing and fine-tuning compensation on a variety of shapes of paddle and mercury troughs to accomplish the desired proportionality. As Wilson has noted, it was the wealthy individuals who purchased or rented Ferranti's meters that had to pay the price of this huge long-term investment of labour and capital.[65] Quite independently of such problems of proportionality, Ferranti was also most concerned to find that this prototypical meter would not register the consumption of any currents below 20 A – a feature that might allow cunningly low-consuming customers to enjoy a virtually free supply! Advice solicited from Sir William Thomson enabled him to overcome the surface tension problems of mercury so that the meter's starting current was reduced to 0.3 A – a sensitivity unrivalled by other extant meters.

Although not documented by Ferranti in print, he also agonized over a problem that had defeated Edison's attempt at a user-readable electrolytic meter and Ayrton's and Perry's attempt at a motor meter. This was the scope and persistency of errors produced by mechanical friction of the gearing for the dials. Indeed, Ferranti once told Ayrton in a moment of humorous irony that his meters would have been 'perfect' if only they had not had to be fitted with a registering mechanism. The compromise solution adopted by Ferranti to this problem was to reduce the magnitude of the friction by making the dials very small – indeed significantly smaller, as Ayrton noted, than on Aron's clock meter.[66] Ferranti's design compromise turned out to be an effective gamble for, although the readings of these meters were judged to have an accuracy to within only 5% – compared with the 1% accuracy claimed for contemporary gas meters – Ferranti found many customers for his devices. Indeed, when Ferranti first advertised his meter in July 1885 he

---

[64] Sebastian Ziani de Ferranti, 'On the Ferranti meter and its evolution', *Transactions of the Royal Scottish Society of Arts*, 14 (1898), pp. 52–66, quote on p. 57.
[65] A meter for a circuit of 100 lights cost £19 to purchase; by comparison, a gas meter normally rented quarterly: a meter for 100 lights would cost 12/6. Sugg, *The domestic uses of coal gas*, pp. 23–6. Note: A gas meter for three lights would cost 3d a quarter to rent. For the Ferranti data see Ferranti and Ince, *Life and letters*, p. 128.
[66] Ayrton in discussion to Forbes, 'Electric meters for central stations', p. 437. Ayrton noted in 1892 that friction was a recurrent problem for proponents of motor meters: 'rise of temperature, and also wear, and the accumulation of dirt would diminish the speed of the motor, which might account for the fact that motor meters were not in favour with the electrical supply companies'; William E. Ayrton, 'Electric meters, motors, and money matters', p. 641.

even had the temerity to declare that it was 'the only commercial meter up to the present time'.

This faith in his product was shared by others who awarded him a silver medal at the International Inventions Exhibition in 1885, as well as a Diploma of Honour at the Antwerp Exhibition; contracts in France and Belgium soon followed.[67] Sales in London were helped by advertising that used elegant iconography[68] and by the spectacle of the Ferranti ac lighting installation at Sir Coutts Lindsay's Grovesnor Gallery installed on New Bond Street in 1885. Lindsay soon found many willing householders in the neighbourhood to whom he could both supply electricity and rent out Ferranti meters to monitor their consumption.[69] Whilst ordinary consumers acquiesced in the familiar design of the Ferranti meter, taking their own readings in a way that was denied to them in the case of the Edison meter, the Board of Trade acting on their behalf were not, however, quite so quickly impressed by Ferranti's dc meters. Only four years after the Board of Trade first awarded Shallenberger's meter the accolade of its state approval in 1892 did the Ferranti dc meter win this accolade, in kinship with the Hookham meter. Even then the Ferranti meter was approved for currents only up to half its maximum load[70] (Figure 14).

Well before Ferranti's meter was accorded official Board of Trade approval in 1896, professional colleagues in electrical engineering pronounced themselves impressed by the 'beautiful' simplicity of its design and especially by his dexterity in overcoming the problems of friction and the impurity of mercury. In his lecture on electric meters at the Society of Arts in January 1889, George Forbes described Ferranti's accomplishment as all the more impressive, given that both Edison and Edward Weston in the USA had attempted to make similar mercury-motor devices without success.[71] To spontaneous applause, Forbes reported that these two inventors had been 'astonished' to hear of Ferranti's success in maintaining the purity of mercury. It had hitherto been assumed to be a 'universal fact' that mercury was unsuitable for permanent contacts in electrical devices because it soon attracted impurities and dirt that interfered with efficient working (see Chapter 3). The reason for the Ferranti company's success with mercury, Forbes averred, was that it had taken 'enormous care' to purify the mercury in its meters so that no

---

[67] Wilson, *Ferranti and the British electrical industry*, pp. 26–7.
[68] See picture in Ridding, *S.Z. de Ferranti*, p. 8.
[69] Wilson, *Ferranti and the British electrical industry*, pp. 27–8.
[70] Editorial note, '"Approved" electric meters', *Electrician*, 38 (1896–7), p. 37. Wilson, ibid.
[71] For Edison's mercury-motor meter, see Jehl, *Menlo Park reminiscences*, Vol. 2, pp. 661–7, and Jenks, 'Six year's practical experience with the Edison chemical meter', pp. 29–30. Jenks commented: 'Probably if mercury did not have such a chronic inability to behave itself in practical continuous work, and if it really possessed the ideal character of a liquid which it commonly gets credit for, we should have seen this meter put into practical use by now'; ibid., pp. 29–30.

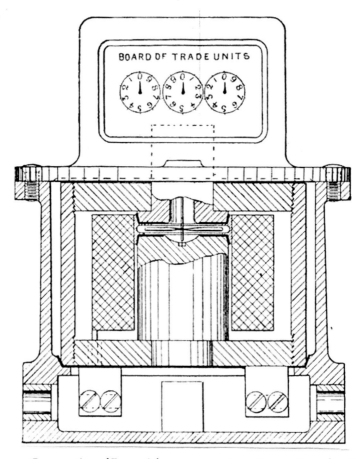

Figure 14. Cross section of Ferranti dc mercury-motor meter ca. 1895, showing paddle rotating in conducting mercury vat, linked to a three-dial counting mechanism almost identical to that cited by Sugg for the gas meter. (*Source*: From Sebastian Ziani de Ferranti, 'On the Ferranti Electricity Meter and its Evolution', *Transactions of the Royal Scottish Society of Arts*, 14(1898), p. 64 (paper originally read to the Society on 14 January 1895)).

interference was found. Indeed, users of Ferranti meters were instructed to fill their meters only with mercury prepared by the Ferranti company itself.[72]

Whilst Forbes considered the Ferranti meter to be beautifully simple in its principle, he did not regard it as perfect. In his lecture, Forbes listed fifteen desiderata for ideal electricity meters. Although he reckoned the Edison,

[72] Forbes, 'Electric meters for central stations', p. 374. Ferranti company catalogue 'Continuous Current Meters', undated ca. 1900–1, p. 17, in S. P. Thompson collection, IEE (dating based on the citation of the Notting Hill Supply Company as a user of Ferranti meters, a company that commenced generation only in 1900 – see data in Jones, op. cit.).

Ferranti, Shallenberger, and Lowrie–Hall models to meet most of them, no meter had yet met them all. Forbes' principal desideratum was that such meters should be 'at least' as accurate as ordinary gas meters, a rather 'low' standard for 'modern' requirements he contended – to the palpable approval of his audience. By accuracy here he meant two things: first, that the readings should faithfully reflect both the strength and duration of currents passing in a suitably linear fashion, so that ten lamps in operation for one hour registered the same consumption on the meter as one lamp operating for ten hours; second, that each meter should be *consistent* with others of its type made by the same manufacturer. Significantly, Forbes offered no quantitative criterion for what would count as accuracy, probably leaving his audience with the impression that this accuracy could be not be any more than agreement within a few per cent. Forbes anyway judged both the Ferranti meter and Edison meter to meet this condition to his satisfaction, and also the correlated condition that the operation of meters should consume less than half a per cent of what customers took from the supply.[73]

None of the meters Forbes discussed met the desideratum of yielding equally accurate results over the full range of its operations from just one lamp lit to a full load. The remarkably low starting current of the Ferranti meter, though, meant that it did not pose serious problems to the supplier that other meters did, especially motor meters. This was the awkward way that, for very low currents, they registered no current at all if just one lamp were left switched on. Like all other contemporary meters except the Forbes thermal-windmill meter (see next section), the Ferranti dc meter was not able also to register ac currents, and the Ferranti meter fell far short of Forbes' desideratum of not being 'very costly'. Unlike the Edison meter, though, the Ferranti meter did not require special 'attention' from company officials to enact a reading, and indeed the latter gave 'convenient' registrations like a gas meter, so that consumers could readily check its accuracy for themselves. And indeed all other meters discussed by Forbes followed the Ferranti meter's use of the gas paradigm in having at least three visible dials for consumers to 'read' their consumption of electricity.

One might see this acquiescence of suppliers in the consumer expectation of readable gas dials as evidence of the power of electricity consumers to insist that the domestic meter be shaped to their concerns. Yet when one considers the issue of what these meters registered and thus what consumers were actually charged *for*, such a naive social-constructivist claim about the power of the consumer in shaping the technology of domestic metering to meet their interests becomes less plausible. Both the Edison and Ferranti meters had one important feature in common with gas supply: Both had mechanisms designed to register the *quantity* of electricity as the subject of consumption just as gas meters registered the volumetric quantity of gas being delivered

[73] Forbes, 'Electric meters for central stations', p. 372.

to householders. Even though later versions of the Ferranti meters were calibrated in Board of Trade units of energy, this calibration was simply a recalibration using the dynamic formula 'energy = current × potential difference'. The fidelity of this calibration hinged on the assumption that the potential difference (hereafter pd) or 'pressure' at which the current supplied was a constant, and, as we shall see in the next section, this was as readily contested as the claim that gas supply was delivered at a constant pressure. There was indeed a considerable irony in the conservatism of customers acquiescing by analogy with paying for a quantity of gas, that they could fairly be charged per quantity of electricity passing through their meter.

6.5. FAIRNESS VERSUS EXPEDIENCY: RIVAL INTERPRETATIONS OF ELECTRICAL CONSUMPTION

With some meters the process of cheating is easier than with others. I speak chiefly of cheating on the part of the consumer (laughter). I notice that generally when I have been speaking about meters with people the other possible way of cheating is not considered of much importance (renewed laughter).

George Forbes, 'Electric Meters for Central Stations', Address at the Society of Arts, January 1889[74]

The design, construction, operation, and reading of early electric meters posed a number of important challenges for all concerned. A major instance of this concerned the *fairness* of financial charging for both supplier and consumer: Forbes' tenth desideratum for a good meter was that that it should pre-empt the possibilities of any form of cheating. Suppliers might cheat consumers with a substandard quality of supply if meters were designed so as not to register this malfeasance; or consumers might manipulate their meter to diminish its readings so as to defraud the supplier. Then again both parties might lose out without overt fraudulence if meters did not register consumption proportionately: They could unfairly underread in the customer's favour at some currents or overread in the supplier's favour at others. In his Society of Arts lecture, Forbes jested that electrical suppliers tended to draw attention to the devious householder or the underregistering meter rather than to other threats to trustworthiness of the encounter between consumer and supplier.

As it was, Forbes considered that a robust meter installation could obviate both the well-known risk of a consumer's short-circuiting a meter to prevent currents passing through it and of tilting it out of level to slow down its registrations. However, he offered no cure for meters which could be retarded by localized heating or application of external magnets. Wily customers who understood the mechanism of clock meters had apparently used the latter technique to oppose the retarding effect of the electromagnetic brake on

[74] Ibid.

the pendulum. At this same meeting Ayrton said that he and Perry thought they had overcome this fraudulent practice. They had redesigned their clock meter so that its current direction was occasionally reversed: Anyone trying to cheat by applying magnets was 'as likely to raise his monthly bill as to lower it'.[75]

Whilst much was at stake in designing meters to be impregnable to consumer meddling, designers of electricity meters devoted at least as much attention to ensuring that the construction of meters did not give biased registrations. Because the interests of consumers in this regard were at least in theory defended by the Board of Trade and its meter inspectors, the electricity suppliers' primary concern was inevitably to prevent injustices against themselves – just as gas suppliers and their allies had long been concerned to do. Sugg's *Domestic Uses of Coal Gas*, for example, explicitly promoted the 'dry' gas meter over what he condemned as the 'barbarously' contrived archimedean screw mechanism of the 'wet' gas meter. His reasoning reveals his sympathy for gas suppliers: Although the dry meter was prone to 2%–3% inaccuracies either way because of friction from dust or dirt, at least it was not as prone as the wet meter to underestimate readings or even to stop altogether, allowing customers the pleasure of unmetered supply. Whilst the known infidelities of the dry meter might compromise consumer rights, Sugg pointed out that the terms of the 1859 Sale of Gas Act would provide adequate protection for customers faced with such problems. If householders were concerned about a dry meter overreading, he emphasized that government inspectors would test it for a small fee, and if the meter had run fast customers would receive appropriate compensation from the company.[76]

The efforts of electricity meter designers to make meters equally accurate for all levels of current in its range were directly analogous. What particularly drove such efforts was a concern to avoid financial loss for suppliers on the one hand, and disapprobation from meter inspectors and consumers on the other. Yet designers repeatedly found that the linearity (proportionality) of operation required for meeting this goal was very hard to accomplish. Swinburne noted in 1892 that supply companies had often simply contented themselves with meters that merely went 'faster' as the load on them increased. Many were distinctly non-linear in performance, underregistering at the lower end of the range and overregistering at the upper end. Board of Trade inspectors were, however, likely to take much more stringent

[75] Ayrton in discussion to Forbes, 'Electric meters for central stations', p. 437, with audience laughter recorded. Ayrton and his students also studied the possibilities of consumer fraud in regard to the Frager 'intermittent' *mechanical integrating* meter. By switching off lights just before the meter registered every hundred seconds it was possible to use 300–400 times the supply that had been recorded. Given the inconvenience of this mode of cheating, Ayrton considered this meter unlikely to worry supply companies. Ayrton, 'Meters, motors, and money matters', p. 641; Brown,'Charging for electricity', pp. 521–2.
[76] Sugg, *The domestic uses of coal gas*, pp. 23–6.

action against the latter than against the former; this is presumably why it approved Ferranti's meter in 1896 up to only half of its full working range. Although Swinburne opposed the contrivance of proportional mechanisms in ammeters and voltmeters (Chapter 4), he performed a striking *volte face* in regard to the domestic meter. He argued to the ICE that extending the linear operation of a domestic meter down to the lowest end of its scale could turn a commercial failure in lighting into a 'brilliant success'. In other words, if a meter could record even the low consumption of the cunning or miserly householder who used only one light at a time, a supply business might have a chance of profitability.[77] It must also be said that suppliers also needed corrective attention for the errant meters that occasionally operated in *reverse* when drawing no current, thus decreasing customers' bills! The consultant engineer Gisbert Kapp reported such an incident with his clock meter in 1892, volunteering this information to the suppliers and inspectors so that the meter could be adjusted to prevent an 'unfair' loss of income to the supplier.[78]

As far as the electrical consumer was concerned, the most significant and pervasive form of 'cheating' was supplier's diminution of the quantity of supply in ways that domestic meters underregistered or did not register at all. This sort of practice was already notorious among gas suppliers whom customers often suspected of fraudulently supplying them with gas at less than the contractually specified pressure. Avaricious gas companies might occasionally decrease the supply pressure with the consequence that customers received less than the agreed amount of combustible material per unit volume. More cynically, such companies might permanently adulterate the coal-gas with less combustible substances such as air or sulphurous by-products of the extraction process from coal. Although consumers might detect the effects of these in a drop in illumination levels, meter readings based on volumetric flow would not show their full loss that was due to pressure drops and would not register adulteration at all. Complaints on these matters proliferated as ordinary householders increasingly became users of gas lighting in the 1870s.[79] So in his 1884 apologetic, William Sugg tried to reassure his readers that the purity and light-giving power of gas were guaranteed by the actions of gas examiners such as John Tyndall, who implemented the protective measures of the Sale of Gas Act (1859) and the Gas Works Clauses (1871).[80] Yet there was much derision on the effectiveness

---

[77] Swinburne, 'Electrical measuring instruments', p. 20.
[78] Kapp in discussion to Swinburne, 'Electrical measuring instruments', pp. 47–8.
[79] Carol Jones, 'Coal, gas and electricity', pp. 81–3; John D. Poulter, *An early history of electricity supply: A study of the electric light in Victorian Leeds*, London: Peregrinus/IEE, 1986. One correspondent, 'Benighted', complained to the *Yorkshire Post* in December 1885 that the gas street lamps of Leeds shed forth 'visible darkness'; ibid, p. 43.
[80] Sugg, *The domestic uses of coal gas*, pp. 38–9. The gas referees appointed by the Board of Trade at the time that Suggs wrote were apparently John Tyndall, A. Vernon Harcourt,

Measurement at a Distance 247

of these measures among proponents of electricity. William Ayrton maintained in his Royal Institution lecture in 1892, for example, that the quality of gas supply varied 'greatly' in different parts of England. At least, Ayrton wryly added, no one had yet found a comparable means of 'adulterating electricity'.[81]

Electricity suppliers could indeed not cheat customers by substitution of any surrogate. Consumers of ac supply were nevertheless vulnerable to a unique form of supply degradation if generating companies increased the frequency of alternation. Swinburne noted that a dishonest ac company could turn a heavy loss into a 'splendid profit' just by increasing its alternator speeds: Owing to increased inductive effects (see Chapter 5), customers would receive less energy than their meter appeared to register.[82] Forbes thus recommended that ac consumers use only those meters that could fully register such changes by giving readings that were entirely independent of frequency – a difficult technological challenge for designers to meet. More generally, however, there was a single major problem for ac and dc customers which had major ramifications for the fairness of charging. By analogy with corrupt gas companies, there was the risk that electricity suppliers would accidentally or deliberately lower the electrical 'pressure' – the pd – of supply below the contractually agreed levels. These were typically a few hundred volts for dc supply and up to a few thousand for ac, and suppliers typically took more care to prevent rises above these levels than they did against drops beneath them – see subsequent discussion.

This matter was very significant in regard to the fairness of different schemes and technologies of metering consumption. The effect of pressure drops on meter readings depended strongly on what form of consumption the meter was designed to register. Edison, Aron, and Ferranti meters all registered the passage of electrical quantity (current × time, i.e., pd/resistance × time), whereas Thomson–Houston and Shallenberger meters registered the passage of electrical energy (current × potential difference × time, i.e., pd^2/resistance × time). So long as the system pressure – pd – was constant, these forms of metering could produce equivalent results. However, any *drop* in voltage would produce a differential effect: Energy meters would register less by a 'squared' factor, whereas quantity meters would register only a smaller linear decrease. Thus suppliers saw a financial advantage in customers' use of quantity meters: Not only were these simpler and thus cheaper to make than energy meters, but underregistrations that were due to pressure

---

and Dr William Pole. although there were also other examiners for the City and metropolis. Charles Vernon Boys later became a gas referee after leaving the Royal College of Science in 1895; see Graeme Gooday, 'Charles Vernon Boys' in *New dictionary of national biography*, forthcoming, 2004. For a general discussion of gas referee activities, see Forbes, 'Electric meters for central stations', p. 372.

[81] Ayrton, 'Electric meters, motors, and money matters', p. 641.
[82] Swinburne, 'Electrical measuring instruments', p. 29.

drops would always be comparatively in their favour. Of course, following the example of gas suppliers' propaganda, electricity suppliers typically insisted that consumers would not risk loss from such fluctuations.

To protect the consumer from the fraudulence of under-pressure supply, Board of Trade inspectors made periodic checks on supply voltages; from 1892 they required these to deviate not more than 4% contractually stipulated levels. Nevertheless, major fluctuations did persistently recur and customers became more sensitive to such variation. Alexander Kennedy of the Westminster Supply Company told the ICE in 1890, for instance, that when he asked for power-station voltmeters to be correct to within one volt, colleagues revealingly 'scoffed' at this unattainable demand. Yet he claimed he was 'sure to hear' from customers if their supply was even two volts adrift, and such complaints about supply levels were evidently not rare.[83]

Whilst consumers also complained from time to time about the trustworthiness of individual meters (see previous discussion), it is much harder to find evidence of consumers complaining about the fairness of metering by quantity vis-à-vis metering by energy supply. To protect householders against unwitting exploitation in this regard, William Ayrton mounted a moralistic campaign to draw their attention to this matter. This ran from 1889 to 1892, the three-year period when electrical supply was starting to grow in scale in the wake of new supplier-friendly legislation in 1888 and in which new customs and practices were starting to be settled. Effectively barred from working directly for the supply industry by the terms of his professorial contract at the City & Guilds Central Institution in South Kensington 1884, Ayrton fashioned himself as a defender of consumer interests. Indeed, it is arguable that this was yet another means of attempting to overcome the credibility problems of technical teaching and research that he and his colleagues suffered during the first decade of this college's operations.[84]

As a defender of consumer interests, Ayrton persistently challenged the claims of certain dc suppliers (and ac suppliers which used adapted electrolytic Lowrie–Hall meters) that it was reasonable and even *fairer* to charge consumers by using 'quantity' meters rather than 'energy' meters. For Ayrton, such claims simply reflected suppliers' financial self-interest in adopting the form of meter that brought them greater income (see previous discussion). More disingenuously, in his view, such suppliers evaded the crucial

---

[83] See Kennedy's comments in discussion of Rookes E. B. Crompton, 'The cost of the generation and distribution of electrical energy', *Minutes of the Proceedings of the ICE*, 106 (1890–1), pp. 1–32, discussion on pp. 33–123, comments on p. 85. Until ca. 1886, Kennedy had been Professor of Mechanical and Civil Engineering at University College, London, where he had run the UK's first academic engineering laboratory and specialized in the testing of material strengths.

[84] Graeme Gooday, 'The premisses of premises: Spatial issues in the historical construction of laboratory credibility' in C. W. Smith and J. Agar (eds.), *Making space for science: Territorial themes in the making of knowledge*, Basingstoke: Macmillan, 1998, pp. 216–45.

question about what was actually *consumed* by householders: Was it 'quantity' of electricity, was it energy, or was it lighting? Some early companies charged customers per hour of illumination or per number of lights, and Ayrton maintained that this was still the fairest principle to follow. Customers should therefore be charged for the amount of light they received and accordingly this was *ideally* what their meters ought to be able to measure. In 1886 he and his Finsbury collaborator John Perry had shown experimentally that the intensity of light from an electric lamp varied as (at least) the *third* power of potential difference across it.[85] Given the problems of supply-pressure decrease, Ayrton was particularly dissatisfied with meters that used 'quantity' as a surrogate for lighting consumption. He noted that if supply voltage dropped by 4%, customers would get approximately 12% less light, but quantity meters would register only a 4% drop. Energy meters by contrast would indicate roughly an 8% drop, and this he considered to be a fairer – if still imperfect – solution to supplier exploitation of customers.

Much of the morals of metering thus hinged on what kind of parameter *ought* to be registered by meters operating in a truly fair system. Early on in his lecture on meters to the Society of Arts in 1889, the Scottish consultant engineer George Forbes moved quickly to resolve the question of 'what it is we want to measure'. According to his account, the critical matter was *who* supplied lamps and thus also who had to replace lamps that were damaged or broken by ageing or overvoltage. If it were the supplier, as was typically the case in France, then for both gas and electric lighting, it would be 'most desirable' to measure the light 'given off'. That way, customers would not lose out if undersupply occurred, and suppliers would be inhibited from oversupply by being liable to pay for lamps damaged thereby. Forbes considered, however, such a mode of practice was not likely to be adopted by British suppliers for his audience, because, for whatever reason, suppliers devolved to consumers the responsibility of choosing and installing their own lamps. Accordingly for British consumers he proposed instead that meters should register the 'light-giving' power of electricity. This he argued should be represented by *quantity* of electricity rather than by energy.[86] The justice of this claim was based on an asymmetry in the effects of negative and positive voltage fluctuations. When voltage levels dropped below the standard level, no harm was done to light bulbs. Yet when voltages were in 'excess', lamps would burn too brightly and the resulting damage would shorten their life. Because replacing the bulbs more often placed a financial penalty on consumers, Forbes reasoned that it would be 'objectionable' for consumers to

---

[85] See William E. Ayrton and John Perry, 'The most economical potential difference to employ with incandescent lamps', *Proceedings of the Physical Society of London*, 7 (1886), pp. 40–50. They claimed a cubic empirical formula linked the light emission of a bulb, θ to the voltage across it, v, as $\theta(v) = a(v-b)^3$, where a and b are constants obtained empirically by curve-fitting techniques; ibid., p. 46.
[86] Forbes, 'Electric meters for central stations', p. 372.

be charged both for replacing bulbs *and* for the extra energy supplied to them during voltage surges. It thus would not be 'fair' for customers to be metered for energy. Because quantity meters always *underregistered* any voltage increase, Forbes argued that such meters would accordingly be the fairest available means of compensating customers for damage to their bulbs. In understanding the basis of his argument, it should noted that a considerable part of Forbes' lecture was devoted to publicizing his own new 'windmill' meter, and this itself was a quantity-measuring device.[87]

Forbes' argument for the greatest fairness of charging by quantity relied on the premise that electrical inspectors could and would prevent voltage drops from occurring. He stipulated that variations should be no more than 4%–5%, and for 'perfect' lighting no more than 2%–3%. Even so, he admitted to having seen much larger variations in the past and hoped – with ingenuous optimism – that new central generating stations would give 'less ground for complaint' against the electric-light companies.[88] It was precisely on this point that his critics differed in the ensuing debate at the Society of Arts. As the self-fashioned defender of the consumer, Ayrton replied that Forbes' admission of irregularity of supply undermined the plausibility that fairness would in fact be accomplished by Forbes' proposed solution. Ayrton thus sought to 'break a lance' with Forbes over this subject, doing so in the tones of relatively benign badinage.

Ayrton dryly observed that generating companies naturally preferred their meters to measure whatever was easiest to measure (quantity of electricity) rather than what customers really 'ought to pay for'. After the carefully recorded audience mirth subsided, he argued that it was in fact *light* that he and other customers wished to purchase from the companies, and thus what they should pay for. Citing his 1886 study with Perry, he pointed out that drops in pd even within Forbes' tolerable margin of 2.5%, would have a very considerable effect on the light they consumed. For example, at a Saturday afternoon concert in St James Hall recently he had noticed that a mundane fluctuation in voltage had so dimmed the light bulbs that the management had been forced to revert to switching on the older gas lighting too. Ayrton thus recommended to his audience that they should not use the quantity meters advocated by Forbes, but ask for meters that registered a 'higher power' of the voltage. Only these gave consumers a prospect that their meters could fairly register the energy consumed (which varied as the square power of voltage) or even the light they received (which varied as the third power or higher of voltage).[89]

---

[87] Forbes had developed this novel and little-used thermal windmill device two years previously: Uniquely, it registered both ac and dc supply. Brown, 'Charging for electricity', pp. 518–19.
[88] Forbes, 'Electric meters for central stations', p. 372.
[89] Ayrton discussion to Forbes, 'Electric meters for central stations', pp. 436–7. See also the account of the debate on 'quantity vs. energy' in Brown, 'Charging for electricity', pp. 514–15.

It is hard to find any evidence of domestic consumers' adopting Ayrton's zealous opposition to quantity meters. Nevertheless many companies soon began to calibrate domestic meters in the Board of Trade unit of energy. This was calibrated to the energy carried by a coulomb of electricity across a potential of 1 V × 360,000 (the number of seconds in 1000 h).[90] Nevertheless, Ayrton told his Royal Institution audience at a Friday Evening Discourse in April 1892 that many meters which now *appeared* to register energy units were still just quantity meters. And he re-emphasized that their calibration would accordingly mislead consumers if the supply voltage were not constant. Again attacking the dishonesty of some supply companies, Ayrton offered this argument:

The meters employed by the electric supply companies simply measured the number of coulombs passed through them, and should therefore be called coulomb meters. The name was not employed by the companies, who usually called these instruments electric supply meters, possibly because they had no desire to call attention to the fact that what was registered by the meter and what was charged to the consumer were two entirely different things.

What especially perturbed Ayrton was a concurrently introduced Board of Trade regulation that required electricity suppliers to maintain the potential difference of their systems to within ±4% of the specified value. Whatever the intentions motivating this limit, it still effectively countenanced a gross misrepresentation of lighting consumption by quantity meters. Whilst Ayrton noted that it was 'greatly to the credit' of supply companies that they had generally been able to keep within these ±4% voltage variations, he and his South Kensington students had found that even within these extremes of the allowed variation, a standard 16-cp Edison–Swan bulb might alter its illumination by ±25% between 12 and 20 cp. More disturbingly, there were places not far from Albermarle Street in which the variation greatly exceeded these limits, and he advised his audience to seek their own independent tests at regular intervals and to seek legal redress as necessary.[91]

It is important not to interpret Ayrton's views simply as those of a sardonic commentator on the margins of the electrical industry. In James Swinburne's lecture on electrical instruments to the ICE in the same month, he concurred with Ayrton that the system pressure was 'seldom' what it ought to be. Indeed, they shared the view that, on those very grounds, it was preferable from the customers' point of view to supply them with energy meters rather than coulomb meters. More strikingly still, he agreed with Ayrton that, to make customer charges 'really fair', the meter should read according to the light given by the lamp rather than the energy consumed. Thus, following

---

[90] The commercial rate for this was between 7¼d and 8d; Mrs Gordon, *Decorative electricity*, p. 14.
[91] Ayrton, 'Electric meters, motors, and money matters', pp. 640–1.

Ayrton's quantitative prescription, he proposed that if the voltage were to run 1% below its prescribed level, the meter should run 5% 'slow' to compensate them duly for loss of illumination. But following the force of Forbes' arguments, Swinburne also maintained that if the main pressure were to rise above the requisite level, meters should run slow or even backwards to compensate customers for the serious reduction in the life of their lamps. Yet at this point Swinburne's commercial instincts as an engineer took precedence over his scruples about what was 'fair' to the customer. Meters which justly compensated consumers for the supplier's inability to maintain the recommended constant pressure were, alas, 'impracticable' – that is to say too difficult and expensive for any engineer to bother developing. So all other things being equal, the joule meter, he concluded, was the 'best' form of meter to use – although both supplier and consumer might both lose out equally because of the rather larger power consumed by these meters in comparison with that consumed by coulomb meters.[92]

The public debate effectively ended at that point, and energy meters have remained the standard technology of measurement at a distance since then. Even so, the question of the fairest meter as one that measured light consumption rather than energy did not entirely disappear. The issue briefly surfaced again following an incident in Paris in late 1893. The Compagnie Parisienne de L'Air Comprimé, or Popp Company, which had a monopoly in supplying a district of Paris, was found on inspection to be supplying the 110-V lamps of its customers at only 95 V. Thus, whereas meters recorded about 70% of the normal energy consumption, customers received a mere 34% of the light that they would have received at 110 V. Although the Popp company soon lost the majority of its clientele for its actions (some presumably went back to gas lighting), the municipal authorities further punished the Popp company by depriving it of its monopoly so that consumers could find a more honest supplier. Discussing this story in December that year, the *Electrician* reminded its readers in the electrical supply industry of Ayrton's demand several years earlier that customers should be metered by light levels. It noted that that this standard should indeed be the *ultima thule* of the meter-maker. The journal's comment was that if Popp's system of 'extra-low' voltage distribution should become general, the demand for a more rigorous, consumer-friendly 'payment-by-result' meter would 'not long be confined to its originator'. This was a cautionary tale warning suppliers to avoid negligence or fraud in voltage control lest consumers rebel and demand a much more expensive form of metering.[93] Given that, to this day, electricity metering is still undertaken on the basis of energy supply, it is evident that customers never came to distrust supply companies sufficiently to demand this change of metering. Rather, they literally bought into an

---

[92] Swinburne, 'Electrical measuring instruments', p. 26.
[93] Notes, '*The Electrician*', 32 (1893), p. 182.

account of consumption of lighting as the consumption of energy that supply companies found most technologically and financially convenient.

6.6. METERS AND THE GENDERED CONSUMPTION OF ELECTRIC LIGHTING

Downstairs is the electric meter, on which is recorded monthly the number of electric units burnt [sic] in the house. The master can himself calculate and check each month the expense of his electric light; and where the servants are not to be trusted, it might be well to have a separate meter for all their rooms.

Mrs J. E. H. Gordon, *Decorative Electricity*, 1891[94]

Previous sections identified some of those involved in the commercial world of meters: designers, vendors, electricity suppliers who bought them for their customers, and 'experts' who claimed to represent the interests of suppliers, consumers, or both. But who were the household consumers whose use of electrical lighting was registered by early electric meters? More specifically, which householder was accorded the responsibility of taking readings from the meter, and why?

Given the sheer expense of electric light in the first decades of supply, early domestic consumers inevitably belonged to the wealthiest sections of society. Some estimates in the 1890s put the cost of installing electrical light in a town house in the region of £300–£400. This was roughly the entire annual income of a moderately affluent middle-class household; and even this outlay might be accompanied by annual bills totalling £40–£50 – roughly the annual income of a poor factory worker. Unsurprisingly, few British cities outside the capital could count on attracting enough customers to risk setting up electrical supply after the new supplier-friendly legislation of 1888: Exeter and Bradford began in 1889, Bath and Newcastle-Upon-Tyne the following year.[95] Nevertheless, with typical metropolitan parochialism, *The Times* reported in August 1891 that the new 'light of luxury' could be found only in the most prosperous districts of London.[96]

Later that month, Lindley Sambourne presented a satirical response to this newspaper article in a cartoon for *Punch*. A boyish sprite holding a representative 'Electric Light' sits in the street outside the front door of a plush London residence, asking cheekily to be allowed in. Faced with such an appeal from this self-styled 'dear little chap', a substantial 'Paterfamilias'

---

[94] Mrs Gordon, *Decorative electricity*, p. 14.
[95] Jones, 'Coal, gas and electricity', p. 88. For the 1888 legislation, see Hughes, *Networks of power*, p. 230.
[96] Figures from Crompton catalogue, 1896, S. P. Thompson Collection, IEE Archives; 'On the growth of electric lighting in London', *The Times*, 19 August 1891, p. 5, column a.

punned in reply 'Ah! You're a little too dear for me – at present' (Figure 15).[97] As demand increased and prices began to drop (for reasons subsequently discussed), most cities around Britain did come to accommodate the boisterous electrical sprite in the ensuing five years. Even so, the cost of electric lighting did not fall to a par with gas until the National Grid was set up more than 50 years after Edison's pronouncements of imminent cheap electricity first panicked British gas shareholders in 1879.[98] Accordingly, to promote domestic sales the electrical lobby had to resort to alternative strategies. One was presenting electric ovens as capable of cheaper and safer cooking than by gas methods and offering a cheaper supply rate for daytime usage of those otherwise costly novelties, the electric kettle and frying pan.[99] Another tactic, as we shall shortly see, was to present the consumption of electric light as a matter of domestic aesthetics at least as much as household economics.

For the period covered by this book, the electric light unquestionably remained a *luxury* commodity beyond the pocket of large sectors of the population. At the same time it was a contentious technology, and some women held strong views on its worthiness as a household commodity that diverged from those of their menfolk. Thus in retrieving the identity of the electrical consumer, and the reasons for their growing number during the 1890s, a gender analysis of domestic illumination practices among the upper social echelons can provide more useful insights than a purely class-based study. It was evident, after all, in earlier sections, that electrical engineers and supply companies represented the domestic consumer of electricity as the financially responsible *male* householder. Not only was this representation reiterated in Sambourne's *Punch* cartoon, but it also was enacted in his domestic arrangements: In 1896 Lindley enthusiastically arranged the installation of electric light in the marital home, evidently paying little heed to the indifference and later disgust of his spouse Marion.[100] Even female enthusiasts for the electric light such as Mrs J. E. H. Gordon in 1891 (see the epigraph

---

[97] 'At the door; or paterfamilias and the young spark', *Punch*, 101 (1891), pp. 98–9. See subsequent discussion on the cartoonist Linley Sambourne and his spouse Marion.

[98] See Byatt, *The British electrical industry, 1875–1914*, and Leslie Hannah, *Electricity before nationalization*, pp. 1–53.

[99] Anonymous, 'Electricity in the kitchen', *Electrician*, 333 (1894), pp. 208–9. The writer commented 'Careful experiments in one instance have shown that for general cooking, other than boiling, the electrical system may actually be cheaper; and it is not unlikely that with careful management the electric oven may prove to be of considerable value. Since "science is measurement," and electrical engineers profess to be more or less scientific, some data as to cost are liable to be forthcoming before long'. Ibid., p. 209. For further discussion of the role of the Crompton Company in developing electrical kitchen equipment and the problems thereof see Bob Gordon, *Early electrical appliances*, 2nd edition, Princes Risborough: Shire Publications, 1998, pp. 11–19.

[100] Shirley Nicholson, *A Victorian household*, Stroud: Sutton, 1998, p. 156. I thank Sophie Forgan for drawing my attention to this source.

AT THE DOOR; OR, PATERFAMILIAS AND THE YOUNG SPARK.

Electric Light. "WHAT, WON'T YOU LET ME IN—A DEAR LITTLE CHAP LIKE ME?"
Householder. "AH! YOU'RE A LITTLE TOO DEAR FOR ME—AT PRESENT."

Figure 15. Linley Sambourne's punning cartoon of an affluent London male householder in August 1891 who was still disinclined to take on the financial burden of the new luxury electric light. (*Source*: Linley Sambourne, 'At the Door; or Paterfamilias and the Young Spark', *Punch*, 101 (1891), pp. 98–99).

that opens this section) represented ownership of domestic electric light as a male phenomenon and the checking of the meter as a specifically masculine prerogative. We need not, however, take this representation of electric-light consumption at face value. Elsewhere in her book, *Domestic Electricity*, Alice

Gordon shows herself to be well informed about both the technical and the aesthetic issues of domestic electrical lighting from her own daily usage of it. And in what follows I shall explore the use and valuation of electric lighting and show that its early consumption was of a more complex gendered character than perhaps was the case for other domestic electrical technologies later introduced to the home in the early twentieth century.[101] In so doing I shall challenge the claim by Sue Bowden and Avner Offer that the early consumption of electric lighting was gender neutral[102] and highlight the gender issues somewhat understated in Schivelbusch's account of the asethetic resistance among many European consumers to the advent of the electric light.[103]

Like a significant number of later Victorian women married to electrical engineers, Mrs Gordon was well acquainted with the issues of electrical metering. Gertrude Ferranti had familiarized herself with her husband Sebastian's activities in meter development after their marriage in 1888 (see previous discussion), and in the same year Hertha Ayrton developed her own form of meter that gauged electrical supply from the thermal evaporation of water.[104] Three years later Mrs Gordon presented a rather more traditional 'womanly' expertise on matters of meters and household economy. In the February 1891 issue of the *Fortnightly Review*, a piece under the name of Alice M. Gordon appeared with the title 'The Development of Decorative Electricity'. This piece emphasized the luxurious and comfortable nature of the well-presented electric light for the solitary sybaritic male. She nevertheless specifically reminded her readers that the 'quantity of electricity' used was to be 'paid for by meter' – as if to imply very clearly that billing was not based on connection time or on simply the number of lights switched on. Given the strictness of such metering, she emphasized how the installation of small corner lights would enable the solitary user to effect greater economy than a householder who skimped on the initial costs by installing only a single central chandelier. Better still, such corner lights

---

[101] Ruth Schwartz Cowan, 'The "Industrial Revolution" in the home household technology and social change in the 20th century', *Technology and Culture*, 17 (1976), pp. 1–23, and Gerrylyn Roberts, 'Electrification', in Colin Chant (ed.), *Science, technology and everyday life' 1870–1950*, London : Routledge, 1989, pp. 68–112.

[102] Sue Bowden and Avner Offer, 'The technological revolution that never was: Gender, class, and the diffusion of household appliances in interwar England', in Victoria de Grazia and E. Furlough (eds.), *The six of things*, Berkeley/London: University of California Press, 1996, pp. 244–74, quote on p. 245. My account of gender and technology follows that of Cynthia Cockburn and Susan Ormrod, *Gender and technology in the making*, London: Sage, 1993, who show that a technology can be gendered from the ways in which consumers use a technology and construct representations of that usage *irrespective* of whether its makers had gendered or non-gendered intentions about its usage.

[103] Schivelbusch, *Disenchanted night*, pp. 157–87.

[104] See Ferranti and Ince, *Life and letters*, pp. 137–8; for Hertha Ayrton's meter, see William Ayrton's comments in discussion on Forbes, 'Electric meters for central stations', p. 437.

could be installed near curtains without risk of fire, and electric lights could be fitted to decorated ceilings without the destruction created by the corrosive vapours of burning gas lights. Most tellingly of all, Alice encouraged her readers to adopt highly decorated forms of light fittings that used only indirect or shielded illumination.[105] The gendered rationale for this 'decorative' approach only became apparent, however, in her next publication.

In the early spring of 1891, Alice published a fully illustrated and populist handbook, *Decorative Electricity*. This time she drew very heavily on her husband's name and professional credentials, fashioning herself explicitly on the title page as Mrs J. E. H. Gordon. And her book was evidently written to support the business of Mr Gordon – 'Director and Consulting Engineer to the Metropolitan Electric Supply Company' – in promoting the take-up of electricity among a distinctly sceptical public. She even interpolated a chapter by him rebutting the oft-repeated allegations among the press and gas fraternity that electric installations posed major forms of fire hazard. Addressed to a mixed audience, it aimed to overcome the image of electric light not only as being economically disadvantaged in comparison with ever-expanding domain of gas lighting, but also unsightly. Her strategy in so doing was to glamorize the electric light as integral to the aesthetic decoration of the home in ways that could overcome women's palpable dislike of the inelegant glare of bare unadorned lamps.[106]

In the very first chapter of *Decorative Electricity*, Alice first sought to persuade economy-conscious readers that electric light need not be *much* more expensive than gas light. Without this reassurance it is not at all obvious that they would have bothered to continue further with any ensuing chapters. Mrs Gordon argued that by very disciplined practices in switching electrical lights off when they were not needed and careful comparative evaluation of all the costs involved, the bill derived from the electrical meter might be nearly as low as that of gas lighting:

In calculating the expenses of the electric light, they should not only be compared with the former gas bills only, but with the gas, paraffin, and wax candles, and a small amount added for tapers and matches. If this is done, I feel sure from my

---

[105] Alice M. Gordon, [Mrs J. E. H. Gordon], 'The development of decorative electricity', *Fortnightly Review*, (new series) 49 (1891), pp. 278–84, quotes from p. 282.

[106] Mrs Gordon's *Decorative electricity* was reviewed widely by both ladies' journals and electrical engineering periodicals. See endpiece of second edition, 1892, for reviews of the first edition, as discussed in Graeme Gooday, '"I never will have the electric light in my house": Alice Gordon, *Decorative Electricity* and the gendered periodical representation of a contentious new technology', in Louise Henson et al. (eds.), *Culture and science in the nineteenth-century media*, Burlington (VT): Ashgate, forthcoming 2004. I thank Sophie Forgan for sharing with me her many insights on this book as part of our joint research on women's work in electrical engineering.

own experience that the electric light bills will compare most favourably with the former bills for illumination, without taking into account the saving to health and decoration.

Although at a cost of a farthing per lamp per hour, electric light was, she conceded, about 20% more expensive than gas, it was possible to overcome this by using the 'far greater facilities for switching off' electric lights when they were not wanted, whereas gas would have to be left burning. With 'ordinary care', she thus contended that the cost per lamp per year could even be less than that of gas, with the added advantages of avoiding the noxious and corrosive vapours produced by unshielded gas lights.[107]

Having dealt thus with the issue of domestic economy that was of concern to both male and female readers, most of the ensuing chapters appear to have been addressed to the aesthetic sensibilities specifically of women readers. This is not surprising since suggestion of women's dislike of electric light were quite widespread. The *Punch* cartoonist, Lindley Sambourne, had treated this subject to unsympathetic satire in 1889, and an *Electrician* editorial published contemporaneously with Mrs Gordon's *Fortnightly* article noted that electric light has been condemned freely and 'not unreasonably' as '"unbecoming" to dresses and complexions'.[108] Accordingly, the major part of *Decorative Electricity* was devoted to overcoming the harshness of electric lighting and turning it into an aesthetic feature of domestic furnishing and entertainment that would overcome some women's conspicuous dislike of it. Mrs Gordon reported that the electric light often found in dining rooms was 'very glaring and disagreeable'. This, she maintained, fully justified the remark she had so often heard made by ladies that 'I never will have the electric light in my house, as it gives me a headache whenever I dine by it.' Recently, reported Alice, she had herself been at a dinner party at which

---

[107] Mrs Gordon, *Decorative electricity*, pp. 14–15. She did not mention the Welsbach gas mantle developed in 1885 to overcome these traditional objections to gas illumination. See Schivelbusch, *Disenchanted night*, on this, pp. 47–8. Some reviewers were highly sceptical of Mrs Gordon's claims concerning the possibility of achieving greater economy with electric lighting: 'No; holders of gas shares may still preserve their equanimity.' See 'Decorative electricity', *Black and White*, 1 (1891), p. 575, and further discussion of this in Gooday, 'I never will have the electric light in my house'.

[108] Linley Sambourne, 'Happy thought', *Punch*, 97 (1889), p. 30, presents electrical illumination as 'not always becoming to the female complexion', obliging women to use decorative Japanese sunshades for protection, thus allegedly delighting male observers. My thanks to Laurie Brewer for drawing this cartoon to my attention – for reproduction and further discussion see Gooday, 'I never will have the electric light in my house'. The anonymous editorial, 'Problems connected with indoor Illumination', *Electrician*, 26 (1891), pp. 480–1, concluded with the comment 'Engineers will do all they can to make electric light safe and cheap; let some artistic genius make it comfortable and beautiful.' A clear allusion to this piece and the correspondence which it provoked (ibid., pp. 501–2, 521) was made in this journal's extremely favourable review of Mrs Gordon's book six weeks later – see *Electrician*, 26 (1891), p. 670.

each guest had a lamp with a lemon-yellow shade hung just above his or her eyes, so that that light had shown up 'every wrinkle and line' on the face. No one, she contended, over the age of eighteen should be asked to sit beneath such a light.[109] In its highly sympathetic review of Mrs Gordon's book, *The Electrician* reported these and other claims verbatim, hoping that by giving them the 'utmost publicity', efforts might be made to prevent more of 'such abuse'.[110]

That this sort of problem was one most acutely problematic for women was apparent from Mrs Gordon's account of women's experiences of overhead electric lighting in public buildings:

How trying and unbecoming it can be, to even the very youngest and prettiest among women can be studied when a lecture or concert is given... Wherever there are large ceiling lights alone with reflectors *over them*, women must be content to look their worst. Ceiling lighting is most unbecoming to a woman's age, and causes dark shadows under the eyes, which accounts for the haggard and worn look of most people at concerts.[111]

Mrs Gordon's solution to the harshness of overhead lighting in the home and concert hall alike was spatial and material. She recommended that various kinds of decorative wrappings should be used to diffuse and tint the illumination in a manner less invidious to the sensibility of ladies; also that a number of sidelights be disposed discreetly on walls, to supplement the more subdued ceiling lamps. Yet this strategy is precisely what seems to have been the source of gender conflict in the home. It appears that some husbands who paid the metered electricity bill disliked this strategy: Covering lights with cloths and installing more than just one central lighting greatly increased the costs of installation and running and decreased the amount of illumination they got for their money. Mrs Gordon herself relates at the start of her book a putatively familiar tale of a married couple's planning and electric installation in which they could 'seldom agree between themselves' as to what they wanted[112]:

The master wishes to get all the light possible, and the mistress to have the light as becoming and pleasant as possible. It is rather difficult to reconcile these two wishes, and after much discussion the master testily exclaims: "My dear, what is the good of going to all this expense if you will tie the light up in bags?"[113]

---

[109] Mrs Gordon, *Decorative electricity*, pp. 59–60.
[110] Anonymous, 'Reviews: Decorative electricity', *Electrician*, 26 (1891), p. 670.
[111] Mrs Gordon, *Decorative electricity*, p. 146.
[112] As a further example of subtly gendered prerogative, Mrs Gordon notes 'In the case of a decorated house, the master should himself go into the details of the proposed wiring, that he may understand how it is to be done' – while also implying that women should consider themselves perfectly competent to undertake the wiring work by themselves too; Mrs Gordon, *Decorative electricity*, p. 7.
[113] Ibid., pp. 3–4.

However, Mrs Gordon hinted at an overall scheme for resolving the use of aesthetically shielded electric light so as not to exacerbate the cost differential between it and gas lighting. She proposed a gendered division of labour in which the lighting of the house was divided into two 'classes': 'decorative' lights for occasional use versus 'practical' lights for daily usage. It was the former of these which was a distinctly womanly prerogative. With regard to the traditionally 'feminine' domain of the drawing room, for example, Mrs Gordon suggested that switches for the larger ceiling and decorative lights should be placed inside an ornamented wooden cupboard next to the door: The key for this should be kept by the mistress of the house. By that means she could ensure that decorative lights were only used when required: She considered that a great deal of electricity was wasted by those householders – male and female – who turned on the light just to 'show their friends'.[114]

Having devoted so much of her book to the multifaceted womanly prerogative in costing, arranging, and prudently managing domestic electrical lighting, why did Mrs Gordon adopt a heavily *masculinized* representation of the 'ownership' of the electric light and the responsibility for checking the meter as indicated in the epigraph that opens this section? This was probably more than just a strategic deference to male readers' notions of domestic control. Research by Erika Rappaport on payment for luxury goods in late Victorian Britain indicates a great degree of complexity in the gendering of marital financial responsibility. Recognized legally as property holders by the late 1880s, married women were increasingly targeted as potential consumers by retailers of luxury goods. Yet whilst their husbands were still legally liable for their debts on agreed expenditure, husbands could and often did refuse to pay the bills on luxury goods for which wives had exceeded agreed budgets. A spate of litigation ensued in which shopkeepers tried to make husbands legally liable for their wives' debts – but generally found court judgements against them. The implication was that vendors had been irresponsible in allowing women credit for such luxury consumption. Whilst this was evidently a serious problem for the unfortunate vendors, Rappaport concludes that the status of luxury-good consumption was highly problematic insofar as it could generate embarrassingly public manifestations of domestic conflict over financial liability.[115]

With this insight from Rappaport we can begin to understand the tensions within Mrs Gordon's account. Clearly she wrote to encourage women to be more active and enthusiastic consumers of electricity, especially of the most luxurious 'decorative' occasional lighting, aiming to stimulate a more

---

[114] Ibid., p. 10.
[115] Erika Rappaport, '"A Husband and his Wife's Dresses": Consumer Credit and the Debtor Family in England, 1864–1914', in V. de Grazia & E. Furlough (eds.) *The Sex of Things: gender and consumption in historical perspective*, Berkeley/London: University of California Press, 1996, pp. 163–87.

Measurement at a Distance 261

extensive installation of electric lighting – to assist her husband's supply business – and a more intensive consumption of it on special occasions. Then again, in claiming that decorative lighting used with *prudence* would not be much more expensive than gas, she took her argument in a quite different direction. In identifying the male householder as the person who *owned* the luxurious electric light and who was responsible for checking the meter regularly, she signalled to readers of both sexes how to avoid financial embarrassment and domestic conflict over women's unauthorized excessive consumption of this luxury. Linking this back to our earlier discussion about the problematic nature of the Edison meter, we can see why male householders – and female observers such as Mrs Gordon – might have insisted on having a directly readable meter. It gave the financially accountable male scope for monitoring both his own and his spouse's consumption of the expensive indulgence of electric lighting.

6.7. CONCLUSION

The problems of trust in using the domestic meter were rather more sharply evident than for the measurement technologies discussed in previous chapters. The requirements for making meters that satisfied either consumers or suppliers were not simply technical, but also were imbued with moral judgements of fairness and accountability. The problems were heightened by the fact that both sides did not entirely trust the other not to try to manipulate the form and operation of the meter's reading for their own interests. Thus any internal pathologies, fallibilities, or opportunities for manipulation or interference were treated with some suspicion or anxiety as a possible source of error or outright fraud. The customer and supplier were almost inevitably in conflict over what was the most convenient and economically desirable form for them, and there was a whole range of issues over which suspicion could be roused as well as more opportunities for dishonest behaviour on both sides.

This story has been one in contrast to the standard social-constructivist account of technology in which consumers and designers figure as cooperative collaborators in contriving technologies into forms that bring mutual benefit for all concerned parties.[116] In the case of the domestic meter, consumers appear – superficially at least – to have won the battles of trust over whose interests should be embodied in the meter. They secured the construction of

---

[116] Trevor Pinch and Wiebe Bijker, 'The social construction of facts and artifacts: Or how the sociology of science and the sociology of technology might benefit each other', in Wiebe Bijker, Trevor Pinch, and Thomas P. Hughes (eds.), *The social construction of technological systems*, Cambridge, MA: MIT Press, 1987, pp. 17–50. Wiebe Bijker and John Law, *Shaping technology/Building society: Studies in sociotechnical change*, London/Cambridge, MA: MIT Press, 1992. Wiebe Bijker, *Of Bicycles, Bakelite and Bulbs*, Cambridge, MA/London: MIT Press, 1995.

meters that were readable by *both* consumer and supplier, and also meters that registered energy consumption rather than quantity. But in regard to the adoption of gas-meter-like dials for the reading technology of electric meters, this is better explained in terms of the conservatism of customers who prefer familiar patterns of meter reading rather than the social-constructivist perspective that consumers' well-articulated interests are explicitly accommodated into a new technology. Indeed, if one considers the extra expense of making meters with readable dials and the inevitability of energy loss through the friction of its operation, it is not obvious that the adoption of such meter-reading technology was in the financial interests of such consumers. The relationship between the measurement technology and relevant social 'interest' thus remains ambiguous.

In some respects it is easier to see consumer conservatism as the more important factor than consumer 'interest'. Many customers were conservatively content to acquiesce in the gas paradigm of being metered simply for the 'quantity' of supply even though this was not the theoretically best means of meeting their financial interests. Only by the intervention of characters such as Ayrton and the stipulations of the Board of Trade were these interests of consumers met in the process of charging for energy rather than quantity of electricity. Then again, no meters were made that registered what perhaps was in the greatest interest of consumers to have registered: the quantity of light that they consumed. The sheer technical difficulties and expense of that evidently determined that customers were metered in the way that suppliers and meter manufacturers found most convenient, rather than what was necessarily 'fairest' for customers to be charged for consuming. Social-constructivist accounts of technology perhaps should be more open to the possibility that the kind of consumer 'interests' which are instantiated in new technologies might generally be those that designers and manufacturers find most convenient to address. By highlighting differences in the meaning of consuming electrical quantity, energy, and lighting, I thus criticize the political glibness of some forms of social constructivism by emphasizing that consumers did not have an autonomous interest-driven choice about *what* they were charged for consuming. The units in which their meters were calibrated were not chosen as a consensual compromise that accommodated the interests of all relevant parties. Historians should not underestimate the power of commerce to dictate which technologies would be available for customers to have to learn to trust.[117]

---

[117] For more on this perspective, see Judy Wajcman and Donald Mackenzie (eds.), *The social shaping of technology*, 2nd edition, London: Routledge, 1999.

Conclusion

It seems, indeed, as if the commercial requirements of the application of electricity to lighting, and other uses of every-day life, were destined to cause an advance of the practical science of electric measurement, not less important and valuable in the higher region of scientific investigation than that which, from twenty to thirty years ago, was brought about by the practical requirements of submarine telegraphy.
Sir William Thomson, 'Electric Units of Measurement', ICE, 1883[1]

The colourful episodes discussed in preceding pages enable us better to understand the changing and contested practices for *measuring* electrical performance in the latter part of the nineteenth century. In his 1883 ICE lecture, Sir William Thomson predicted that the practice of electric lighting would transform electrical-measurement techniques just as telegraphy had done since the early 1860s (Chapter 3). Chapters 4–6 of this book describe how the new technological enterprise did indeed stimulate practitioners to develop new kinds of instruments that embodied new techniques and new understandings of what constituted measurement – both in the general case and in the specific case of electrical practice. Yet in this process they brought lingering unresolved problems into the foreground about what measurement actually was, who counted as a measurer, and what and whom should be trusted or otherwise in the measurement process. Moreover, it raised awkward questions about whether accuracy was an easily identifiable and readily quantifiable attribute of a measurement.

As we saw in Chapter 2, such developments served to highlight the extent to which different conceptions of measurement – the comparative and the absolute – had long coexisted in a sometimes perplexing dualism for practitioners and students. By the late-1880s a lighting engineer could plausibly contend that an accurate measurement could be accomplished by a mere glance at a dial or scale to read a deflection directly representing what was hitherto measurable only indirectly by comparative or absolute methods. This move away from time-consuming and labour-intensive methods to technologically simplified techniques of reading instruments certainly constituted

---

[1] William Thomson, 'Electrical units of measurement', *Popular Lectures*, London, 1891, Vol. 1, p. 86.

a radical change in measurement practice. Let me recap what this meant for the apparently familiar case of measuring electrical resistance.

A variety of techniques were employed in the 1860s that would pass muster under the pluralist orthodoxy concerning electrical measurement: These allowed measurement to be either a matter of determining length and or mass or a matter of direct comparison with a standard unit. Both Werner von Siemens' and Augustus Matthiessen's initial approaches to constructing a comparative standard of resistance required the delicate *weighing* of the amount of metal required (mercury or metal alloy) to fulfil a specification framed in terms of the linear dimensions of a conductor. The BAAS Committee on which Matthiessen worked from 1861 to 1870 sought to measure absolute resistance by gauging the deflection of a needle on a Weber dynamometer galvanometer, converting this to a value of resistance by time-consuming multiple trials and algebraic calculations. Once this had all been done, with all due care taken to pre-empt error, the resistance units calibrated to this value could subsequently be trusted by a relatively unskilled practitioner using a Wheatstone bridge to undertake a direct comparative null measurement from a carefully adjusted bridge. By the mid-1880s, however, it was possible for a busy engineer to measure resistance with a direct-reading ohmmeter. Once the instrument had been set up and checked, an engineer could in a matter of seconds use it to take readings of resistance by glancing at a needle deflected against a precalibrated dial.[2]

The ohmmeter, like its direct-reading siblings the ammeter, voltmeter, wattmeter, and secohmmeter, embodied and thus further perpetuated the industrial imperatives for labour-saving and time-saving techniques. The bodily skill and labour of workers in measurement had largely been displaced by sophisticated machinery, so that technological skill, labour, and attention could be deployed to other tasks. In the wider cultural domain, this industrialization of measurement is what enabled twentieth-century car drivers to read road speed or engine speed by glancing at dials on a car dashboard. As I have indicated, though, this transition was not one, however, to which all late Victorian practitioners uniformly and easily subscribed. For those who identified with older traditions of natural philosophy and telegraphy, measurement was an activity that *ought* to involve skilled bodily labour enacted with self-discipline, patience, self-reliance, and care. For such practitioners, the loss of such virtuous corporeal conduct in the new industrialized machinery of measurement also entailed a further loss: A loss of older means of knowing how to judge the trustworthiness of numerical outcomes of measurement practice. With the new automated machinery, the trustworthiness and accuracy of a measurement could no longer be guaranteed by reference to the character of the measurer or the degree of care taken in measurement. These were instead replaced with a requirement to trust the designer and maker of the instrument's mechanism – not a straightforward or risk-free

[2] James Swinburne, *Practical electrical measurement*, London: 1888, pp. 16–22.

business. Quite apart from this there were, as I showed in Chapter 2, new statistical protocols of error and precision imported to Britain from German and American sources that enabled trustworthiness and accuracy to be quantified by means that were more algorithmic and or mechanistic in kind. These required no particular knowledge of the character of material or human participants in the measurement process and thus helped to circumvent traditional needs for morally laden judgements to be executed in measurement work.

Looking back on Thomson's 1883 prediction from some years, it is thus not obvious that all instrument-users in the electrical world would have regarded changes in measurement practice wrought in the ensuing decade as being self-evidently an advance. And it was not just the end-users who might have been concerned about this. The development of the direct-reading measurement instrument was facilitated by a more complex division of labour that devolved much of the obligation for virtuosity and concern for care in measurement from the end-user to those who had earlier designed and constructed such instruments. Yet the continued association of the authorship of a measurement with this end user of conveniently pre-calibrated and error-proofed devices tended to obscure the distributed and communal nature of the labour that made such easy measurement possible. As Charles Boys, Carey Foster, and, later, Norman Campbell pointed out after some decades of working in this new practice, it was simply a confusion for those who took mere visual readings from such instruments to believe that they had undertaken a measurement at all.

Given such opposition, how did the change in measurement practice nevertheless take place? The relevant factors, I suggest, were the growing time pressure of industrialization and the growing scope of technical practices within the electrical industry. The demand for electric light from the late 1880s grew in tandem with a community of electrical practitioners with so many diverse tasks to attend to within a given working day, they were too busy to engage in the more fastidious practice of measurement or worry about who should be ascribed authorship of a measurement. The correlated mass production of direct-reading instruments made it easier and more affordable for others to adopt this new measurement technology in the laboratory and workshop, especially teachers in the universities, colleges, and schools of the early twentieth century. Preoccupied with the exciting new areas of X-rays, electrons, radio activity, and wireless telegraphy, and thus with an ever fuller curriculum to handle, such teachers and their students paid less attention to measurement as a virtuosic and virtuous activity of such intrinsic importance as to properly preoccupy individuals.[3] Although

---

[3] See preface and Graeme Gooday, 'The morals of energy metering: Constructing and deconstructing the precision of the electrical engineer's ammeter and voltmeter', in M. N. Wise (ed.), *The values of precision*, Princeton, NJ: Princeton University Press, 1995, pp. 239–82, esp. pp. 274–5.

Norman Campbell, among others, developed a new theoretical understanding of measurement in the wake of these developments, the authorship of a measurement has evidently little troubled scientists and engineers in the later twentieth and early twenty-first centuries. Many people have acquiesced in what Campbell excoriated: a solipsistic delusion about the distributed nature of measurement practice in a highly technologized society. The moral for the historian is that the identification of authorship for a measurement in the past is not a morally neutral matter. It is a potentially fraught task that cannot evade the need to examine the social–political assumptions involved in making either individualist or collectivist attributions of authorial agency in bringing quantification to bear on the material world.

The role of trust in measurement work and its relation to the hardware of measurement relates closely to this issue of authorship. The distributed nature of labour in electrical measurement discussed in Chapters 3–6 was not problematic so long as all were able to trust the others involved in the measurement process, such as those who worked *within* either the Siemens Company or the BAAS Electrical Standards Committee. This matter could become highly contentious, though, if a lapse in trust or breach of trustworthiness occurred; generally speaking when such distrust arose it was most typically between unseen persons who were not personally acquainted with each other. From the inside of a community of close-knit workers, sharing a common sense of mutual obligation and immediate experience of their working practices, it was easy to judge who was likely to be honest and open about their results and whose techniques were prone to errors or artefactual results. Without such a knowledge of personal character and integrity, distrust about measurement practices could be precipitated by relatively small-scale contingencies within rivalries of reputation, project, and commerce.

For example, as we saw in Chapter 3 Matthiessen and Siemens were not mutually acquainted before they began their acrimonious dispute about the relative trustworthiness of their labours, techniques, and preferred metals for constructing resistance standards. Their subsequent distrust was not obviously an *inevitable* outcome of differences in their personalities, respective research agendas, metrological commitments, national identities, geographical locations, or laboratory training in chemical analysis. Rather, such distrust emerged from an acrimonious spat in 1860–1 over undiplomatic comments each published about each other's expertise in metals in important journals. As each side brought in further supportive allies, the conflict escalated into major socio–industrial antagonism in which attributions of untrustworthiness were the major explanatory device used to attach and dismiss the results of the other as erroneous. There were, however, frequent ambiguities and shifting claims as to the sources of this untrustworthiness: the flawed moral character of the measurer; the treacherous nature of the metals involved; the unreliable techniques employed or deceitful representational practices used in publication. The long-term and widespread ramifications of this absence of trust and its independence from considerations of metrology (absolute vs. arbitrary) have hitherto been somewhat underestimated by historians of resistance standards. Without a study of these multifarious issues of trust for the case of the Siemens–Matthiessen dispute, it is not possible to fully understand why decisions over the use of such metals as mercury and alloys in constructing standards could be such a disputed matter for several decades in the late nineteenth century.

Overt distrust of people and instruments was not, however, obviously the norm for late nineteenth-century measurement practices; had it been otherwise, as Steven Shapin might remind us, it could not have become a *collective* practice at all. Some degree of trust in at least some other people, instruments, and reading techniques was, as we have seen a prerequisite of effective measurement practice. Such trust was neither monolithic nor always uncritical in its instantiation, though. In Chapter 4 we saw that different kinds of measurement instruments were employed within different expectations of what and whom could and should be trusted, especially in regard to the reading technology of instruments. Users of the tangent galvanometer learned largely to trust their own labours, albeit with many silent or suppressed caveats about the need to borrow theoretical analysis and geomagnetic data from other workers. By contrast, the various forms of pre-calibrated ammeter and voltmeter of the mid-1880s were far less labour intensive to use, but rather more difficult to trust – *unless* one knew who had made it, how they had calibrated it, and could judge from experience how much one should trust their testimony on its reliability.

Accordingly one might qualify Shapin's suggestion that, unlike the early modern era, trust in the industrial era became a matter of trusting aloof technological experts.[4] At least some late Victorians still preferred to judge the trustworthiness of others from judging their character in face-to-face contact, and establishing trust by this means was hardly ever entirely replaced with the unknown and impersonal 'expert' in the nineteenth century. How else could one explain why Peter Willans found that so many dynamo manufacturers in the early 1890s tended, initially at least, to prefer to use their own brand of ammeters rather than trusting anybody else's in gauging the efficiency of their machines? This problem of trust was solved only by strenuous efforts among Willans' staff to develop universal measurement techniques that enabled the multiple and instantaneous witnessing of instrument readings by all parties hitherto prone to sceptical distrust. Contemporary domestic electricity customers also had, initially at least, a comparable anxiety about the integrity of suppliers. We saw in Chapter 6 that they could be very reluctant to trust the Edison Company's use of faceless laboratory experts to adjudicate monthly domestic consumption. And just as one might have expected from application of Ted Porter's account of expert public relations

---

[4] Steven Shapin, *A social history of truth*, Chicago: University of Chicago Press, 1985, p. 412.

in *Trust in Numbers*,[5] it was consumer distrust that effectively forced the Edison Company to produce a more readily readable technology and unlock its black-box meters. The company indeed adopted a much more familiar and trusted yet less efficient dial interface that allowed customers to share in the meter reading with the visiting meter man and thus render Edison's staff more accountable to consumers.

The case of the Edison meter shows us that the relationship between trust and accuracy could be complicated: It was not always the case that greater accuracy of itself inspired greater trust in all constituencies nor that any demand for greater trustworthiness could necessarily be satisfied by greater quantitative authenticity. Instead of a meter that allowed precision laboratory reading methods but entirely excluded non-laboratory witnesses from the meter-reading process, Edison customers preferred a meter that was incontrovertibly less accurate but allowed them to share in the measurement process and thus have greater trust in its outcome. This is just one example of the contingent and audience-dependent way in which a given quantitative threshold might be labelled as accuracy. In the Matthiessen–Siemens metals dispute, once trust had failed between the two protagonists, no amount of increased accuracy in (reporting of) measurements could restore that trust between them. A more extreme version of this audience dependence is apparent when the very *measurability* of an electrical parameter is contested. Whereas Ayrton and Perry claimed that their secohmmeter could with reasonable accuracy measure the self-induction of a moving alternator, Swinburne the sceptic contended that the simplifying approximations embodied in the instrument's design and operation vitiated all claims to accuracy as meaningless.

For those measuring devices that did receive at least some general degree of trust, I would like to suggest how a useful generalization about judgements of their accuracy was made. What counted as accuracy was what constituted a *sufficient* degree of accuracy for a particular purpose to be undertaken within existing contextual constraints of money and time to the satisfactions of relevant audiences. In the 1890s a resistance standard was meant to be reliable to within 0.01% for calibrating other resistance coils. By contrast, an individual stand-alone domestic meter in a contemporary consumer's home was expected only to be true to within around 5% – although often demands could be more strenuous. This was the quantitative threshold to which consumers might have been accustomed from their experience of gas metering and was not one that manufacturers or suppliers were keen to improve on lest consumer pressure forced them to produce expensive high-precision meters. Importantly, such thresholds were also diachronic products: They emerged from prior ongoing processes driven by such factors

---

[5] Theodore Porter, *Trust in numbers: The pursuit of objectivity in science and public life*, Princeton, NJ: Princeton University Press, 1995.

as commercial pressure, professional rivalry, and exploration of new forms of technical virtuosity. For such reasons the accuracies to which resistance standards could be replicated had increased by two orders of magnitude over the preceding three decades and over the 1880s the mean percentage error of readings for several domestic meters had been roughly halved. Accordingly I suggest that the historian should seek to locate judgements about (sufficient) accuracy within particular social–historical contexts if they are to understand better how the credence given to claims about accuracy can be linked to the trust accorded them by specific audiences for their own specific reasons.

I thus draw a further conclusion from studies in this book about how we might see trust working in relation to accuracy in late nineteenth-century physics and engineering. Although trust was, in many respects, as Shapin has noted, construed as a moral category, we should not see trust operating in these domains as an exclusively moral matter. Certainly, both the Siemens–Matthiessen controversy and disputes over Edison meters involved highly moralistic condemnation of certain individuals, organizations, and measurement practices as dishonest and or unfair. As already indicated, though, there was always scope for *more charitable* interpretations of suspect measurement outcomes. These could be attributed instead to the unreliability of the materials or instruments concerned, for example, irredeemable and local idiosyncrasies in the performance of mercury columns or household electricity meters. Alternatively, where the virtuousness of practitioners was not in doubt, unresolved discrepancies could be explained by reference to a lack of due care and precaution in application of materials, design methods, or mathematical technique. Thus, when Swinburne attacked the trustworthiness of results obtained with ammeters and secohmmeters developed by Ayrton and Perry, he did not question their integrity or scruples. Rather, he criticized their imprudent attempts to design and use instruments that were supposed to measure that which simply could not be measured.

From the evidence of physicists, engineers, and electrical consumers in the foregoing chapters it is evident, however, they did not construe the relations between trust and measurement devices in only strategically negative terms. The late Victorians did not regard instruments in the same way as Shapin presents Restoration natural philosophers in seeing their status in asymmetric terms: only as potentially blameworthy in explaining error rather than potentially praiseworthy in explaining success. Certain high-quality measurement devices – especially those with William Thomson's pedigree – were in fact regarded as *positive* bearers of trustworthiness in their own right. Such trustworthiness did not only develop as a result of laboratory socialization, as Steve Woolgar's ethnographic account presents the matter in Chapter 1. For the users of such technologies, trust in an instrument importantly derived from faith in the quality of its designer or maker. Whilst Carey Foster and his generation preferred – or claimed to prefer – to trust only their own labours and skill in measurements, electrical power engineers learned to be more pragmatic in trusting the complex direct-reading devices constructed by instrument-makers. A need to understand how (far) the trustworthiness of instruments was judged to depend on the reputation of their manufacturers adds to the reasons proposed by Mari Williams and others for why historians of science and technology should pay more attention to the work of instrument-makers.

The significance of users' trust or distrust in instruments is an issue that might be generalized to all technologies. Conventional accounts of the social construction of technology focus on how they are fashioned to meet the social interests of a range of actual or potential users. Yet it might be also pertinent to consider what features of technologies such users might expect to be able to *trust* as this was not always co-extensive with their interest in the trustworthiness of technologies. As we saw in Chapter 6, many householders preferred to place trust in meter-reading mechanisms in which it was *not* obviously in their financial self-interest to trust, namely, friction-dependent index dial mechanisms that simply added to the apparent consumption and thus financial billings of householders. Ironically, such householders did not trust the electric meters that had more efficient but less familiar mechanisms which did not give them so transparent a reading of their consumption. They did not trust the usage of the Edison electrolytic meter because they could not trust the company to take disinterested laboratory readings, and Ayrton's and Perry's radically new two-clock meter system required too much complicated calculation for householders to trust themselves to take authentic readings. Such consumers adopted the conservative strategy of embracing technologies they trusted when uncertain of how or whether new technologies might serve their interests. 'Trust' can thus trump 'interest' as an explanatory category for the historian seeking to account for past actions. Moreover, actors' testimony on what and whom they trusted or distrusted is a matter of explicit evidence, and this could not so generally be said either of their putative interests or of their propensity to act to further such interests.

It might also be argued that trust is a more important concept than power in understanding the collective character of quantification practices. The persistent pluralism in many kinds of measurement instrument or reading technique shows that hardly any single authority could wield sufficient power to bring all measurers to practice in the same way. There were no unequivocal 'centres of calculation' that controlled their choice of measurement equipment or that held them accountable in the accuracy of measurements. Insofar as any power to persuade operated among networks of electrical practitioners, this power followed the contours either of trust or distrust. Those who had to choose between Siemens or BA resistances, moving-coil or moving-iron ammeter, secohmmeters or induction bridges, and Edison or Ferranti domestic meters were not necessarily all coerced into their choice nor aimed solely to win favour with the authority. Rather, they had reasons for deciding positively to trust one alternative and to distrust the other. A similar comment can be made regarding the quantitative outcomes claimed by users

Conclusion

of such measurement instruments. Epistemic judgement on claims for accurate measurement was not simply determined by centralized authorities or institutional sanction. Rather, it was the outcome of discretionary practitioners' using informed judgement of the complexities of measurement practice to decide whether to trust or distrust the measurer and the measurement techniques involved to produce the accuracy claimed.

But trust – moral or otherwise – is by no means the only interesting category through which to understand the social history of measurement. As I said in my introduction, there are limits to all interpretive enterprises, and there are certain real limits to any attempt to reduce the history of measurement practice merely to an account of who trusted whom and what and why. Other themes that I touched on in the latter chapters – gender, materiality, and corporeality – point to fruitful directions for future research on measurement practice that takes discussion beyond the topic of trust. Given that I have written elsewhere on the latter topic,[6] I shall comment only on issues of gender and materiality in relation to measurement. It need hardly be said that the gendering of late Victorian electrical-measurement practice is an underresearched topic, nor indeed need it be pointed that such an issue does not reduce to the demographics of inclusion or exclusion of either sex. Certainly in contrast to the significant proportion of women involved in menial (and secluded) aspects of instrument manufacture,[7] comparatively few women were ever regular longer-term participants in publicly visible measurement enterprises, for example, Eleanor Sidgwick and Hertha Ayrton. More telling is the way that Mrs J. E. H. (Alice) Gordon explicitly encouraged female readers of *Decorative Electricity* to see their menfolk as having the prerogative for reading the new-fangled domestic electrical meter that the latter were also explicitly presented as owning. To understand this and other related phenomena we need rather to understand what grounded differential expectations of, and opportunities for, men's and women's participation in the increasingly distributed labour of electrical measurement. In addition to understand how practitioners – whether men or women – came to be trusted as measurers, we need to grasp how late Victorian contemporaries judged their trustworthiness in terms of gendered expectations of masculine or feminine character. That is as historiographically important as recovering how women were mostly given opportunities to undertake those tasks in the workshop or laboratory that men chose (or preferred) not to do. And clearly this needs to be related to men's and women's differential access to training, access to instrumental resources, and time to devote to such purposes – let

---

[6] Graeme Gooday, 'Spot-watching, bodily postures and the 'practised eye': the material practice of instrument reading in late Victorian electrical life', in Iwan Morus (ed.), *Bodies/Machines*, Oxford: Berg, 2002, pp. 165–195.

[7] Alison Morrison-Low, 'Women in the nineteenth-century scientific instrument trade', in Marina Benjamin (ed.), *Science and sensibility: Gender and scientific enquiry, 1780–1945*, Oxford: Blackwell, 1991, pp. 89–117.

alone grasping why they were interested at all in devoting themselves to the esoteric practice of quantifying electrical performance anyway.

Throughout this book I have presented the material culture of measurement as a recurrent source of anxiety over what or whom to trust in quantification. By this I do not just mean the instruments and associated paraphernalia of measurement, but also the *metals* used in the construction of electrical apparatus. It is clear that late Victorian physicists, engineers, and instrument-makers had a number of reasons to deliberate cautiously on the use of several important metals in their work. Mercury, iron, and a wide range of alloys were particularly invaluable as they were readily moulded to a large number of different purposes in the conduction and construction of measurement equipment. But they were also problematic because their long-term performance was not easily made subject to facile guarantees of consistency or control. Practitioners indeed presented such metals as having a distinctive kind of congenial or uncongenial character and showing distinctive degrees of trustworthiness or untrustworthiness. The use of mercury to define the parameters of resistance measurement was as contentious as the use of iron magnets to provide a controlling force in current measuring instruments. And these metals could no more be entirely subordinated to the technocracy of industrialized measurement than could the recalcitrantly autonomous worker in the late Victorian factory or workshop. Such problems in using metals cannot be interpreted reductively simply as disputes about the theoretical scheme of standards to which they should be adapted – that is the metrological fallacy against which I argued in Chapter 1. Instead, to grasp how persistently metallic conduct (and conductance) was a problem for electrical technologists of the nineteenth century, future accounts need to treat metals as having an interesting social history in their own right – however implausible it might *prima facie* be to do so.

By undertaking a wide-ranging study of the history of measurement, we can see why measurement *cannot* be dismissed as merely a dreary if essential part of technoscience in the past and present. Moreover, we can address the concern that so troubled Daniel Yankelovich: that of establishing who loses as well as who gains when measurement techniques come to eclipse other concerns – especially when such techniques are distractingly innovative. Revisiting the famous pronouncement of William Thomson, the irony of this story is that without knowledge of how measurement practices are informed by value-laden decisions and judgements of trustworthiness, our understanding of how such measurement practices were constituted in past forms of technoscience can at best be 'meagre and unsatisfactory'. The business of trusting people, instruments, and materials is in many ways a subject beyond measurement.

Index

Abel, Frederick, 120
Accuracy (and Inaccuracy), xix, 42
   degree of, xix, 42, 62, 76, 94, 100, 107, 144, 150 (*see also* Precision)
   judgement of, by qualitative means, 73
   method-dependence of, 77, 78
   qualitative meanings of, 62
     as closeness to true/expected value, 57
     as consistency between (manufactured) instruments, 243
     as economic parameter, 59
     as fidelity of registration, 243
     as product of 'care' in experiment, 62
     as product of error avoidance, 144
   quantitative meanings of, 62, 243, 268
   relationship to precision, 57
   relationship to sensitivity, 144
   relationship to truth, 170
   relation to labour invested, 71
   responsibility for, 66, 144, 158
   sufficient, 12, 59, 93, 167, 231
   technology-dependence of, 77
Acme Company, 161
Adams, William Grylls, 189, 203
Airy, George Biddell, 74
Alloy
   German-silver, 86
   gold/silver, 89, 100, 105
   standard (*see* resistance)
Alternate current generation, 173 (*see also* Battle of the Systems)
Alternator (alternate current generator)
   De Meritens, 189
   Elwell-Parker, 212
   Ferranti, 179, 200, 202
   Hopkinson, 214
   Lowrie-Parker, 204
   Siemens, 188
   Westinghouse, 179, 202, 203, 204, 212
   Zipernowski, 179, 204
   parallel operation (paralleling) of, 213 (*see also* Secohmmeter)
   armature reaction in, 192, 206, 207, 212, 214
   'care' in effecting, 189
   condition of maximum work available in, 192
   difficulties in accomplishing, 177
   Electrician, role in controversy over, 209, 212
   explanations of success in, 212, 213
   historiographical neglect of, 176
   Hopkinson's theory of, 175, 186, 190, 205
     criticism by Ayrton, 191
   'hunting' in, 178
   Kapp's theory of, 175, 201, 202, 203, 207
   Mordey's account of success in, 205, 206, 212
   self-induction in, 203
Ammeters and voltmeters, 51, 52, 156
Ampere, André Marie, theory of magnet-current interaction, 46, 131
Andersen (*see* Sankey)
Anderson, Olive, 118
Angell, John, 5
Anglo-American Brush Company, 204 (*see also* Mordey, William)
Anthony, William, 147, 159
Approximation, 36, 192, 197, 199, 206, 268 (*see also* Error)
Arago, Dominique François, 131
Armature reaction – *see* Alternator
Ashmore, Malcolm, xv
Assistants, 3, 29
Assumptions, simplification/falsity of, 37
Atkinson, Edmund, 5
Atkinson, Llewellyn, 191, 211

Atwood's Machine, 43
Australia, 109
Authority, 60, 98
Authorship, xxi, *see also* Measurement
Ayrton, Hertha (*née* Marks), 256, 271
Ayrton, William, xxi, xxii, 122, 124 (*see also*, Ammeter, Voltmeter Patents, Scales, Secohmmeter)
    as acolyte of William Thomson, 153
    collaboration with Perry, John, xxi, 7, 46, 153, 173, 192, 195, 228, 249
    direct (instantaneous) reading instruments, 46, 158
    as orthodox practice electrical engineering, 152
    controversial use in determining mechanical equivalent, 51
    electric tricycle, 156, 157
    end of (collaboration), 209
    equal-spaced (linear) scales, advocacy of, 38
    education of, 153
    as expert commentator on electrical measurement, 167
    moralistic campaign on domestic meters, 248
    on Maxwell's *Treatise*, 182
    Practical Electricity (1887), xi
    research with students in London (South Kensington), 203, 208, 215
    teaching in London (Finsbury), 20, 21

Bachelard, Gaston, 129
Baier, Annette, 24
Balance
    current ('weighing'), 3, 47, 51, 163
    null methods (*see* Measurement)
    resistance (Wheatstone bridge), 48, 193, 195, 196
Barrett, William F., 69
Basalla, George, 136, 137
'Battle of the Systems', 176, 177
Bell, Alexander Graham, 65
Bennett, Stuart, 218
Berg, Maxine, xix
Berlin, 17, 102
Besant, Walter, 210
Bias, 245 (*see also* Moral, Fairness)
Bijker, Wiebe (*see* Social Constructivism, Law)
Bloor, David, xv
Board of Trade (UK Government)
    approval of domestic electric meters, 222, 241
    Committee on Electrical Standards (1891), 84, 86, 122
    inspectors, 245, 248, 250
    requirements for accuracy in metering apparatus, 170
    rules on maximum allowable variation in electrical supply, 251
    standardising laboratory, Whitehall, 18, 40, 124
    unit of energy, 244, 251
Body
    bodily deportment/posture, xxi, 143, 168
    bodily senses, xi
    bodily skills/capacities/competences, xv, 19, 113, 128, 151, 197
    body as electrical conductor, 33, 155
    gestural knowledge, 11 (*see also* Sibum, Otto)
    'instinctive' understanding, 195
Bok, Sissela, 23
Bowden, Sue (*see* Offer, Avner)
Boys, Charles Vernon, 5, 51, 70, 71, 146
Bridge (*see* Balance)
Britain
    colonies of, 16, 116
    rejection of mercury for resistance standards in
British Association for the Advancement of Science (B.A.A.S.)
    annual meetings of
        Leeds, 1890, 120
        Manchester, 1861, 66
        Montreal, 1884, 120
    Committee on Electrical Standards, 20, 47, 77, 82, 87, 103
    adoption of 'absolute' scheme, 103
    promotion of 'absolute' scheme, 110
    unit of Electrical Resistance, 10, 56, 60, 173 (*see also* Resistance, electrical)
    sale of coils calibrated to the 1865 unit, 112
    Section A, 77
Broad, William and Wade, Nicholas, xvi
Brock, William H., 5
Brown, C. Neil, 222
Buckingham Palace, 164
Bunsen, Robert, 95
Byatt, Ian C.R., 224

Cables (*see* Telegraph)
Cahan, David (*see* Physikalische-Technische Reichsanstalt)

Index

Calibration, xx, 33 (see also Instruments)
  linear (proportional), 145, 153, 160
  non-linear, 160, 168 (c.f. Mechanism, non-linear)
  standards (precalibration of), 48
Campbell, Norman, xxii, 20, 30, 266
  career of, 43
  critique of accounts of measurement, 43, 44, 50
Cannadine, David, 26
Cardew, Major Philip, 44, 156
Cardwell, Donald, xiv
Care, 42, 89, 125 (see also Trustworthiness)
  feminist analyses of, 63
  in using instruments (ordinary, absolute, great), 34, 107, 241, 258
  as masculine preoccupation in technological management, 63
  as prerequisite for eliminating error, 76, 93, 97, 107, 168
  in reading of meters, 231
  reputation as a guarantee of, 114
Cartwright, Nancy, 35, 192
Charging methods (see Meters)
Cheating (see Fraudulence)
Chemical Society, 118
Christie, John, 89
Chrystal, George, 138
Churchill Lord and Lady Randolph, 220
Clark, Latimer, xxi, 54, 136
  and Robert Sabine, *Electrical Tables and Formulae*, (1871), 44, 118
Class, socio-political (see Cannadine, Shapin, Swinburne)
Clifford, William Kingdom, 42
Coal gas (see Gas)
Cockburn, Cynthia & Susan Ormrod, 256
Collins, Harry, 11, 46
Cookson, Gillian (see Colin Hempstead)
Congress, Electrical
  Chicago, (1893), 83, 123
  Paris, (1881), 84 (see also Resistance, Electrical)
  St Louis (1905), 124
Contamination (see Material, impurity of)
Copper, 33 (see also Matthiessen)
  purity of, 120
  use in resistance standards, 14
  use in telegraph cables, 4
Cornell University, 147
Crompton, Rookes E.B, 65, 68, 123, 161, 164, 167, 248

Crookes, William, 167
Current, electrical, 54
  measurement of, 129
Current balance (see Balance, current)
Customers/consumers, xviii, 53, 80, 219 (see also Meters)
  confidence/good will of, 177, 226, 229
  conservatism of, 226, 230, 239, 262
  dissatisfaction of, 176, 179, 248
  identity of, 253
  readings of meters, 60

Darwin, Charles, 69
Daston, Lorraine, 16 – see also Moral economy
De Morgan, Augustus, 63
Deprez, Marcel (see Galvanometer, Fishbone)
Diez, José, 40
Direct-reading (see Ayrton)
Distrust, 12, 26, 69, 125, 126 (see also Trust, Meters)
Disturbance, 56, 142, 147, 156, 161
Domestic economy, 258, 260
Dörries, Matthias, xix, 47
Dugan, Sally, 13
Duhem-Quine thesis, 21, 22
Dunsheath, Percy (see Historiography, internalist)
Duty – see morals

Edgcumbe, Kenelm, and Franklin Punga, 38, 68, 79, 171
Edinburgh University, 55
Edison, Thomas Alva, xxii, 176, 241
  Company (English), 186
  customer relations, 220
  Edison-Swan light bulb, 7
  gas-supply model, 225
  patents, 234, 238
  technology for measuring at a distance, 232
Einsteins, Albert, Hermann & Jakob, 224
Electricians vs electrical engineers, 168
Electricity
  central generating stations for, 164
  cost of installation, 253
  cost of supply, compared to gas, 224, 226, 257
  intangibility of, 220
  nature/ontology of, xiii, 7, 54, 56
  measurement of (see Measurement)
  methods of charging for (see Meters)
  Supply Act (1882), 223
  Supply Act (1888), 253

Electrodynamometer, 163, 201
  Gordon's criticisms of, 152
  Siemens, 131, 133, 134
  Weber's development of, 133
Ellis, Keith, 13
Energy (*see also* Board of Trade)
  absolute measurement, relation to (*see also* Thomson, William), 10
  conservation of, 82
  Maxwell on, 184
  preoccupation of British physicists with, 87
  standard parameter measured by domestic meters
Environmental conditions, 49
Error, 42 (*see also*, Self-induction)
  calculational, 102
  caused by complexity, 232
  caused by inaccurate reading, 231
  caused by simplifying assumptions, 191
  changeability of, 165
  construction-induced, 115
  cumulative, 230
  friction-induced, 240
  Germanic approach to, 73
  habitual expectation of, 12
  intrinsic, 79
  least squares analysis of, 89, 98
  (maximum) limits of, 79
  minimization by design, 227
  pre-emptive elimination of, 76, 96
  probable, 73, 75, 76
  reporting of, 89, 94
  responsibility for (causing/preventing), 142, 167
  theory of, Legendre-Gauss, 73
  toleration of, 13
  visual, 72
Esselbach, Ernst, 94, 103
Experimentation
  effort and ingenuity in, 100
  incompetence in, 114
  precautions in, 105
Europe, 9, 111
Evershed, Sydney, 148, 160, 168
Ewing, James Alfred, 164
Exhibition, Paris, International Electrical, (1881), 15, 83, 233
Extra-mural laboratory (*see* Latour, Bruno)

Fairbairn, William, 66
Fairness (*see* Moral)
Faith (in instruments), 55, 128, 166, 170, 232
Faraday, Michael
  electrolytic techniques, 48, 233
  on electromagnetic induction, 180
  Hopkinson on, 186, 191
Feinstein, Alvan, xvii
Ferranti, Gertrude (*see* Ferranti, Sebastian)
Ferranti Sebastian Ziani de, 35, 178, 224, 232 (*see also* Alternators, Meters)
  advice from Thomson, William, 240
Feyerabend, Paul, xvi
Figure of merit (*see* Galvanometers)
[Figure 1, 93]
[Figure 2, 134]
[Figure 3, 138]
[Figure 4, 143]
[Figure 5, 157]
[Figure 6, 162]
[Figure 7, 163]
[Figure 8, 187]
[Figure 9, 190]
[Figure 10, 198]
[Figure 11, 213]
[Figure 12, 228]
[Figure 13, 235]
[Figure 14, 242]
[Figure 15, 255]
Fleming, John Ambrose, 106, 182, 192, 215, 216, 236
Forbes, George, 65, 205, 223, 229, 232, 233, 239, 241, 244, 249 (*see also* Meter)
Forgan, Sophie, 57, 72, 227
Foster, George Carey, 40, 60, 69, 87, 269
Fox, Robert & Ann Guagnini, 133
France, 15, 110, 249
Frankland, Edward, 69
Fraudulence (dishonesty), 58, 244, 251
  by supply companies, 237, 246

Galison, Peter, 18
  trading zones, 18
Galvanometer
  dead-beat, 151, 156, 161
  Deprez' D'Arsonval, 133, 161, 163, 164
  figure of merit, 149
  fishbone (Deprez), 152, 155
  mirror
    as archetypal 'linear' instrument, 153
    extension to electrical engineering, 7, 161, 168

Index 277

integrated form, 138
marine (iron-clad) form, 137, 142
patent-laden device, 146
Poggendorf's original proposal for, 137
William Thomson's development of (see Thomson, William)
needle vs. lightspot reading of, 169
robustness of, 144, 156
tangent-, 45, 46, 132, 146, 219
Ganot, Adolphe, 239
Gas (Coal-gas), 230
  Act (1859), Sale of, 245, 246
  cost of (see Electricity)
  engines, 214
  examiners (inspectors), 246
  meter, 227 (see also Meter)
  share-holders, 254
  supply of, 220
  Welsbach gas mantle, 258
Genauigkeit (truthfulness), 74
Gender, 253, 271
Germany (see also Prussia), 10, 15, 58, 83, 91, 109
  approach to error (see Error)
  educational system for physics and mathematics, 74
German-silver (see Alloy)
Gitelman, Lisa, 130
Glasgow, University of, 3, 18
Glazebrook, Richard Tetley, 120, 121
  & W.N. Shaw, *Practical Physics*, 19, 42, 45, 48, 76
  teaching at Cavendish laboratory, Cambridge, 20, 43
Goldstein, Jack, xvii
Golinski, Jan, 138
Gooday, Graeme, 17, 32, 47, 54, 63, 70, 83, 86, 129, 132, 137, 149, 189, 196, 248, 257, 271
Gooding, David, 5
Goodman, George, xiii
Gordon, Alice M. (Mrs J.E.H.), 155, 176, 253, 254, 271
  'The Development of Decorative Electricity' (*Fortnightly Review*), 256
  the Development of Decorative Electricity, 257
  reports women's complaints about electric lighting, 258
Gordon, James Edward Henry (J.E.H.), 134, 152, 236

on fire risks in electric lighting, 226
Metropolitan Supply Company, 257
Mrs J.E.H. Gordon (see Gordon, Alice)
Paddington a.c. installation, 177
Telegraph Construction and Maintenance Company, 178
Gravitational constant 'g', 43
Gray, Andrew, 8
  teaching at Bangor, 20
Grovesnor Gallery, 241
Guthrie, Frederick, 5, 38, 69, 94
Gutta Percha, 4, 78, 87, 91

Hacking, Ian, xvii, 41, 74
Hague, Brian, 216
Hankins, Thomas & Van Helden Albert, 139
Hannah, Leslie, 223
Harman, Peter, 36, 58, 183
Harré, Rom, 1
Hempstead, Colin & Gill Cookson, 88
Heaviside, Oliver, 181, 182, 194
Helmholtz, Hermann von, 132
Historiography, xxi, 11, 61, 136, 222, 225
  internalist, 131
Hockin, Charles, 107, 110
Hollis, Martin, 23
Honesty (see Moral)
Hong, Sungook, xiv
Hopkinson, John, xxii, 122
  Ayrton's analysis of career of, 214
  education at Cambridge University, 26
  lecture on electrical lighting to ICE, 1883, 186, 187
  practice drawn from Cambridge Mathematics Tripos, 37, 206
  theory of alternator operation (see Alternators)
Hopkinson, Edward, collaboration with in formulating (d.c.) dynamo theory, 201
Hughes, David E., 193
  Ayrton's interpretation of, 194, 195
  criticisms by Lord Rayleigh, 193
  'skin effect' interpretation, 194
  Thompson's response to, 194
Hughes, Thomas Parkes, 176, 225
Hunt, Bruce, xiv, 6, 9, 86, 91, 116
Hunting (see Alternators, paralleling)
Huxley, Thomas H., 69, 70
Hysteresis (see Iron)

Iliffe, Rob, 219
Impurity (see Materials)
Induction, self (electromagnetic), 182 (see also Alternator, parallel operation of)
   analogy/disanalogy with inertia/momentum, 37, 173, 175, 180, 181, 183, 188
   (in)constant co-efficient of, 175, 189, 201
   iron, effect of, 181, 202
   Maxwell's mechanical demonstration model of, 185, 187, 188
   measurement of, 193 – see also Secohmmeter
   (non) measurability of, 37, 174, 196, 217
   as source of error, 173
   spatial-dependence of, 180, 181, 183
   Swinburne's critique of, 175, 207
   units of (henry/secohm), 196, 215
Induction, mutual (electromagnetic), 175, 189
Induction balance, 193
India, 107, 109, 153
Indicators (instruments as mere), 46
Industrialization, xix, 6
Institution of Civil Engineers, 164 (see also Swinburne, Thomson)
Instrument (see also Ammeter, Galvanometer, Meter, Reading, Voltmeter etc.), xiv
   absolute, 144
   as 'reified theory', 129 (see also Bachelard)
   character of, 32
   end-users of, 29, 50, 71, 72, 79, 109, 229
   internal mechanisms of, 130, 222
      moving coil, 132
      moving iron, 133, 155
      moving magnet, 132
      proportional spring, 158, 159
      non-linearity (non-proportional performance), 245
   precalibrated, 52, 71
      calibration chart, 155
      direct reading, 128
   proportionality of, 163, 197, 239
Instrument making companies, 30, 50, 125 (see also Acme, Nalder Brothers, Siemens & Halske, Johnson & Phillips)
   engineers' mistrust of, 171

Interests (self-interest), 172, 224, 248, 262
   trust vs. interest in historical explanation, 270
Interference (see Disturbance)
Interpretive flexibility, 22 (see also Social constructivism)
Iron, 33
   time-dependent behaviour of (hysteresis), 33, 35, 158
   use in resistance standards, 14
Irony, xviii, 4, 6, 11, 31, 41, 69, 71, 87, 90, 93, 105, 109, 119, 154, 185, 199, 206, 216, 230, 234, 240, 244

Japan, 153
Jehl, Francis, 233, 236
Jenkin, Fleeming, 55, 88, 103, 110
   as former assistant to Werner von Siemens, 111
   critique of Siemens 1866 publication, 114
Jenks, William, 233
Johns, Adrian (see Authorship)
Johnson & Phillips hotwire ammeter, 162
Jordan, Dominic, 174, 181, 193
Joule, James Prescott, 47, 87, 119
Jungnickel, Christa (& McCormmach, Russell), 53, 181

Kapp Gisbert, 202, 246 (see also Alternator, theory)
Kelvin, Lord (see Thomson, William)
Kempe, Harry, xxi, 44, 48, 78, 142, 145
Kennedy, Alexander B.W., 248
Kennedy, Rankin, 162, 163
King's College London, xxii
Kirchhoff, Gustav, 95, 103, 181
Kjaergaard, Peter, 227
Kline, Ronald, xiv
Knorr-Cetina, Karin, 33
Knowledge (Epistemic)-quantification relationship
   problems in, 41, 50
   (sufficient) accuracy as issue in, 59, 63, 80
   William Thomson's view of, xvii, 3
Kohlrausch, Friedrich, 18, 74, 146, 159
Leitfaden der Praktischen Physik, 19
Krige, John & Pestre, Dominique, xvi
Kuhn, Thomas, 12
   on reasonable agreement, 58
Kula, Witold, 13

Index

Laboratory, 10, 92
  Edison Company, 48, 233
  moral order of, 31
  National Physical Laboratory (UK), 18, 124
  standardizing, 48
  training, as a means of establishing uniform practice, 19, 20
  vs workshop methods of measurement, 154, 168, 169
Labour, xx
  division of, 23, 42, 68, 71 (see also Measurement, Trustworthiness)
  in calibrating, 50, 79
  intensity of, 238
Labour-saving devices, 52
Lagerstrom, Larry, 15, 84
Latour, Bruno
  criticisms of, 37
  extra-mural laboratory, 3, 140
  immutable mobiles, 17–34
  metrology (centres of calculation for), 16, 17
  on instruments, 139
  papyocentrism of, 130, 139
Law, John, 261 (see also Power, social)
Lawrence, Chris (& Steve Shapin), 96, 139
Least squares analysis (see Error)
Ledeboer & Maneuvrier, 195
Lenz, H.F. Emil, 55
Lighting, electric
  aesthetic issues, 256, 257, 258 (see also Gordon, Alice)
  disciplined usage of in the home, 257
  gendered responses, 259
  luxury of, 253
  street-lighting, 246
  value of money of, 259
  women's resistance to, 258
Localization, Locality, 10, 117, 215
  in instrument-making, relationships, 38
  of standards in electricity, 86
  of standards of length and weight, 14
  persistence of in measurement practices, 137
  in value of geomagnetic field, 146
Lodge, Oliver, 122
London, 1862 Exhibition, 105
Lynch, Arnold, 9, 82
Lupton, Sydney, 75

McCormmach Russell (see Jungnickel, Christa)
Machine-tools, 66
Macintyre, Alisdair, 27 (see also Practice)
Mackenzie, Donald, 2, 61, 223
Magnello, Eileen, 18, 124
Magnet, permanent, 155, 158, 168
Magnetic Field, Strength of Earth's, 146
Manchester, 61, 66, 75, 154
Marsden, Ben, 89
Marvin, Carolyn, 220
Marx, Karl, 66
Material culture, 30
Materials
  conductivity of, 33
  historicity of, 33
  metals, variability in performance of, 83, 92, 99, 100, 272
  purity and impurity, 33, 97, 99, 104, 121
  stability, 33, 85, 89
  'transparency' of performance, 34
  trustworthiness of, 2, 23
Mathematics
  Cambridge analytic methods, xxii, 184, 189 (see also Hopkinson)
  graphical/engineering, 188
  Mathematicians (see also Shapin)
  vs practical men, 26
Mather, Thomas, 163
Matthiessen, Augustus, xxi, 84
  adoption of absolute measurement principles, 100
  disability, 95
  disputes with Werner Siemens, 7, 12, 60 (see also Metals controversy)
  early rejection of absolute measurement, 95
  education of, 95
  proposal of solid standard, 94
  suicide of, 96, 117
  Thomson, William, dealings with, 7, 96
Maudslay, Henry, 66
Mayer, Anna, xvi
Maxwell, James Clerk, xxi
  analysis of electromagnetic self-induction in *Treatise*, 175, 182, 184, 195
  Cavendish Laboratory, Cambridge, 18, 19, 21, 37, 187
  determination of absolute unit of resistance, 110
  displacement current, 4

Maxwell, James Clerk (*cont.*)
  dynamical theory of the electromagnetic
    field, 183
  electromagnetic theory of light, 4, 58
  Hopkinson on, 186
  on instruments, 132, 142
  lecture on Thomson's galvanometer, 149
  Maxwellian theory, 218
  Treatise on Electricity and Magnetism,
    xiii, xxi, 44, 54, 76, 117, 118,
    144, 180, 196
Measurement
  attributes of measurements
    agreement between, 12
    authenticity of, 49, 52
    authorship of, 29, 42, 50, 53, 72, 265
    constitution of, xiv, 6, 43
    credibility of, 56
    division of labour in, 3, 49, 50, 265
    identity of, 57
    meaning(s) of, 41
    performance of, xvi
    presuppositions of, xvi
    process of, 45, 79
    purpose of, 80
    ramifications of, xvi
    reporting of, xvi
    rhetoric of, 42
    temporality of, 148, 196, 197
    trustworthiness of (*see* Trustworthiness)
  kinds/methods of measurement, 263
    absolute, 47, 48, 49, 82, 85, 133
      legal implementation of, in Britain
        (1894), 83
      Wilhelm Weber on, 98
    at a distance (*see* Meter)
    changing approaches to, 263
    comparative, 44
    direct-reading, 149
    laboratory, 55 (*see also* Laboratory)
    mensurational (as reducible to
      determination of length), 43, 46
    null, 48, 133, 152, 169, 193
  theory of, 40, 54
Mechanical engineering, 46, 154
Mechanical Engineers, Institution of, 61
Mendenhall, Thomas, 62
Mercury, 33
  De la Rue & Marié-Davy's use of in
    resistance measurement, 92
  Ferranti domestic motor meter, use in,
    239, 241
  Matthiessen's denigration (*see* Metals
    Controversy)
  resistance standards, use in, 9
  secohmmeter, use in, 199
  Siemens' advocacy of, 10
  source of corruption
  specific gravity (density) of, errors in
    values of, 93, 109, 111, 112, 113
  'ticklishness' of, 89
  untrustworthiness of, 102
Merriman, Mansfield, 62
Merton, Robert, xvi
'Metals controversy', 83, 85
  care in handling metals as key issue, 89,
    114, 115
  impurities as an issue in, 99
  Maxwell's comment upon, 117
  temperature management as an issue in,
    101, 113
Meters, domestic electrical, 53 (*see also*
  Power, electrical)
  ampere (*see* ammeters)
  charging methods, 223, 237, 244
  clock (Aron, Ayrton & Perry), 222, 229,
    230
  compared to gas meter paradigm, 234,
    239, 243, 262
  consumer-readable dial, 228, 239
  debate what meters should register, 221,
    250, 252
  demand for trustworthiness in, 48, 220,
    261
  desiderata in design, 232
  electrolytic (Edison), 48, 222
    accuracy of, 237
    criticisms of, 237
    not readable by householders, 234,
      235
    polarization effects in, 234
  electrolytic (Lowrie-Hall), 248
  energy, as standard parameter measured
    by, 252
  gendered responsibility for
  induction meter (Shallenberger), 222
  integrating meters (Thomson etc), 223
  overcharging householders for, 233
  mistakes in reading, 231
  motor meter (Ferranti), 222, 239, 242
  revenue from metered consumption, 219
  temperature-dependence of performance,
    237
  thermal convection (Forbes), 223

Index

recording wattmeter (Thomson-Houston etc), 238
volt (see voltmeters)
Metrological fallacy, 9, 11, 272
Metrology, 2, 11, 88
Metropolitan Supply Company, 178, 223
Mirowski, Philip, 41
Mirror galvanometer – See Galvanometer
Misa, Thomas, 89
Moral(s), xv, 20, 52 (see also Fraudulence)
   as lesson, xiv
   character, 118
   compared with 'social', 31
   fairness, 221, 244, 250, 252
   honesty, 24, 101, 172
   integrity, 24, 28
   responsibility, 38, 46, 53, 65
   rights, 1
   virtue(s), virtuous conduct, 28, 84
Moral economy, 23, 27
Mordey, William, 175, 204, 210 (see also Alternator)
Morrell, Jack, 116
Morrisson-Low, Alison, 29, 271
Morus, Iwan, xviii
Muirhead, Alexander, 64, 84, 86, 123
Muirhead, Elizabeth, 64

Nalder Brothers, 199, 216
Nasmyth, James, 66
Nationalism, nationalistic rivalry, 88, 126
'Nature', constancy of, 32
Neumann, Franz, 74
Newall & Co., 91
Nicholas 1, Czar, 91
Nicholson, Shirley, 254

Obach, Eugen, 156
Ohm, 83 – see also Resistance, electrical
   defined in absolute terms in 1881, 15
Ohmmeter, 264
Ohm's Law, 55, 129
O'Connell, Joseph, 15, 219
Offer, Avner & Sue Bowden, 256
Olesko, Kathryn, 9, 10, 28, 58, 74, 94
Olsen, Marvin & Marger, Martin, 20, see Power, theory of social
Ormrod, Susan (see Cockburn, Cynthia)

Pedagogy – see Training
Parsons, Charles, 171

Parsons, Robert H., xviii, 164, 177, 212, 214
Patents, 136, 138, 228 (see also Ayrton, Edison, Swinburne, mirror-galvanometer)
Paterfamilias (see Sambourne, Linley)
Paul, Robert, 163
Pels, Dick, xv
Pepper, Henry, 5
Performance (see Measurement)
Perry, John (see Ayrton, William)
Physical Society of London, 157, 163
Physics, Modern, 8, 265
Physikalische-Technische Reichanstalt, 17, 124
Pickering, Andrew, 137
Pinch, Trevor (& Bijker Wiebe.), xvi, 135
Poggendorf (see Galvanometer, mirror)
Porter, Theodore (Ted), 1, 22, 27, 238
Pouillet, Claude, 132
Power, electrical, ac vs dc – (see also Battle of the Systems)
   financial/economic/profit issues in, 178, 223, 247 (see also Metering)
Power, social
   as circumscribed by social resistance (Marger & Olsen), 20, 21
   impetus theory of, 21
   relation to metrology, 17
   relation to trust, 2, 16, 21
   Law, John, critique of essentialist theories of power
Power stations
   conditions in, 57
   development of, 79
   switchboard instruments, 248
'Practical men', 26, 56, 173
Practice
   as constitutive feature of everyday life, xi
   Alisdair Macintyre on, xi
   representations of, 31
Precision,
   as educational topic, 62
   relation to accuracy (see Accuracy)
   relation to delicacy/beauty, 64
   relation to 'degree of accuracy' (see Accuracy)
   relation to objectivity, 28
   relation to linguistic 'refinement', 65
   Schärfe, Präzision, 73
   as 'sharpness' of data set, 73
   telephone as instrument of, 66

Proctor, Robert, xvi
Preece, William H, 59, 68, 148, 181
Prussia, 88 (*see also* Germany)
Punch magazine, 227, 253
Pycior, Helena, 29

Quantification, xi, 1, 2
  Quantifying 'spirit', xiv

Rappaport, Erika, 260
Ravetz, Jerome, 43
Rayleigh, Lord, 13, 77
  at Cavendish Laboratory, 18
  collaborating with Eleanor Sidgwick, 48 (*see also* Sidgwick)
  on experimental determination of mercury ohm, 121
  on self-induction, 77, 173, 175, 196, 216
Reading instruments, 72, 139, 174
  as heterogeneous act, 199
  counter-rotating dials in meters, 231
  distinctive power station practices, 178
Reading technologies (kinds of), 129, 130, 172
  spot-watching, 131, 151, 169
Reasonable agreement – 13, 57, 99, 124, 125
  see also Kuhn, Thomas
Reflexivity, xix
Resistance, Electrical (*see also* Ohm)
  commercial precalibrated sets (and problems of), 49, 52, 105
  devices for measuring, 116
  standard of resistance, material construction of
    BA alloy standard, 110
    later British proposals for solid standard, 122
    Matthiessen's proposal of gold-silver alloy, 95, 97
    rejection at paris congress 1881, (*see* Congress)
    absolute form adopted at Paris Congress in 1881, 120
    calculation/recalculation of, 106
    continued British ambivalence towards, 122
    proposal for absolute form of, 106
    revised determination of absolute form, 121
    Siemens' proposal for metre column (*see* Siemens)
  temperature-dependence of, 97, 98, 120
  ubiquity of, 4
unit of resistance
  British Association absolute unit determination and redetermination of, 119
  dimensions of 'velocity', 10
  Jacobi copper-wire unit, 91
  Jacobi iron unit, 86
  Siemens' mercury unit
    conformity with Prussian metric system
    official abandonment in 1881, 83
    proposal in spatial form., 92
    sales of coils based on 1864 and subsequent units, 109, 115, 116, 119
Resistance, Social (*see* Power, Social)
Responsibility (*see* Accuracy, Error, Morals, Meters)
Richards, Eveleen, xv
Rimington, E.C., 200
Roberts, Richard, 67
Rorty, Richard, xviii, xix
Rouse, Joseph (*see* Practice)
Rowland, Henry, 121
Russia, 91
Rutherford, Ernest, 72

Sabine, Robert, xxi, 88, 102, 106, 117 (*see also* Clark, Latimer)
Sambourne, Linley, 253, 255
Sambourne, Marion, 254
Sankey, Captain H.R., 161
Scales (Instrument reading)
  'care' in preparation, 171
  disagreement between Ayrton and Swinburne on, 38
  equally divided scales, untrustworthiness of, 169
  minimum discernible scale division, 150
Schaffer, Simon, 34, 126 (*see also* Materials, transparency of)
  on Cavendish Laboratory as Manufactory of Ohms, 16 (*see also* Maxwell)
  metrology, 9, 83
  socio-political administration of experimentation, 19
  & Shapin, Steven (*see* Shapin)
Schivelbusch, Wolfgang, xix, 226, 256
Schuster, Arthur, 13
Secohm (*see* Induction, self)
Secohmmeter, 7, 173, 198 (*see also* Ayrton, Induction, Self)

Index 283

alternators, use in analysing, 203, 206
  as bridge-based device, 49
  Ayrton and Perry's presentation to STEE in 1887, 55, 194, 199
  criticisms of, 200
  initial reception of, 200
  limitations of, 195
  marginal career of, 215
  versatility of, 216
Self-help, self-reliance, self-discipline, 67, 71
Self-induction (inductance) (*see* Induction, self-)
Senior, John, 33
Senses – *see* Body
Sensitivity, 64, 141, 144, 149 (*see also* Accuracy)
Shapin, Steven
  and Schaffer, Simon, xv, 236
  Social History of Truth, xvi, 1, 23
    Catholics in Restoration natural philosophy, 26
    mathematicians, trustworthiness of in Restoration natural philosophy, 26
    Robert Hooke at the Royal Society socio-political class, 25
    trust and trustworthiness in Restoration experimental philosophy, 30
Shaw, W.N. (*see* Glazebrook)
Shoolbred, James, 46, 156, 233
Sibum, Otto, 11, 139
Sidgwick, Eleanor, 271
  collaboration with Lord Rayleigh, 13 (*see also* Rayleigh)
Siemens & Halske Co.,
  cable-laying activities, 91
  formation in 1847, 91
  London location, 88
  production of provisional (1863) BA resistance standards, 107
  production of commercial resistance coils (*see* Resistance, Electrical).
Siemens, Carl (Charles), 88
Siemens dynamometer, *see* Electrodynamometer
Siemens, Werner von, xxi, 10, 12 (*see also* Metals controversy, Resistance, electrical)
  collaboration with Carl (Charles) von Siemens, 87, 90
  education and early career, 90
  international reputation, 114

Lebenserinnerungen (1892), 84
  proposal of mercury unit, 1860, 90
  vindication by Rayleigh *et al.*, 13
Silva, Elizabeth B. da, 63
Skill (*see* Body)
Smiles, Samuel, 61, 66, 67, 68
Smith, A[dam] (*see* Goodman, George, J.W.)
Smith, Crosbie (& M. Norton Wise), 4, 64, 83
Smith, Willoughby
  & Oliver Smith, on board Great Eastern
  President of STEE, 1883, 219, 221
Social constructivism, 136, 221, 243, 261, 262
Society of Telegraph Engineers (and Electricians), 46, 55, 186
South Kensington, xxi, 146, 248
Spot-watching (*see* Reading)
Spouses, 256 (*see also* Women, Assistants)
Stability (*see* Material)
Standard, metrological, (*see also* Metrology, O'Connell, Olesko)
  reproducibility of, 100, 108, 115
  Urmaass, 9
Steam engines, 164, 213
Steam gauge, 46, 154
Stewart, Balfour, 44
Stokes, George Gabriel, 63
Strange, Phillip, 169
Students (*see* training)
Sturgeon, William, 132
Sugg, Jennie Marie, 230
Sugg, William, 230, 245
Sumpner William E., 163, 199
Surrogate, 36, 46, 48, 197, 249
Swinburne, James, xxii (*see also* Induction, Scales)
  advocacy of alternate current, 179
  'armature reaction' (*see* Alternators)
  as expert in matters of electrical measurement, 168, 200
  criticisms of Ayrton (& Perry)'s instruments, 20, 210, 269
  ICE lecture on electrical instruments (1892), 30, 152, 171, 219, 251
  mirror-reading techniques, objections to, 20
  Practical Electrical Measurement (1887), 44, 79, 136, 147
  views on trustworthiness of instruments, 35, 159
  views on trustworthiness of theories, 37

Sydenham, Phillip, 41
Sylvester, John Joseph, 63

Tait, Peter Guthrie (*see* Thomson, William)
Teaching of electrical physics/engineering, 42, 55
Technicians (*see* Assistants)
Technology (*see* Telegraph, Power, Lighting)
Telegraph industry
   Atlantic cables, 59, 96
   fault-finding, 59, 150
   signalling, theory of, 181
   stimulus to develop of electrical measurement, 3, 10
   testing rooms, 19
   use of electrical resistance standards in, 14, 83, 86
Telephone (*see* Precision)
Tennyson, Alfred Lord, 149
Theory
   embodied theory, instruments as (*see* Bachelard)
   theory laden, theory laden-ness, 45, 128
   trustworthiness of, 2 (*see also* Trust)
Thompson, Edward P., 27
Thomson, John Joseph, 32, 122, 185, 194
Thompson, Silvanus P., 51, 123, 124, 137, 191, 200, 214
Thomson, William (Lord Kelvin), xvii, xxi, 122, 123 (*see also* Knowledge-quantification relationship, and Precision)
   absolute measurement, advocacy of, 10
   as authority on measurement, 2, 3
   Atlantic cables, involvement in, 4, 137
   British Association Committee on Electrical Standards, involvement in, 103
   collaboration with James White, 3, 38
   collaboration with P.G. Tait on *Treatise on Natural Philosophy* (1867), 44
   'curse of Kelvin', xvii
   direct reading instruments, support for, 46
   efficiency and economy, concern with, 6
   elevation to peerage as Lord Kelvin 1892, 3
   IEE President in 1888-9, 207
   lecture to ICE on electrical units, 1883, 3, 14, 263
   mirror galvanometer, 64, 135, 137
   metrology, involvement in, 9

   portable marine voltmeter, 47
   self-induction, treatment of, 181
   Siemens mercury unit, support for, 13
Thomson-Houston company, 231
Training, 26, 70, 89
Trial and error methods (of instrument design), 46
Tricycle, electric – see Ayrton
Trotter, Alexander P., 169
Trotter, Coutts, 32
Trust, xix, 1, 52, *see also* Distrust
   in construction of instruments, 37, 50
   interpersonal/social basis of, 23, 125
   in theories, 35
   normative vs predictive, 24
   power, contrast with (*see* Power, social)
   problems generated by new technologies, 130
   problems of distrust in others' instruments, 165
   relation to division of labour, xx, 2, 28, 49, 52, 130
   relation to personal labour, 46
Trustworthiness
   accuracy as prerequisite for, 67
   amnesia concerning involvement of prior labour, 50, 51
   care in experimental procedure, relation to, 34, 114
   context-dependence of, 127
   customer, 166
   human vs non-human bearers of, 2
   of instrument readings, 12, 128, 220
   of material substances, 83
   of measurements, 28
   of methods, 119
Tyndall, John, 5

Universality, universalization (*see also* Localization, Locality)
   epistemological, 10
   in relation to metrology, 16
   of electrical units, 15
   of electrical practice, 20
   of devices, 130, 131, 136, 162, 171, 172
   of truth-telling, 23
United States of America, 9, 62, 109
Unwin, Alfred, 165

Value(s), 19
Value-ladenness, xv, 6, 63

Index 285

Van Helden, Albert., 138 (see also Hankins)
Varley, Cromwell, 115, 132, 135
Victoria, H.R.H. Queen, 83, 123
Vienna telegraph Congress (1868), 120
Vignoles, E.B., 47
Viriamu Jones, John, 121
Virtue (see Morals)
Voltmeter (see Ammeter, Thomson)

Wajcman, Judy (& Donald Mackenzie), 262
Warwick, Andrew, 26, 54, 182, 192
Watt, James, 61
Weber, Heinrich, 160
Weber, Wilhelm, 10 (see also Electrodynamometer)
Wells, Herbert George, xxii, 69
Westminster Supply Company, 248
Weston, Edward, 132, 137, 241
Wheatstone, Charles, 103, 115 (see also Balance, resistance)
White, James (see William Thomson)
White, Morton, xvi
Whitehouse, 'Wildman', 6
Whitworth, Joseph, 61, 66, 169
Wilde, Henry, 189
Wilhelm II, Kaiser, 83
Willans, Peter, 128, 130, 135, 169
  economy of steam engines produced by, 171
  reputation for high accuracy in testing steam-consumption, 165
Willans & Robinson Co, 164, 212 (see also Sankey and Andersen)
  defence of employees instrumental practices (Sankey and Andersen), 170
  specifications for ideal electrical testing instruments, 165
Williams, Mari, 133, 270
Williamson, Alexander, 103
Wilson, John F., 178, 240
Wise, M Norton, xix, 42
  industrial cultures of precision, 28
Witnessing
  conjoint, 170
  virtual, 236
Women, xxiii (see also Ayrton, Ferranti, Gordon, Muirhead, Sidgwick)
Woolgar, Steve, 1, 30, 269
Wynne, Brian, xv

Yankelovich, Daniel, xiii, xvii
Yavetz, Ido, 174, 182

Zacharias, Jerrold, xvii

Lightning Source UK Ltd.
Milton Keynes UK
UKOW02f0823170317
296853UK00001B/54/P